Statistics for Social and Behavioral Sciences

Series editor
Stephen E. Fienberg
Carnegie Mellon University
Pittsburgh
Pennsylvania
USA

Statistics for Social and Behavioral Sciences (SSBS) includes monographs and advanced textbooks relating to education, psychology, sociology, political science, public policy, and law.

More information about this series at http://www.springernature.com/series/3463

Linda M. Collins

Optimization of Behavioral, Biobehavioral, and Biomedical Interventions

The Multiphase Optimization Strategy (MOST)

Springer

Linda M. Collins
The Pennsylvania State University
The Methodology Center
University Park, PA, USA

ISSN 2199-7357 ISSN 2199-7365 (electronic)
Statistics for Social and Behavioral Sciences
ISBN 978-3-319-72205-4 ISBN 978-3-319-72206-1 (eBook)
https://doi.org/10.1007/978-3-319-72206-1

Library of Congress Control Number: 2017960298

© Springer International Publishing AG 2018

This work is subject to copyright. All rights are reserved by the Publisher, whether the whole or part of the material is concerned, specifically the rights of translation, reprinting, reuse of illustrations, recitation, broadcasting, reproduction on microfilms or in any other physical way, and transmission or information storage and retrieval, electronic adaptation, computer software, or by similar or dissimilar methodology now known or hereafter developed.

The use of general descriptive names, registered names, trademarks, service marks, etc. in this publication does not imply, even in the absence of a specific statement, that such names are exempt from the relevant protective laws and regulations and therefore free for general use.

The publisher, the authors and the editors are safe to assume that the advice and information in this book are believed to be true and accurate at the date of publication. Neither the publisher nor the authors or the editors give a warranty, express or implied, with respect to the material contained herein or for any errors or omissions that may have been made. The publisher remains neutral with regard to jurisdictional claims in published maps and institutional affiliations.

Printed on acid-free paper

This Springer imprint is published by Springer Nature
The registered company is Springer International Publishing AG
The registered company address is: Gewerbestrasse 11, 6330 Cham, Switzerland

To John, Matthew, Joy, and Jazeya

Preface

In the United States and worldwide, billions of dollars have been spent to develop behavioral, biobehavioral, and biomedical interventions (hereafter referred to simply as interventions) to prevent and treat health problems, promote health and well-being, prevent violence, improve learning, promote academic achievement, and generally improve the human condition. Numerous interventions are in use that are successful in the sense that they have demonstrated a statistically and clinically significant effect in a randomized controlled trial (RCT). However, many are less successful in terms of progress toward solving problems. In fact, after decades of research, as a society we continue to struggle with the very issues these interventions have been designed to ameliorate. Only very slow progress is being made in many areas; in some, the problem continues to worsen. Let us consider two examples in the public health domain, both from the Healthy People goals set every ten years by the United States Centers for Disease Control and Prevention (CDC).

The first example concerns adult obesity. One of the CDC's Healthy People 2010 goals was to reduce the prevalence of adult obesity from the 2000 baseline of 23% down to 15%. Unfortunately, by 2010 adult obesity had *increased* to 34%. This is a serious issue for American society; according to Finkelstein, Trogdon, Cohen, and Dietz (2009), the medical care costs of obesity in the United States are $147 billion per year (in 2008 dollars). The healthy people 2020 goal is to reduce adult obesity to 30.5%. On the one hand, this would be a 10% reduction from the 2010 prevalence; on the other hand, it is higher than the 2000 base rate.

The second example concerns cigarette smoking. Cigarette smoking remains the leading preventable cause of death in the United States and worldwide (CDC, 2016b). In 2015, the prevalence of adult smoking in the United States was about 15% (CDC, 2016a). This is a marked improvement over the 2000 base rate of 21%. The goal for Healthy People 2020 is to reduce the prevalence of adult smoking to 12%, and there is optimism that this goal can be met. However, there are startling disparities in the prevalence of smoking. For example, in 2015 the prevalence of smoking among adults with a general equivalency degree (GED) certificate (a credential equivalent to completion of high school, earned by taking an examination) was more than 30%, which is nearly five times the prevalence among adults with a college degree and more than nine times the prevalence among those with a graduate degree.

In my view, there are three primary reasons why progress in areas like obesity and tobacco use, as well as many other areas in which interventions could potentially make a huge difference, has been slower than it ideally could have been.

First, *intervention science is making little progress toward accumulating a coherent base of scientific knowledge* about which intervention components work and which components do not work, which ones work particularly well together, and which components work well for subgroups of individuals with particular characteristics. If an ever-expanding base of scientific knowledge in this area were established, scientists could draw on this established knowledge when developing an intervention, and add to it by investigating new components. Such an approach would enable scientists to keep improving interventions. Instead, most of today's interventions are "black boxes" – it is unclear what the active ingredients are, so the mechanisms that produce any observed effects are poorly understood. This makes it difficult to see the way forward to improve the intervention or to build on what previous studies have accomplished.

Second, *there has not been sufficient emphasis on steady, programmatic, incremental, and measureable improvement of interventions.* It would be helpful if every new intervention were an improvement over its predecessors in specific ways and by specific amounts; in this way interventions would become better and better over time. Think of how much improvement has been made in intervention science in the past 40 years, and compare this to the steady progress made in consumer products, such as the automobile, during this time. It is easy to point to a variety of metrics that can be used to express improvements in the automobile. For example, compared to the cars of 40 years ago, today's cars are more fuel efficient, more comfortable, more reliable, and safer. By contrast, even though new interventions appear regularly in the scientific literature, it is difficult to point to similar metrics to express clear progress over time. Compared to 40 years ago, or last year, are today's interventions measurably more effective? Less expensive? Easier to implement? Less burdensome?

Third, *many evidence-based interventions are too expensive, demanding, or complex to be scalable.* Today there is little expectation that efficiency, economy (in terms of both immediate monetary outlay, and time and other resource demands), and scalability will be given serious consideration while an intervention is being developed and evaluated. It follows that there is little disincentive to develop interventions by including any and all components the investigator believes may boost the treatment effect and produce a significant hypothesis test in the RCT. The thinking is that the first order of business is to demonstrate a significant effect in an RCT; once an effect has been demonstrated, there will be an opportunity to consider the matters of efficiency, economy, and scalability before the intervention is implemented in the intended setting.

This perspective has led to the development of many interventions that have earned the prized designation "evidence-based" but at the same time are difficult or impractical to implement because they are prohibitively expensive, dauntingly complex, or too time-consuming for staff. Some end up never being implemented broadly; others are implemented broadly only after removal or revision of components to reduce cost or complexity. Because, as mentioned above, the active ingredients of most interventions are unknown, there are no guidelines for which components are essential and which may be removed without a huge sacrifice in effectiveness. As a result, any revisions to the intervention are ad hoc and run the risk of removing what may be an essential element, thereby reducing or even nullifying the intervention's effect. (See Chap. 7 for a more concrete explanation of how ad hoc revisions can backfire.) In any case, once such revisions have been made, the revised intervention no longer merits the designation "evidence-based" because it is not the same as the one that was originally evaluated.

A Vision for the Future

I remain optimistic that society can meet its goals in areas like public health, education, and human well-being. But to accomplish this, I believe better interventions are needed—interventions that are more effective, more cost-effective, more efficient, and more scalable. These interventions will be more practical to implement exactly as designed and evaluated. Imagine a near future in which...

>...only components that positively and demonstrably affect the outcome are eligible for inclusion in an intervention, so no time or money is squandered on useless or counterproductive components.
>
>...interventions are built in a principled manner to meet clearly specified standards of effectiveness, efficiency, economy (including cost-effectiveness), and scalability. It is always clearly reported what these standards are and how a particular intervention measures up to them. The standards may vary depending on the needs of the area, resources available, and the environment in which the intervention is to be applied.
>
>...key constraints expected to affect scalability, such as constraints on implementation costs, are taken into account from the beginning of intervention development. The objective is to develop an intervention that delivers the best outcome achievable within those constraints and is immediately scalable without the need for any ad hoc modifications.
>
>...as findings emerge from basic research, approaches for translating them to components for inclusion in interventions are programmatically tested, and only the approaches that work are added to interventions. Investigators continue to test dramatically new ideas, but must demonstrate specifically how and to what extent they are better than existing alternatives.
>
>...every study adds to the knowledge base about which intervention components do and do not work, even if it does not produce an incrementally and materially improved intervention.
>
>...as time goes on, a knowledge base grows about what works and how. Scientists developing new interventions build on this knowledge base and keep moving forward. This accelerates the pace of intervention science.
>
>...looking back over a period of time, say ten years, it is clearly evident that the interventions in a particular area have steadily become measurably more effective, efficient, cost-effective, and scalable.

How can this future become a reality? To realize the tremendous potential that interventions have to make the world a better place, it is necessary to start taking a very different approach to their development and evaluation.

This book and its companion volume (Collins & Kugler, 2018) are about one possible such approach, the multiphase optimization strategy (MOST). MOST is a framework for development, optimization, and evaluation of behavioral, biobehavioral, and biomedical interventions.

A Brief Comparison of the Classical Approach and MOST

MOST integrates perspectives, concepts, and approaches used in engineering, statistics, biostatistics, and behavioral science. The fundamental idea is that *interventions can and should be optimized to meet specific criteria, and only after an intervention has been optimized should it be evaluated in an RCT*. Optimization of interventions and evaluation of interventions are different phases of research, pose different questions, and require different methodological approaches.

As described in detail in this book, MOST consists of three phases: preparation, optimization, and evaluation, in that order. Activities in the preparation phase include selection of the components that are candidates for inclusion in the intervention, and development of a detailed conceptual model of the process to be intervened on. In the optimization phase of MOST, which occurs before an intervention is evaluated in an RCT, steps are taken to optimize the intervention to meet specific criteria established a priori by the investigator. This optimization may take any of a variety of forms, depending on the type of intervention to be optimized. In the evaluation phase of MOST, the effectiveness of the optimized intervention is evaluated in a standard RCT. For the first seven chapters of this book, the emphasis is on fixed interventions—those in which all participants undergo the same treatment. In Chap. 8 adaptive interventions, in which treatment may vary across participants and over time, are discussed. Optimization of adaptive interventions is discussed further in the companion volume in the chapters by Almirall, Nahum-Shani, Wang, and Kasari and Rivera, Hekler, Savage, and Downs.

The classical approach to intervention research has consisted of identifying the components that are to make up the intervention; possibly pilot testing[1] them to ensure that they are implementable, not toxic, etc.; making any necessary revisions based on the pilot testing; immediately combining the components into an intervention package; and testing the efficacy or effectiveness of the package in an RCT. Typically the RCT will have two "arms" (experimental conditions): a treatment arm in which subjects receive the intervention package and a control arm in which subjects receive a suitable control protocol, such as the current standard of care. After data have been collected on sufficient subjects to afford the desired level of

---

[1] See the definition of pilot test in Chap. 2 and in the glossary.

statistical power, the difference between the outcome for the treatment and control groups is estimated. If this difference is statistically significant, then the intervention is said to be efficacious (if the experiment was conducted in a tightly controlled setting) or effective (if the experiment was conducted in a real-world setting).

From the MOST perspective, one major difficulty with the classical approach is its sole reliance on the RCT, to the exclusion of other approaches, for all phases of research. As mentioned above, optimization and evaluation of an intervention are distinctly different phases of research. Optimization refers to identifying the set of components and component levels that will make up the optimized intervention, and evaluation refers to assessing whether the optimized intervention has a statistically and clinically significant effect. The RCT, although eminently well-suited to the primary research question that motivates evaluation—that is, whether an intervention has a significant effect as a package—is poorly suited to addressing the kinds of research questions that come up during optimization of an intervention. An important part of optimization and, over time, ongoing improvement of an intervention, is assessment of the performance of the individual components under consideration for inclusion in an intervention and whether and how the components affect each other's performance. The RCT does not provide this information. Post-hoc analyses on the data from an RCT, such as mediation analyses, can be helpful for testing theories and generating hypotheses for future research, but they are limited in what they reveal about the performance of individual intervention components. In MOST, an intervention is always optimized before evaluation in an RCT, and this optimization is based on experimentation that provides direct information about the performance of individual components and, ideally, how they affect each other. Thus, experimental designs other than the RCT are needed. These ideas are developed in detail in the book.

The RCT has been, and remains, an excellent way to evaluate interventions once they have been optimized, because it provides a direct, straightforward, and sensible way of determining whether a single treatment, or an array of treatments *as a package*, performs better than a control. Some intervention scientists have expressed concern that using MOST implies abandoning the RCT. My response has always been that the RCT is an integral part of the evaluation phase of MOST; this is discussed in Chap. 1. This book does not include a chapter on the RCT per se, because there are already many excellent books on the topic.

How MOST Has Been Developed So Far

MOST emerged from a conversation that took place in 2003 between Susan Murphy and Vijay Nair in a hallway of the Statistics Department at the University of Michigan. Susan Murphy's area is statistical theory and applied methodology pertaining to development of adaptive interventions, and Vijay Nair's area is theoretical and applied statistics related to engineering research. Susan and I had been collaborating for several years on some conceptual aspects of methodology related to development of adaptive interventions. I had been complaining to Susan repeatedly for some time about how, in my view, intervention science was not making very rapid progress. Susan and Vijay started discussing this perceived lack of progress, and Vijay briefly described to her the way products are developed in engineering, noting the differences from how intervention scientists seemed to develop interventions. Susan, who has always been an outstanding matchmaker of collaborators, immediately phoned me to say I needed to visit Michigan right away to spend a day meeting with Vijay and her, to see how far we could get adapting the engineering approach for use in intervention science. The eventual outcome of that meeting was Collins, Murphy, Nair, and Strecher (2005), the first article on MOST.

Methodological research can be as messy as other areas of science, and there are aspects of research on MOST that have been a bit messy. I would like to take this opportunity to call two messy aspects of research on MOST to your attention. The first is the change in how the phases of MOST are conceptualized. In Collins et al. (2005) we outlined three phases of MOST, namely screening, refining, and confirming. Those phases came straight out of engineering and map very well onto what is done in that field. Numerous publications related to MOST have been structured around those phases. However, after about eight years of working with intervention scientists within the MOST framework, I came to the conclusion that screening, refining, and confirming did not map well enough onto what these scientists needed to do. I then reconceptualized MOST so that it consists of the three phases discussed in this book, namely preparation, optimization, and evaluation, and soon began using the new phases in my speaking and writing. The evaluation phase is simply a renaming of the confirming phase, but the preparation and optimization phases are different from the previous screening and refining phases. For example, the previous screening phase did not include the critically important conceptual model (Chap. 2). The first article using the new framework was Collins, Kugler, and Gwadz (2016). Intervention scientists immediately seemed to be more comfortable with the new framework and to be better able to relate the framework to their own research, so the change had a positive effect overall. However, I regret that the reformulation has created some confusion.

A second aspect of research on MOST that has been messier than I had hoped is the integration of MOST and work related to development of adaptive interventions. After our 2005 publication, Susan Murphy continued to work in the area of adaptive interventions, particularly on the sequential, multiple assignment, randomized trial (SMART), a pioneering experimental design. (More recently, she has been working in the m-health area, pioneering the microrandomized trial.) I continued to develop the MOST framework. Not until some time later did it dawn on us that the SMART is a type of optimization trial, and that Susan's work on adaptive interventions, as well as the work of Daniel Almirall and Inbal (Billie) Nahum-Shani, fit squarely within the MOST framework. I wish we had realized this sooner, partly because now it seems so obvious, and partly because our publications before about 2012 reinforce the mistaken idea that MOST and SMART are two nearly unrelated approaches. I sincerely hope this book and the companion volume help to integrate MOST and SMART in the minds of readers. More work remains to be done on this integration.

I do not claim that all of the ideas in this book are new or original, although I hope some of them will be new to some readers. Previous authors, for example Yeaton and Sechrest (1981), have called for a better understanding of the effectiveness of individual intervention components. The work of Stephen West and Leona Aiken (e.g., West & Aiken, 1997; West, Aiken, & Todd, 1993), including an excellent presentation by Steve West I attended in the mid-1990s, has stimulated my thinking. In particular, West et al. (1993) is the first time I know of that the idea of optimizing behavioral interventions appears in the scientific literature. The MOST framework is adapted from standard operating procedures in wide use in engineering. I have cited Wu and Hamada (2011) many times in this book and other writing, but these citations are an inadequate reflection of how much their book, which I consider a modern classic, has influenced me.

Some readers may find a few of the ideas in this book controversial—for example, the position taken on hypothesis testing when selecting components and component levels for the optimized intervention. My view is that none of the ideas in this book are written in stone, and I would be extremely gratified if in some small way this book helps to stimulate discussion that ultimately will move the science of optimization of interventions forward.

Some Examples of Implementations of MOST

This book will demonstrate that the MOST approach can be implemented without the need for much of an increase in the level of resources that is devoted today to the classical approach. However, a realignment in how those resources are spent, as compared to how resources are typically spent in the classical approach, will be necessary in most cases. The ideas presented in this book and the companion volume have been successfully implemented to develop interventions in a variety of areas and in a variety of settings. In fact, the list of applications of MOST and related ideas to optimize interventions is growing rapidly. I will list a few examples here of studies that have used or are using the MOST framework to develop an intervention.

A team of investigators led by Michael Fiore and Timothy Baker at the University of Wisconsin, and including Megan Piper, Robin Mermelstein (University of Illinois, Chicago), and me, conducted several optimization trials examining components of a clinic-based intervention for adult smoking cessation. More about this project, which was funded by the National Cancer Institute (part of the United States National Institutes of Health), can be found in Baker et al. (2011, 2016), Collins et al. (2011), Cook et al. (2016), Piper et al. (2016), and Schlam et al. (2016).

Connie Kasari at the University of California, Los Angeles, and her collaborators developed an adaptive intervention aimed at improving communication skills in non-verbal schoolchildren. More about this project, which was funded by Autism Speaks, can be found in Kasari et al. (2014).

Linda Caldwell and Edward Smith at Penn State, along with their collaborators, including me, investigated three factors hypothesized to affect the fidelity of delivery of a school-based intervention aimed at preventing drug abuse and HIV in South Africa. More about this project, which was funded by the National Institute on Drug Abuse (part of the United States National Institutes of Health), can be found in Caldwell et al. (2012).

Bonnie Spring at Northwestern University and I, along with a team of collaborators, conducted an optimization trial examining several components of a weight-loss intervention for overweight adults. More about this project, which was funded by the National Institute of Diabetes and Digestive and Kidney Diseases (part of the United States National Institutes of Health), can be found in Pellegrini et al. (2014, 2015).

Amy Kilbourne, Daniel Almirall, and their collaborators at the University of Michigan developed an adaptive strategy to enhance the implementation of an evidence-based mental health intervention. More about this project, which was funded by the National Institute on Mental Health (part of the United States National Institutes of Health), can be found in Kilbourne et al. (2014).

At this writing, a team of investigators including Kari Kugler, Kate Guastaferro, and me at Penn State, and David Wyrick, Jeffrey Milroy, and Amanda Tanner at the University of North Carolina, Greensboro, are conducting a series of optimization trials to examine components of an online intervention to prevent excessive alcohol use and risky sex in college students. More about this project can be found in the Kugler, Wyrick, Tanner, Milroy, Chambers, Ma, Guastaferro, and Collins chapter in the companion volume.

Also at this writing, a team led by Dr. Marya Gwadz at New York University and I are conducting an optimization trial examining several components of an intervention to persuade HIV-positive individuals who are not currently on antiretroviral therapy (ART) to engage in the healthcare system, start taking ART, and reduce their viral loads. More about this project can be found in Gwadz and colleagues (2017).

These are just a few examples chosen to illustrate several points. First, it is possible to obtain funding to conduct research to optimize interventions, and this funding can be obtained from both government and private sources. Second, interventions in any domain can be optimized. The above examples are in the areas of smoking cessation, weight loss, drug abuse and HIV prevention, mental health, learning disabilities, and health care services; there are projects in many other areas. Third, optimization can be aimed not only at the content of the intervention, but also at participant involvement and adherence or implementation quality and fidelity.

Objective and Chapters

The objective of this book and the companion volume is to provide readers with the background needed to use MOST to develop and evaluate optimized interventions. The present book offers a comprehensive introduction to MOST. It also provides an orientation to what might be called the MOST mindset, which is a perspective on, and approach to, intervention research that is different from how most of today's intervention scientists have been trained.

Chapter 1 provides a conceptual overview of MOST. In writing this book, my biggest struggle was determining the order in which to present the material. For example, as you may see if you read the book from start to finish (which I recommend), selection of an experimental design for the optimization trial can seem abstract without a sense of how the results are to be used later in decision-making. Yet, decision-making cannot be covered before experimental design, because to discuss decision-making it is necessary to understand what kind of experimental results the decision-making is to be based on. Chapter 1 is my attempt to deal with this dilemma by helping the reader see how the topics covered in this book are pieces that fit together and form a coherent whole.

Chapter 2 covers the first of the three phases of MOST, the preparation phase. This chapter emphasizes the importance of a well-specified conceptual model and discusses how the investigator can specify such a model. This model then guides subsequent decisions that are made in MOST. This chapter also covers the role of pilot testing in MOST and introduces the concept of the optimization criterion.

Chapters 3, 4, 5, 6, and 7 are concerned with the optimization phase of MOST. Chapters 3 and 4 introduce the factorial experiment. Particular care is taken with this introduction, because factorial designs and their close relatives are important tools in the optimization phase of MOST. An appropriately designed factorial experiment can produce a high yield of scientific information in the optimization phase. Yet it has been my experience that there are pervasive misunderstandings among intervention scientists about factorial experiments. For example, some years ago I gave a presentation about MOST at a highly-regarded medical school in the United States, in which I discussed the idea of using a factorial experimental design for the optimization trial. I was invited to have lunch with their junior intervention scientists after the presentation, and gladly accepted. During lunch one of the junior scientists said she was very taken with the idea of MOST and could readily see what it offered her research area, but she would probably never use it. When I asked why, she said her department head had told them all that he would never permit a grant proposal using a factorial experiment to be submitted by anyone in the department, because in his view it is not practical to power factorial experiments. If, like that eminent scientist, you believe this mistaken notion about factorial experiments now, I hope after reading Chap. 3 you are convinced that factorial experiments can be highly efficient when properly conducted and analyzed.

Chapter 4 contains an extensive discussion of the interaction, which is an important feature of factorial experiments. Why include a separate chapter on interactions in this book? Interactions are a complex topic, and in my experience, there is considerable confusion about them. This confusion concerns interpretation of interactions, whether main effects can be interpreted if interactions are present, and whether it is possible to power an experiment so that it has a reasonable chance of detecting an interaction if it is present. Perhaps due to this confusion, some investigators may be reluctant to undertake a factorial experiment and prefer simpler designs that do not involve estimation of interactions. This book takes the opposite perspective and proposes that to understand what works, why, and for whom, it is necessary to examine interactions. Thus, interactions are critically important in science; in fact, it can be argued that behavioral, biobehavioral, and biomedical science can advance only so far without incorporating the concept of interactions into theory and gathering reliable empirical information on interactions. From this perspective, interactions merit more attention than they have typically received in empirical investigations, particularly in intervention science.

Chapter 5, which is probably the most technical in the book, discusses reduced factorial designs, highlighting the fractional factorial design. Fractional factorial designs can be very economical and efficient in situations where it is desired to reduce the number of experimental conditions that must be implemented. Fractional factorial designs make exactly the same overall sample size requirements as complete factorial designs, but require implementation of fewer experimental conditions–usually half or fewer. The choice of which conditions to include in the design is made solely on statistical grounds, to preserve important properties of the factorial experiment. Fractional factorial designs require certain assumptions that may or may not be plausible in a given situation; these are reviewed in the chapter. Fractional factorial designs are not for every situation, but under the right circumstances they can enable intervention scientists to do more with less, and so they merit consideration alongside other design options.

Chapter 6 discusses how to be a good manager of research resources by selecting the most efficient experimental design that provides the desired scientific information. Different experimental designs make different resource demands. Some require more experimental subjects but fewer experimental conditions; others require fewer experimental subjects but more experimental conditions. Depending on which is more costly, adding subjects or adding experimental conditions, different designs may cost very different amounts. Chapter 6 also covers some practical issues related to implementation of factorial experiments in field settings.

Chapter 7 starts with the premise that an investigator has conducted an optimization trial using a factorial or fractional factorial design, properly analyzed the data, and obtained results. These results are to form the basis for making the necessary decisions about the composition of the optimized intervention. Chapter 7 walks through a suggested decision-making process that starts with experimental results and the optimization criterion, and ends with the optimized intervention.

Chapters 1, 2, 3, 4, 5, 6, and 7 emphasize fixed interventions, that is, interventions in which the design calls for providing all participants with the same intervention. Chapter 8 discusses adaptive interventions. In an adaptive intervention, the intervention is altered at critical decision points, according to pre-specified decision rules. This is done to adapt the intervention so that it responds to characteristics of the individual or setting, or to what amount and kind of progress the individual is making over time. The purpose of Chap. 8 is primarily to introduce optimization of adaptive interventions. As mentioned below, two chapters in the companion volume provide a more advanced and detailed treatment of optimization of adaptive interventions.

Because individual chapters of this book are available for downloading, I tried to make each chapter as self-contained as possible. This resulted in some unavoidable redundancy; for example, most chapters begin with a summary of the hypothetical example that is threaded through the book. I apologize if those who read the book from start to finish find this tedious.

Intended Audiences and How to Use This Book

The intended audience for this book includes scientists who develop and evaluate behavioral, biobehavioral, and biomedical interventions; statisticians, biostatisticians, quantitative psychologists, and other methodologists working in intervention science; and trainees preparing for careers in these areas.

I have tried to keep this book relatively non-technical. This preface, Chap. 1, and much of Chap. 2 have been written for a general scientific audience. The rest of the book has been written to be understandable to anyone who has had graduate training in statistics up through multiple regression. Those with more technical backgrounds, such as statisticians, may find the treatment in this book incomplete and wish to do some additional reading on some topics. Examples include fractional factorial designs and the details of multivariate analysis of data gathered via a factorial experiment. A good starting point would be the reference lists in each chapter.

This book is a suitable textbook for an advanced graduate course, provided students have had the necessary training in multiple regression. Instructors also may wish to assign some or all of the chapters in the companion volume.

Some readers may be wondering whether it is necessary to read every chapter. Of course, I recommend that everyone read the entire book! I particularly make this recommendation to investigators who are planning to write a grant proposal featuring MOST or to lead the optimization of an intervention using MOST. However, it is realistic to assume that, depending on an individual's role in intervention science, some chapters may be of more interest than others. Some scientists work in a team where responsibilities are divided among team members. Those who are primarily responsible for development of interventions from a conceptual perspective may be particularly interested in Chaps. 1, 2, and 8. Team members primarily responsible for the methodological aspects of research may be particularly interested in Chaps. 1, 3, 4, 5, 6, and 7.

Some readers may primarily be looking for an overview of MOST. Examples are program officials at granting agencies, such as the National Institutes of Health in the United States, who field inquiries from prospective grantees who wish to include MOST in their proposals; and senior scientists responsible for mentoring junior scientists who are considering using MOST in their work. These individuals will probably find Chaps. 1 and 2 particularly helpful.

Interventions in Different Domains

There is a single hypothetical example threaded through this book for pedagogical purposes, concerning an intervention aimed at improving adherence to antiretroviral therapy (ART) in HIV+ individuals. This would be classified as a behavioral or perhaps biobehavioral intervention (see definitions in Chap. 1 and the glossary). I selected this example because of my background in these types of interventions. Unfortunately, this means two other types of interventions receive less emphasis in the book. One is biomedical interventions, which consist of pharmaceuticals, surgery, physical therapy, and the like. The other is educational interventions, which are an important subset of behavioral interventions. Everything said in this book also applies directly to optimization and evaluation of both biomedical and educational interventions. Factorial experimentation when cluster randomization is necessary is an important topic for optimization of educational interventions. This is not covered in detail in the present volume, but it is covered in the Nahum-Shani and Dziak chapter in the companion volume.

Additional Resources

Each chapter in the companion volume offers a treatment of an advanced topic, written by experts in those areas. The material in the present book provides a necessary foundation for these chapters. A very brief description of the chapters in the companion volume is provided at the end of Chap. 8.

Additional resources can be found at http://methodology.psu.edu/ra/MOST. These resources include material that is supplementary to this book or the companion volume; brief descriptions of applications of MOST in a variety of areas; FAQ in which some of the material in this book is presented in an informal way; suggested readings; and software.

Acknowledgments

In this brief acknowledgments paragraph I wish to thank a few people who were directly instrumental in the preparation of this book. Several cohorts of students in my graduate seminar read previous drafts of many of the chapters. These students, along with hundreds of people who have participated in presentations and workshops I have given, have asked questions that stimulated my thinking or helped me to find a better way to explain something. Donna Coffman, David Conroy, John Dziak, Kate Guastaferro, Kari Kugler, Susan Murphy, and Kelly Rulison gave me extremely helpful comments on an earlier version of the manuscript. John Graham made himself available to talk over some key aspects of the book at critical times. Amanda Applegate's editorial assistance was invaluable throughout the process. Many thanks to all of you.

Concluding Remarks

Above I asked to you imagine a scenario in which interventions are made up exclusively of components of demonstrated effectiveness; interventions are built to meet clearly specified standards and are immediately scalable; translation from basic science into intervention practice is done scientifically and programmatically; every study adds to the knowledge base about what works and how; and interventions become incrementally and steadily more effective, efficient, cost-effective, and scalable over time. In this book I hope to convince you that MOST can help behavioral, biobehavioral, and biomedical scientists make this scenario a reality.

It is my sincere hope that this book and the companion volume will help intervention scientists to think in a new way about development and evaluation of interventions. I hope readers will consider shifting a bit of the focus of their own work away from the evaluation of interventions as a package and toward optimization of interventions to meet specific criteria. I also hope the field as a whole will start to value demonstrable, incremental, and cumulative improvement over time in interventions. My dream is that fifteen years from now, we can all look back to the current state of intervention science and be able to say convincingly, "Today's interventions are much more effective, efficient, economical, and scalable than those were."

Finally, to those who see value in the ideas offered in this book, but are hesitant to implement them because they seem to be too radical a departure from business as usual, I offer this quote from Pythagoras (570 BC – 495 BC; quoted in Edwards, 1891, p. 101):

> *Choose always the way that seems best, however rough it may be, and custom will soon render it easy and agreeable.*

University Park, PA, USA
2017

Linda M. Collins

References

Almirall, D., Nahum-Shani, I., Wang, L., & Kasari, C. (2018). Experimental designs for research on adaptive interventions: Singly and sequentially randomized trials. In L. M. Collins & K. C. Kugler (Eds.), *Optimization of multicomponent behavioral, biobehavioral, and biomedical interventions: Advanced topics* (forthcoming). New York, NY: Springer.

Baker, T. B., Collins, L. M., Mermelstein, R., Piper, M. E., Schlam, T. R., Cook, J. W., . . . Fiore, M. C. (2016). Enhancing the effectiveness of smoking treatment research: Conceptual bases and progress. *Addiction, 111*, 107–116.

Baker, T. B., Mermelstein, R. J., Collins, L. M., Piper, M. E., Jorenby, D. E., Smith, S. S., . . . Fiore, M. C. (2011). New methods for tobacco dependence treatment research. *Annals of Behavioral Medicine, 41*, 192–207.

Caldwell, L. L., Smith, E. A., Collins, L. M., Graham, J. W., Lai, M., Wegner, L., ... Jacobs, J. (2012). Translational research in South Africa: Evaluating implementation quality using a factorial design. *Child and Youth Care Forum, 41,* 119–136.

Centers for Disease Control. (2016b). Retrieved January 23, 2017. https://www.cdc.gov/tobacco/data_statistics/fact_sheets/adult_data/cig_smoking/

Centers for Disease Control and Prevention. (2016a). Retrieved January 23, 2017. https://www.cdc.gov/tobacco/data_statistics/fact_sheets/fast_facts/

Collins, L. M., Baker, T. B., Mermelstein, R. J., Piper, M. E., Jorenby, D. E., Smith, S. S., ... Fiore, M. C. (2011). The multiphase optimization strategy for engineering effective tobacco use interventions. *Annals of Behavioral Medicine, 41,* 208–226.

Collins, L. M., & Kugler, K. C. (Eds.). (2018). *Optimization of multicomponent behavioral, biobehavioral, and biomedical interventions: Advanced topics.* New York, NY: Springer.

Collins, L. M., Kugler, K. C., & Gwadz, M. V. (2016). Optimization of multicomponent behavioral and biobehavioral interventions for the prevention and treatment of HIV/AIDS. *AIDS and Behavior, 20,* 197–214.

Collins, L. M., Murphy, S. A., Nair, V. N., & Strecher, V. J. (2005). A strategy for optimizing and evaluating behavioral interventions. *Annals of Behavioral Medicine, 30*(1), 65–73.

Cook, J. W., Collins, L. M., Fiore, M. C., Smith, S. S., Fraser, D., Bolt, D. M., ... Mermelstein, R. (2016). Comparative effectiveness of motivation phase intervention components for use with smokers unwilling to quit: A factorial screening experiment. *Addiction, 111,* 117–128.

Edwards, T. (1891). *A dictionary of thoughts; being a cyclopedia of laconic quotations from the best authors, both ancient and modern.* New York, NY: Cassell Publishing Company.

Finkelstein, E. A, Trogdon, J. G, Cohen, J. W, & Dietz, W. (2009). Annual medical spending attributable to obesity: Payer- and service-specific estimates. *Health Affairs, 28*(5), w822–w831.

Gwadz, M. V., Collins, L. M., Cleland, C. M., Leonard, N. R., Wilton, L., Gandhi, M., ... Ritchie, A. S. (2017). Using the multiphase optimization strategy (MOST) to optimize an HIV care continuum intervention for vulnerable populations: A study protocol. *BMC Public Health, 17,* 383.

Kasari, C., Kaiser, A., Goods, K., Nietfeld, J., Mathy, P., Landa, R., ... Almirall, D. (2014). Communication interventions for minimally verbal children with autism: A sequential multiple assignment randomized trial. *Journal of the American Academy of Child & Adolescent Psychiatry, 53*(6), 635–646.

Kilbourne, A. M., Almirall, D., Eisenberg, D., Waxmonsky, J., Goodrich, D. E., Fortney, J. C., ... Thomas, M. R. (2014). Adaptive implementation of effective programs trial (ADEPT): Cluster randomized SMART trial comparing a standard vs. enhanced implementation strategy to improve outcomes of a mood disorders program. *Implementation Science, 9,* 132–146.

Kugler, K. C., Wyrick, D. L., Tanner, A. E., Milroy, J. J., Chambers, B., Ma, A., & Collins, L. M. (2018). An iterative approach to building an optimized STI prevention intervention aimed at college students: The important of the conceptual model. In L. M. Collins & K. C. Kugler (Eds.), *Optimization of multicomponent behavioral, biobehavioral, and biomedical interventions: Advanced topics* (forthcoming). New York, NY: Springer.

Nahum-Shani, I., & Dziak, J. J. (2018). Multilevel factorial designs in intervention development. In L. M. Collins & K. C. Kugler (Eds.), *Optimization of multicomponent behavioral, biobehavioral, and biomedical interventions: Advanced topics* (forthcoming). New York, NY: Springer.

Pellegrini, C. A., Hoffman, S. A., Collins, L. M., & Spring, B. (2014). Optimization of remotely delivered intensive lifestyle treatment for obesity using the multiphase optimization strategy: Opt-IN study protocol. *Contemporary Clinical Trials, 38,* 251–259.

Pellegrini, C. A., Hoffman, S. A., Collins, L. M., & Spring, B. (2015). Corrigendum to "Optimization of remotely delivered intensive lifestyle treatment for obesity using the multiphase optimization strategy: Opt-IN study protocol." *Contemporary Clinical Trials, 45,* 468–469.

Piper, M. E., Fiore, M. C., Smith, S. S., Fraser, D., Bolt, D. M., Collins, L. M., ... Baker, T. B. (2016). Identifying effective intervention components for smoking cessation: A factorial screening experiment. *Addiction, 111,* 129–141.

Rivera, D. E., Hekler, E. B., Savage, J. S., & Downs, D. S. (2018). Intensively adaptive interventions using control systems engineering: Two illustrative examples. In L. M. Collins & K. C. Kugler (Eds.), *Optimization of multicomponent behavioral, biobehavioral, and biomedical interventions: Advanced topics* (forthcoming). New York, NY: Springer.

Schlam, T. R., Fiore, M. C., Smith, S. S., Fraser, S., Bolt, D. M., Collins, L. M., ... Baker, T. B. (2016). Comparative effectiveness of intervention components for producing long-term abstinence from smoking: A factorial screening experiment. *Addiction, 111,* 142–155.

West, S. G., & Aiken, L. S. (1997). Toward understanding individual effects in multicomponent prevention programs: Design and analysis strategies. In Bryant, K. J., Windle, M., & West, S. G. (Eds.), *The science of prevention: Methodological advances from alcohol and substance abuse research* (pp. 167–209). Washington, DC: American Psychological Association.

West, S. G., Aiken, L. S., & Todd, M. (1993). Probing the effects of individual components in multiple component prevention programs. *American Journal of Community Psychology, 21,* 571–605.

Wu, C. J., & Hamada, M. S. (2011). *Experiments: Planning, analysis, and optimization.* Hoboken, NJ: Wiley.

Yeaton, W. H., & Sechrest, L. (1981). Critical dimensions in the choice and maintenance of successful treatments: Strength, integrity, and effectiveness. *Journal of Consulting and Clinical Psychology, 49,* 156.

Contents

1 Conceptual Introduction to the Multiphase Optimization Strategy (MOST)... 1
  1.1 Introduction.. 2
    1.1.1 An Approach Inspired by Ideas From Engineering..... 3
    1.1.2 The Objective: Optimized Rather Than Best.......... 4
    1.1.3 The Kind of Information Needed for Optimization.... 5
    1.1.4 Using Research Resources Strategically to Obtain the Needed Information.................. 6
  1.2 Optimizing an Intervention: A Brief Hypothetical Example..... 7
    1.2.1 The MOST Perspective............................... 8
    1.2.2 Gathering the Information Needed to Optimize the Intervention.................................... 10
    1.2.3 From Experimental Results to Optimized Intervention....................................... 11
  1.3 Definition of Optimization of an Intervention................ 12
    1.3.1 Optimization Is a Process: The Continual Optimization Principle............................. 12
    1.3.2 Four Desiderata: Effectiveness, Efficiency, Economy, and Scalability.......................... 14
    1.3.3 Trade-Offs Among the Desiderata................... 15
  1.4 The Resource Management Principle......................... 16
  1.5 Some Differences Between the Classical and MOST Perspectives.. 17
  1.6 Definitions of Some Important Terms........................ 20
    1.6.1 Design.. 20
    1.6.2 Component... 22
    1.6.3 Multicomponent Interventions...................... 23
  1.7 The MOST Framework: The Three Phases....................... 24
    1.7.1 The Preparation Phase............................. 24

xxiii

		1.7.2 The Optimization Phase..........................	25
		1.7.3 The Evaluation Phase............................	28
	1.8	Reasons for Returning to the Preparation Phase...............	28
		1.8.1 Returning to the Preparation Phase Immediately After the Optimization Phase.....................	28
		1.8.2 Returning to the Preparation Phase After the Evaluation Phase.............................	30
	1.9	The Distinction Between Optimization and Evaluation.........	31
	1.10	A Different Way of Thinking About Intervention Research: MOST-Induced Dilemmas...............................	31
	1.11	What's Next..	33
	References..		34
2	**The Preparation Phase of MOST**...........................		35
	2.1	The Conceptual Model: The "Engine" that Drives the Intervention......................................	36
		2.1.1 Specifying the Conceptual Model..................	37
		2.1.2 Targeted Populations and Participant Heterogeneity....	38
	2.2	Using a Figure to Represent a Conceptual Model.............	38
		2.2.1 A Hypothetical Example.........................	39
		2.2.2 Tracing the Causal Chain.........................	42
	2.3	Using the Conceptual Model to Select Outcome Variables for the Optimization and Evaluation Phases of MOST..........	43
		2.3.1 When the Outcome of Primary Interest Is Far in the Future..................................	44
	2.4	The Living Conceptual Model...........................	45
	2.5	Correspondence Between Components and Target Mediators...	46
		2.5.1 The Granularity of Components...................	49
	2.6	Meta-Analysis and the Conceptual Model..................	51
	2.7	Including Moderation (Effect Modification) in a Figure Representing a Conceptual Model........................	51
	2.8	Why Bother With a Conceptual Model?....................	53
	2.9	The Role of Pilot Testing in MOST.......................	54
	2.10	The Optimization Criterion..............................	55
		2.10.1 Reasons for Identifying the Optimization Criterion in the Preparation Phase Rather Than the Optimization Phase..........................	56
		2.10.2 The Role of the Screened-In Set in Optimization.......	56
		2.10.3 All Active Components..........................	57
		2.10.4 Optimization Criteria Involving Specific Constraints....	58
		2.10.5 Overview of Incorporating Constraints Into the Optimization Criterion.........................	58
		2.10.6 Other Kinds of Constraints.......................	60
		2.10.7 Constraints on Multiple Resources.................	61

2.11		Identifying Key Constraints	62
2.12		What's Next	62
References			63

3 Introduction to the Factorial Optimization Trial 67
3.1 Introduction 68
3.2 The Basics of the Factorial Experiment 70
3.2.1 Intervention Components vs. Factors 71
3.3 Scientific Information Provided by the Factorial Experiment: Main Effects and Interactions 72
3.3.1 Main Effects 73
3.3.2 Two-Way Interactions 74
3.3.3 Three-Way and Higher-Order Interactions 75
3.3.4 The Importance of Interactions in Building an Intervention 76
3.4 The Balance Property and the Efficiency of Factorial Experiments 76
3.4.1 Criteria that Define Balance 76
3.4.2 Balance and Efficiency 77
3.5 The Concept of Experimental Control 79
3.5.1 The Concept of Control in the RCT and the Factorial Experiment 79
3.5.2 Different Control for Different Factors in a Factorial Experiment 80
3.6 Including a Constant Component in a Factorial Experiment 81
3.7 Translating Intervention Components Into Factors and Levels of Factors 83
3.7.1 Ensuring the Factors Can Be Fully Crossed 83
3.7.2 Ensuring the Factors Can Be Manipulated Independently 84
3.8 When Subjects are Clustered 85
3.9 A Very Brief Review of Statistical Power 87
3.10 Conclusion-Priority and Decision-Priority Perspectives on Research 88
3.10.1 Conclusion-Priority Perspective vs. Decision-Priority Perspective 88
3.10.2 When to Take Each Perspective 93
3.11 The General Linear Model (GLM) Approach to Classical Factorial Analysis of Variance (ANOVA) 94
3.11.1 Effect Coding 94
3.11.2 Dummy Coding 97
3.11.3 Why Effect Coding Is Preferred for Analysis of Data from Factorial Optimization Trials 97

	3.12	Powering a Factorial Experiment: Main Effects............	99
		3.12.1 Power and Factorial Experiments................	99
		3.12.2 Identifying Effect Sizes for Power Analysis From the Conclusion-Priority and Decision-Priority Perspectives...................................	100
		3.12.3 Comparison of Two Putatively Active Levels of a Component.............................	100
		3.12.4 Demonstration of Power Analysis for a Factorial Experiment.................................	101
		3.12.5 The Negligible Impact on Power of Adding a Factor to a Factorial Experiment.......................	102
		3.12.6 The Large Impact on Power of Adding a Level to a Factor: Why 2^k Factorial Optimization Trials are Recommended.............................	105
		3.12.7 Recommendations Based on the Resource Management Principle.........................	107
		3.12.8 Small N Situations................................	109
	3.13	The Coefficient Correction: Why Is There a 2 in the Denominator of the Two-Way Interaction?...........	109
	3.14	Summary: The Efficiency of the Factorial Experiment and the Efficiency of the RCT...........................	111
	3.15	What's Next?...	112
	References...	113
4	**Interactions Between Components and Moderation of Component Effects**................................	115
	4.1	Introduction...	116
	4.2	Interactions and Moderation............................	117
	4.3	Definition of the Interaction Effect in Factorial ANOVA.......	118
		4.3.1 Review of the Classical Definition of the Interaction....	118
	4.4	Interpreting Interactions by Plotting.....................	119
		4.4.1 Plots of Means Where There Is No Interaction........	119
		4.4.2 Synergistic and Antagonistic Interactions............	123
		4.4.3 Synergistic Two-Way Interactions.................	123
		4.4.4 Antagonistic Two-Way Interactions................	125
		4.4.5 Higher-Order Interactions.......................	128
	4.5	The Role of Main Effects and Interactions in Decision-Making: Effect Hierarchy, Effect Sparsity, and Effect Heredity...................................	130
		4.5.1 A Brief Reminder.............................	131
	4.6	A Decision-Priority Perspective on Interactions.............	132
		4.6.1 Interactions and Decision-Making.................	132
		4.6.2 Implications for Powering Factorial Experiments.......	133
	4.7	What Is the Statistical Power for Detection of Interaction Effects as Compared to Main Effects?.....................	134

Contents

	4.7.1 Power Is Identical for Main Effects and Interactions With Identical Regression Weights in 2^k Experiments, but Not Necessarily in Other Experiments	134
	4.7.2 Cohen and Fleiss Use Different Definitions of "Interaction"	135
	4.7.3 The Ability to Detect Interactions Depends Partly on the Expected Effect Size of the Interaction, but Who Knows What to Expect?	138
4.8	Moderation of the Effects of Factors by Observed Variables (Effect Modification)	138
	4.8.1 Examining Moderation Based on Naturally Varying Observed Moderators	139
	4.8.2 Examining Moderation by Including the Moderator as a Factor in the Experiment	140
	4.8.3 Moderation of Component Effects and the Continual Optimization Principle	142
4.9	What's Next?	142
References		143

5 Balanced and Unbalanced Reduced Factorial Designs 145

5.1	Introduction to Balanced Reduced Designs (i.e., Fractional Factorial Designs)	146
5.2	Hypothetical Example of a Fractional Factorial Design	147
	5.2.1 Some Interesting Aspects of Fractional Factorial Designs	149
	5.2.2 Notation for Fractional Factorial Designs	151
5.3	What's the Catch? Aliasing	151
	5.3.1 A Small Example of Combining of Effects	152
	5.3.2 Confounding and Aliasing	155
5.4	Overview of Rationale and Strategy	155
	5.4.1 Rationale: Targeting Resources to Scientifically Important Effects	155
	5.4.2 Strategy of Selecting a Design That Deliberately Aliases Effects	157
5.5	More About the Strategy of Deliberately Aliasing Effects	158
5.6	Clear and Strongly Clear Effects	159
5.7	Two Key Characteristics of Fractional Factorial Designs	160
5.8	Trade-Offs to Consider When Selecting a Fractional Factorial Design	161
	5.8.1 Resolution and Resource Requirements	162
	5.8.2 Bundling of Effects and Resource Requirements	164
5.9	Aliasing: What to Look For	165
5.10	Selecting a Fractional Factorial Design Using Software	166
	5.10.1 Overview: Different Approaches to Using Software to Select a Fractional Factorial Design	166

		5.10.2	Selecting a Design by Specifying the Desired Number of Experimental Conditions....................	166
		5.10.3	Selecting a Design by Specifying a Minimum Desired Resolution............................	170
		5.10.4	Selecting a Design by Specifying Both the Desired Number of Experimental Conditions and a Desired Resolution....................................	170
		5.10.5	Selecting a Design by Specifying Which Effects Are in Which Categories........................	171
	5.11	The Term "Experimental Design" as Used by Statisticians and as Used by Intervention Scientists...................		172
		5.11.1	Switching Factor Level Labels....................	173
		5.11.2	Permuting the Order in Which Factors Are Listed......	174
		5.11.3	When the Investigator Desires to Omit a Particular Experimental Condition.......................	175
	5.12	Fractional Factorial Designs With Clustered Data............		177
	5.13	Following the Resource Management Principle When Considering a Fractional Factorial Design.................		178
		5.13.1	General Recommendations......................	179
		5.13.2	Some Specific Recommendations..................	181
		5.13.3	Resources for Further Study.....................	181
	5.14	Unbalanced Reduced Factorial Designs, Interactions, and Aliasing.....................................		181
		5.14.1	Familiar Experimental Designs as Unbalanced Reduced Factorial Designs......................	182
		5.14.2	Aliasing in Individual Experiments and MACEs........	184
		5.14.3	What Is the Effect of a Factor?...................	187
	5.15	A Few Points to Remember About Fractional Factorial Designs.......................................		189
	5.16	What's Next.....................................		190
	References...			190
6	**Gathering Information for Decision-Making in the Optimization Phase: Resource Management and Practical Issues**...............			**193**
	6.1	Introduction.....................................		194
	6.2	Selecting an Experimental Design Based on the Resource Management Principle................................		195
		6.2.1	Cost......................................	196
		6.2.2	Scientific Yield..............................	196
	6.3	Number of Experimental Conditions and Number of Subjects Required by Individual Experiments, MACEs, Factorial Experiments, and Fractional Factorial Experiments....		197
		6.3.1	Number of Experimental Conditions................	197
		6.3.2	Sample Size................................	199

	6.4	Costs of Conducting an Experiment	200
		6.4.1 Per-Subject Costs	201
		6.4.2 Per-Condition Overhead Costs	202
		6.4.3 Constant Overhead Costs	202
		6.4.4 Scenarios Illustrating Different Costs	203
	6.5	Identifying the Least Expensive Experimental Design When Exact Costs Are Unknown	207
	6.6	Different Experiments Estimate Different Quantities: The Scientific Yield of an Experimental Design	209
		6.6.1 The Decision-Priority Perspective and the Resource Management Principle	211
	6.7	Type of Outcome Variable	212
	6.8	Conducting Random Assignment	213
		6.8.1 Simple Random Assignment	213
		6.8.2 Stratified Random Assignment	214
	6.9	Avoiding and Dealing with Protocol Deviations	215
		6.9.1 Training and Supervising Staff	216
		6.9.2 Dealing with Unanticipated Disruptions to the Experimental Design	217
		6.9.3 The Robustness of Factorial Experiments	219
		6.9.4 The Aviation Approach to Experimentation	220
	6.10	Avoiding Contamination Across Experimental Conditions	221
	6.11	Registry of Optimization Trials	223
	6.12	What's Next	224
	References		224
7	**The Completion of the Optimization Phase**		**227**
	7.1	Introduction	228
	7.2	Overview of the Decision-Making Process	231
	7.3	Some Fundamentals	233
		7.3.1 Effect Hierarchy, Effect Sparsity, and Effect Heredity Revisited	233
		7.3.2 A Decision-Making Approach Based on Effect Coding	234
	7.4	Step 1: Identify the Important Main Effects and Interactions	234
		7.4.1 Defining What Constitutes an Important Effect: Main Effect and Interaction Criteria	234
		7.4.2 Selecting the Criteria	236
		7.4.3 The Main Effect and Interaction Criteria and Statistical Power	237
		7.4.4 Example of Step 1	237
	7.5	Step 2: Divide the Candidate Components Into the Screened-In and Screened-Out Sets	239
		7.5.1 When Provisional Decisions Are Likely to Be Reversed	240

	7.5.2	Reconsidering Provisional Decisions Based on Interactions Involving PEER	242
	7.5.3	Reconsidering Provisional Decisions Based on Interactions Involving SKILLS	245
	7.5.4	Reconsidering Provisional Decisions Based on Interactions Involving MI	245
	7.5.5	Summary of Selection of the Screened-In and Screened-Out Sets of Components	246
	7.5.6	A Note About the Lower Level of Components	247
	7.5.7	Orphan Interactions	247
7.6	Step 3: Apply the Optimization Criterion	248
	7.6.1	The All Active Components Criterion	248
	7.6.2	The Need for Other Optimization Criteria	250
	7.6.3	Constraints on Money	252
	7.6.4	Constraints on Time	253
	7.6.5	A Note on Cost and the All Active Components Criterion	254
	7.6.6	Constraints on Multiple Resources	254
	7.6.7	Four Different Optimization Criteria, Four Different Interventions...Which One Is Best?	255
	7.6.8	When a Component From the Screened-In Set Must Be Omitted From the Optimized Intervention	257
	7.6.9	The Shortcomings of Ad Hoc Modifications	257
	7.6.10	Reusing the Results of an Optimization Trial With Different Constraints	258
	7.6.11	A Note on the Estimation of Cost	259
	7.6.12	The Possibility of a Bayesian Approach	259
7.7	When There Is More Than One Outcome Variable	260
	7.7.1	When Measures of Mediators Are Used as Short-Term Outcomes	260
	7.7.2	When There Is More Than One Primary Outcome	261
7.8	Why Bother?	262
7.9	After the Decision-Making Is Complete	263
	7.9.1	Secondary Analysis and Qualitative Data	263
	7.9.2	The Evaluation Phase of MOST	263
7.10	What's Next	265
References	265

8 Introduction to Adaptive Interventions 267
8.1	Adaptive Interventions: The Basics	268
	8.1.1	Rationale for Adaptive Interventions	268
	8.1.2	The Anatomy of an Adaptive Intervention	269
8.2	Intensity of Adaptation	272

8.3		Identifying the Components of an Adaptive Intervention and Selecting an Approach for the Optimization Trial	273
	8.3.1	The Sequential, Multiple Assignment, Randomized Trial (SMART)	276
	8.3.2	A Brief Note About Powering a SMART	278
	8.3.3	The Outcome Variable in SMARTs	279
	8.3.4	Two Persistent Sources of Confusion	279
	8.3.5	Optimization Trials for Higher-Intensity Adaptive Interventions	280
8.4		Summary of Selection From the MOST Optimization Phase Toolbox	281
8.5		Some Open Areas and Open Questions	283
	8.5.1	The Preparation Phase of MOST and Adaptive Interventions	283
	8.5.2	The Use of Optimization Criteria With Adaptive Interventions	284
	8.5.3	How Personally Tailored Should an Intervention Be?	284
	8.5.4	Robust Adaptive Interventions	285
8.6		What's Next	285
References			286

Glossary 289

Index 295

Chapter 1
Conceptual Introduction to the Multiphase Optimization Strategy (MOST)

Abstract The multiphase optimization strategy (MOST) is an engineering-inspired framework for development, optimization, and evaluation of behavioral, biobehavioral, and biomedical interventions. This chapter provides a conceptual overview of MOST and discusses how it is different from the classical approach. The focus of the classical approach is on developing an intervention a priori and then evaluating it in a randomized controlled trial (RCT). By contrast, the focus of MOST is on a phased approach: first developing and optimizing an intervention and then evaluating the optimized intervention in an RCT. The optimization is based on carefully conducted and fully powered optimization trials. The objective of MOST is to arrive at an intervention that not only demonstrates effectiveness in an RCT but also is efficient, economical, and scalable. Readers are encouraged to read the preface before this chapter, because it provides a rationale for and orientation to the book.

Contents

1.1	Introduction	2
	1.1.1 An Approach Inspired by Ideas From Engineering	3
	1.1.2 The Objective: Optimized Rather Than Best	4
	1.1.3 The Kind of Information Needed for Optimization	5
	1.1.4 Using Research Resources Strategically to Obtain the Needed Information	6
1.2	Optimizing an Intervention: A Brief Hypothetical Example	7
	1.2.1 The MOST Perspective	8
	1.2.2 Gathering the Information Needed to Optimize the Intervention	10
	1.2.3 From Experimental Results to Optimized Intervention	11
1.3	Definition of Optimization of an Intervention	12
	1.3.1 Optimization Is a Process: The Continual Optimization Principle	12
	1.3.2 Four Desiderata: Effectiveness, Efficiency, Economy, and Scalability	14
	1.3.3 Trade-Offs Among the Desiderata	15
1.4	The Resource Management Principle	16
1.5	Some Differences Between the Classical and MOST Perspectives	17
1.6	Definitions of Some Important Terms	20
	1.6.1 Design	20
	1.6.2 Component	22
	1.6.3 Multicomponent Interventions	23

© Springer International Publishing AG 2018
L. M. Collins, *Optimization of Behavioral, Biobehavioral, and Biomedical Interventions*, Statistics for Social and Behavioral Sciences,
https://doi.org/10.1007/978-3-319-72206-1_1

1.7	The MOST Framework: The Three Phases	24
	1.7.1 The Preparation Phase	24
	1.7.2 The Optimization Phase	25
	1.7.3 The Evaluation Phase	28
1.8	Reasons for Returning to the Preparation Phase	28
	1.8.1 Returning to the Preparation Phase Immediately After the Optimization Phase	28
	1.8.2 Returning to the Preparation Phase After the Evaluation Phase	30
1.9	The Distinction Between Optimization and Evaluation	31
1.10	A Different Way of Thinking About Intervention Research: MOST-Induced Dilemmas	31
1.11	What's Next	33
References		34

1.1 Introduction

This book and its companion volume (Collins & Kugler, 2018) are about MOST, a framework for development, optimization, and evaluation of behavioral, biobehavioral, and biomedical interventions. Here the term *intervention* refers to a program with the objective of improving and/or maintaining human health and well-being, broadly defined. Behavioral interventions use a strategy based on modification of affective, cognitive, or behavioral factors. Biomedical interventions use a strategy based on pharmaceuticals, surgery, and the like. Biobehavioral interventions use a strategy based on both behavioral and biomedical approaches. Interventions may have any of a variety of purposes. Examples include preventing and treating physical and mental health disorders, promoting physical and mental health, improving family functioning, preventing violence, improving learning, and promoting academic achievement. An intervention may be aimed at the individual, family, school, organizational, or community level or at a combination of levels.

The classical approach to developing and evaluating interventions has been about the same for many years. The intervention scientist identifies a set of components that potentially can be included in the intervention; perhaps conducts a pilot study[1] on these components to determine whether they are acceptable to participants,[2] safe, and reasonably practical to implement; and then assembles the components into an intervention package and evaluates the package by means of a randomized controlled trial (RCT). In a typical RCT, subjects are randomly assigned to one of two or more experimental conditions or "arms." In a two-arm RCT, subjects in the

---

[1]This book will use the term pilot study to refer to a study that is aimed at examining feasibility in preparation for a more formal study and not intended for hypothesis testing (Leon, Davis, & Kraemer, 2011).

[2]This book will use the term participant to refer to an individual who is taking part in an implementation of an intervention for clinical, as opposed to research, purposes and the term subject to refer to an individual who is taking part in, and providing data for, a research study.

treatment arm are provided with an intervention package to be evaluated, and subjects in the control arm are provided with a suitable comparison intervention, such as the current standard of care. If the difference between the treatment arm and the control arm is found to be statistically significant, then the intervention package is considered efficacious (if the RCT is conducted under controlled experimental circumstances) or effective (if the RCT is conducted under real-world circumstances). Often the investigator conducts mediation analyses of the data collected during the RCT, with the objective of revealing which variables mediate the effect of the treatment package on the outcome, thereby gaining an understanding of how the intervention operates.

1.1.1 An Approach Inspired by Ideas From Engineering

MOST offers a perspective on intervention development and evaluation that is different from the classical approach, because it has been inspired by ideas from engineering. Let us begin by considering how an industrial engineer might go about developing an improved process for manufacturing vehicle leaf springs, which are a part of the suspension system of most cars and trucks. Manufacturing may appear on the surface to have little relevance to intervention science, but closer examination shows that there are important conceptual similarities between what an industrial engineer sets out to accomplish when developing and evaluating a manufacturing process and what an intervention scientist sets out to accomplish when developing and evaluating an intervention. For now, there is no need to be concerned about details or definitions of terms; that will come later in this chapter and in the remainder of the book. The purpose of this introduction is to impart a conceptual feel for the difference between the way an engineer's training and an intervention scientist's training lead them to approach a problem and to illustrate how and why an engineering-inspired approach may have some value in intervention science.

Suppose the owners of a plant that manufactures leaf springs wish to improve the manufacturing process. There is always some variation in the length of the leaf springs when they come off the assembly line. The owners would like to improve the manufacturing process so that the leaf springs are closer to a particular desired ideal length. An industrial engineer, Dr. E, is brought in to study the problem.

Based on prior research and experience, Dr. E hypothesizes that the following components are critical in the leaf spring manufacturing process: furnace temperature, heating time, time on the conveyor belt, time in the high-pressure press, and range of quench oil temperatures. (This scenario and list of components are very loosely based on Pignatiello and Ramberg, 1985, as described in Wu and Hamada, 2011). In the current process, these components are set to the following levels: low furnace temperature, short heating time, short time on the conveyor belt, short time in the high-pressure press, and lower quench oil temperature range. Dr. E hypothesizes that the manufacturing process would be improved—that is, the leaf springs coming off the assembly line would be closer to the desired length—if the settings were changed so that the furnace temperature was higher; the heating time, time on the conveyor belt, and time in the high-pressure press were longer; and the quench oil temperature range was higher.

If Dr. E had been trained as an intervention scientist, that training might suggest approaching this problem by relying primarily on the RCT. Dr. E would create a new manufacturing process involving higher furnace temperature; longer heating time, time on the conveyor belt, and time in the high-pressure press; and higher quench oil temperature range. Then the new process as a package would be compared to the standard of care, that is, the old process.

However, this is not how Dr. E would proceed. Let us consider why.

1.1.2 The Objective: Optimized Rather Than Best

Imagine Dr. E were to come up with the very best manufacturing process currently possible, one that produces leaf springs so uniform that differences from the target length are virtually undetectable, and proudly presents this to the plant owners. The plant owners ask how much the new manufacturing process costs, and Dr. E replies that the process costs $1 million per spring. From one perspective, this is undeniably an improved manufacturing process. However, from the plant owners' perspective, it is anything but an improvement. They inform Dr. E that to stay competitive in the industry, the company must contain the selling price of the leaf springs, and therefore the manufacturing cost cannot exceed $25 per spring. Thus although the new manufacturing process produces leaf springs that are uniformly close to the desired length, it is useless to the company because of its prohibitive cost. Dr. E would probably be sent packing!

In reality this misunderstanding would be unlikely to happen because, as an industrial engineer, Dr. E knows that the plant owners are primarily interested in a manufacturing process they can actually use. In other words, the objective is to arrive at an *optimized* manufacturing process rather than the best manufacturing process. An optimized manufacturing process represents the best expected manufacturing process that can be obtained *subject to constraints*, for example, constraints on the per spring cost. Soon after being hired, and before starting any research, Dr. E would discuss the key constraints with the plant owners. Constraints may be expressed in terms of money, time, personnel, equipment, or any other quantity. At this time the plant owner would tell Dr. E about the upper bound of $25 per leaf spring in manufacturing costs. Dr. E and the plant owners acknowledge that there may be manufacturing processes that are better in the sense that they produce more uniform leaf springs, but if they cost more than $25 per leaf spring, they are of no more than academic interest in this situation. An optimized manufacturing process will not produce leaf springs that are closest to the ideal that can be obtained in any absolute sense; rather, it will produce leaf springs that are the closest to the ideal that the plant can afford and thus will be genuinely practical.

1.1.3 The Kind of Information Needed for Optimization

Dr. E will conduct experimentation to obtain the information necessary to identify the optimized manufacturing process. In general, only experimental designs that enable Dr. E to address the following questions will be considered:

What are the size and direction of each component's effect?
Which components demonstrate a positive effect? A null effect? A negative effect?
Is the performance of one component affected by the presence or level of one or more other components?

There are several reasons why the answers to these questions are important for development of the optimized manufacturing process.

First, to ensure the efficiency of the new manufacturing process, Dr. E wishes to select only components and component levels that make a positive contribution toward obtaining leaf springs that are as close to the target length as possible. No resources in the new process are to be wasted on components that have very small or null effects, and it is particularly important not to include components and component levels that are counterproductive.

Second, in selecting components and component levels, Dr. E knows it is important to account for how one component may affect the performance of another. The effect of one component may be enhanced or undermined by the presence or level of one or more other components. For example, maybe when the manufacturing process allows more time on the conveyor belt, more time in the high-pressure press has a much larger effect on uniformity in the length of the leaf springs. A component may be effective only when certain other components are included or may be effective *unless* combined with certain other components. For example, maybe if the furnace temperature is set to low, the longer heating time results in leaf spring lengths that are closer to the ideal, but when the furnace temperature is turned to high, the longer heating time does not make any difference.

Third, Dr. E is mindful that different components may be associated with different costs. For example, suppose the plant owners have told Dr. E that the fuel needed to increase the furnace heat is expensive; by contrast, it does not cost much to increase time spent on the conveyor belt. To arrive at the best manufacturing process that can be obtained for a cost of $25 per leaf spring or less, it will be important to take cost differences in components into account. A very expensive component may have to demonstrate a correspondingly large effect on manufacturing precision to be included. This is another reason it will be important to be able to estimate both individual component effects and component interactions. It will also be important to have information on cost.

Fourth, in engineering there is an expectation that improvement is an ongoing process. This means that once the new manufacturing procedure is identified, almost immediately Dr. E will be expected to start working on producing the next one. In this way the manufacturing process will become incrementally better and better over time. To know where to focus efforts to create an even better manufacturing process the next time, one must build on a previously obtained understanding of which components work well, which are weak and could be improved, which are not working at all or are counterproductive and should be abandoned, and which enhance or undermine the effectiveness of others.

1.1.4 Using Research Resources Strategically to Obtain the Needed Information

For all of the reasons listed above, Dr. E needs to be able to estimate the individual effects of components and how the components may affect each other (and, if not already known, the costs associated with each component). Dr. E must take stock of the amount and type of resources the plant owner is willing to spend on research to obtain this information. Resources might be money, time in the plant, technician time, space, materials, etc. Because different experimental designs can make very different resource demands, Dr. E will compare the resource requirements of several different designs, carefully selecting a highly efficient experimental alternative that makes use of whatever resources are available to obtain as much reliable information as possible. The idea is to work strategically to gather the most, and the highest quality, information that can be obtained with the available resources.

In approaching the task of developing the new manufacturing process, Dr. E would be unlikely to select a two-arm RCT or similar experimental design, because it would not yield the needed data. In a two-arm RCT, the components would be manipulated as a set. In the treatment group, all five components would be set to the higher/more intense level, and in the control group, all five would be set to the level specified in the current manufacturing process. In other words, in a two-arm RCT, the components all would be confounded with each other. To obtain the information needed, it is necessary to manipulate the components individually. Rather than using the RCT at this point, Dr. E will look for a highly efficient experimental design that can deliver a lot of scientific information about individual component effects in relation to the resources it requires.

Once Dr. E has selected an experimental design, conducted the experiment, and analyzed the data, the results will provide estimates of the effect of each of the five components and how each component's effects may be impacted by the presence or absence of other components. This information will form the basis for identifying which set of components and component levels produces the best outcome (i.e., leaf springs closest in length to the target) while staying within the limit of $25 per leaf spring in manufacturing costs.

After an optimized manufacturing process has been developed, Dr. E may directly compare its performance to that of the current process, if this is deemed necessary. In this case the final decision about whether or not the new process will replace the current one would be made on the basis of the outcome of this experiment.

1.2 Optimizing an Intervention: A Brief Hypothetical Example

Suppose an intervention scientist, Dr. B, wishes to develop and evaluate an approach for reducing viral load in HIV-positive individuals who drink heavily. Based on prior scientific literature, clinical experience, and a well-specified conceptual model (example presented in Collins, Kugler, & Gwadz, 2016, and reviewed in detail in Chap. 2), Dr. B has identified five intervention components that are hypothesized to be critical in helping HIV-positive individuals to reduce their drinking and improve their adherence to antiretroviral therapy (ART). They are (a) motivational interviewing to engage the participant in the process of examining his or her alcohol use and ART adherence, (b) peer mentoring to provide contact with a positive role model, (c) text messaging to provide access to a support system, (d) mindfulness meditation to improve mental health, and (e) behavioral skills training to improve behavioral skills for managing alcohol use and ART. Suppose there is currently no standard of care for the first four components, but physicians and HIV clinics routinely give HIV-positive individuals who drink heavily and have poor ART adherence a workbook to complete to help improve their behavioral skills.

The training of most of today's intervention scientists would suggest taking the classical approach. The five components would first be pilot tested and then assembled into an intervention package that would be evaluated in a two-arm RCT. Experimental subjects would be randomly assigned to either the treatment arm or the control arm. The treatment arm would receive motivational interviewing, peer mentoring, text messaging, mindfulness meditation, and behavioral skills training. The control arm would receive only the workbook. For the purposes of this example, the primary outcome variable is number of days of adherence to ART in the 30-day period following the conclusion of the intervention. This will be abbreviated *adhere*. After the RCT has been completed and the primary test of intervention efficacy or effectiveness has been performed, mediation analyses would probably be undertaken to provide a better sense of how the intervention worked.

The premise of this book and the companion volume is that the pace of progress toward better interventions can be accelerated by using MOST to develop, optimize, and evaluate interventions. MOST is similar to the approach taken by Dr. E to arrive at an optimized leaf spring manufacturing process but adapted for use in intervention science. Instead of using the classical approach, Dr. B can use MOST to develop, optimize, and evaluate the ART adherence intervention. The next section will illustrate how Dr. B's thinking parallels that of Dr. E. The discussion provides merely a thumbnail sketch; it is superficial and intended only to provide the reader with a conceptual sense of how MOST works. Moreover, it describes only *one way* that an investigator might apply MOST. It is important to realize that because MOST is a framework, not a specific procedure, different applications of MOST can proceed very differently. In particular, they can involve different approaches to experimentation, depending on what is most appropriate and efficient for that application. Much more detail on various approaches that can be used within the

MOST framework is provided in the subsequent chapters in this book and in the companion volume.

1.2.1 The MOST Perspective

Recall that Dr. B has identified five intervention components. Rather than immediately including all of these components in the intervention, Dr. B views the five components as a set of *candidates* to be selected for inclusion in the intervention. Like Dr. E, Dr. B will conduct experimentation to obtain the information necessary to identify the optimized manufacturing process.

Dr. B starts by taking stock of what resources will be available to implement the optimized intervention. Suppose Dr. B contacts several government-funded HIV clinics and engages in extensive discussions about what kind of intervention would be immediately scalable. It is determined that an intervention aimed at HIV-positive individuals who drink heavily that effectively improves adherence to ART and, thereby, reduces viral load would be immediately scalable if it cost no more than $500 per individual to implement. Thus, the clear goal is to develop the most effective intervention that can be implemented for $500 per individual or less. This will serve as Dr. B's definition of the optimized intervention, in other words, the *optimization criterion*.

Next, Dr. B turns to planning the experimentation that will be done to obtain the information needed for optimization. Dr. B is interested primarily in experimental designs that enable estimation of the size and direction of individual component effects and also assessment of whether the effect of one component is affected by the presence or level of any of the other components, for the following reasons.

First, to increase the efficiency of the new intervention, Dr. B wishes to select only components and component levels that are making a positive contribution toward increasing *adhere* (West, Aiken, & Todd, 1993). No resources in the new intervention are to be wasted on components that have very small or null effects, and it is particularly important not to waste resources by including components that appear to decrease *adhere*. The term resources here is broadly defined to include money, staff time or expertise, participant time or burden, space, equipment, or any other quantifiable resource.

Second, in selecting components and component levels, Dr. B knows that it is important to account for how one component may affect the performance of another. The effect of one component may be enhanced or undermined by the presence of one or more other components. For example, maybe use of mindfulness meditation enhances the effect of behavioral skills training because the meditation practice helps the individual be more open to the training.

Third, Dr. B knows that different components may be associated with different resource demands. Therefore, to increase the economy and scalability of the new intervention, it may be important to take these resource demands into account in relation to the contribution a component makes. For example, text messaging may be relatively inexpensive, whereas motivational interviewing by a trained interviewer may be very expensive.

Fourth, like Dr. E, Dr. B knows that improvement is an ongoing process. Once the new ART adherence intervention has been identified, almost immediately Dr. B will start working on producing the next, even better one. To know where to focus efforts to create an even better intervention, Dr. B needs to build on a previously obtained understanding of which components work well, which are weak and could be improved, which do not work at all or are counterproductive and should be abandoned, and which enhance or undermine the effectiveness of others. In this way a coherent base of knowledge about what works and what does not work will be accumulated, and the ART adherence intervention will become incrementally better and better over time with repeated efforts to improve it.

Thus, to obtain the information needed in this phase of research, that is, the information needed to select what components/component levels will make up the optimized intervention, it is necessary to conduct experimentation that enables manipulation of the components individually. Dr. B takes stock of what resources are available to conduct this experimentation. Dr. B is in a somewhat different situation from Dr. E, in that it is necessary to write a grant proposal to obtain funding for the project. The proposal will be subject to whatever limitations the funding source may place on how much money can be requested. Rather than proposing use of the RCT design at this point, Dr. B will look for a highly efficient experimental design that can deliver a lot of scientific information for the resources it requires. Dr. B knows the grant proposal will be viewed as stronger by the review committee if it includes a clear rationale for why the experimental design selected is the one that will make the best use of the funds requested. Therefore, the proposal will contain a comparison of the resource requirements of a few experimental designs that could be used (comparison of resource requirements of different experimental designs is discussed in Chap. 6).

Once Dr. B has selected an experimental design, obtained the necessary funding, conducted the experiment, and analyzed the data, the results will provide estimates of the effect of each of the five components and how each component's effects may be impacted by the presence or absence of other components. This information will form the basis for identifying (a) which components should be eliminated from consideration and (b) out of those that remain, which set of components and component levels meets the optimization criterion, in other words, produces the best outcome on *adhere* without exceeding an implementation cost of $500.

Dr. B remains committed to evaluating the effectiveness of the optimized intervention as a package. Once the optimized intervention has been identified, Dr. B will directly compare its performance to that of a suitable control treatment in an RCT. The results will form the basis for the final decision about whether or not the new intervention is effective.

1.2.2 Gathering the Information Needed to Optimize the Intervention

To obtain the information required to decide what set of components and component levels should be selected to comprise the optimized intervention, Dr. B needs to gather several different kinds of information. Information is needed on the individual performance of each component. Information is also needed on how components perform together. This is represented statistically in interaction terms. Are there synergistic interactions between components—in other words, does one component enhance the effect of another? Are there antagonistic interactions—in other words, does one component undermine the effect of another?

An experiment designed to collect the information needed to optimize an intervention is called an *optimization trial*. In these two companion volumes, many pages are devoted to various approaches to experimentation that are appropriate for an optimization trial. In the example being discussed here, Dr. B needs to screen out components that do not show sufficient effectiveness and then, from the set of components remaining, select which components and component levels should be included in the optimized intervention. Dr. B needs to find an efficient experimental design that will provide the information needed to accomplish this.

For now, let us say that Dr. B decides to conduct a factorial experiment. Dr. B selects this experimental design because, as will be discussed at length in subsequent chapters in this volume, the factorial experiment is a highly efficient and economical way of assessing the performance of individual intervention components and determining how components may affect each other's performance. (The idea that factorial experiments can be highly efficient and economical may be counterintuitive to some intervention scientists. An introduction to factorial experiments, along with an explanation of their efficiency, is provided in Chap. 3). The experiment will include a factor corresponding to each component. Each factor will have two levels. The factors corresponding to the first four components can be set to no, that is, not included in the intervention, or yes, that is, included. The factor corresponding to the behavioral skills component can be at either a low level, consisting of the workbook only (i.e., standard of care), or a high level, consisting of the workbook plus training delivered by a behavioral skills counselor. Dr. B has selected *adhere* as the primary outcome variable.

In the course of conducting the experiment, Dr. B obtains information not only on the primary outcome and any secondary outcomes but also on implementation costs, because cost will be an important consideration when the final components and levels are selected. As explained above, Dr. B seeks an intervention that can be implemented for no more than $500 per person. Data will be collected on the cost of implementation of each component and whether there are economies of scale or additional expenses when certain components are combined.

1.2.3 From Experimental Results to Optimized Intervention

After conducting the factorial experiment, Dr. B analyzes the data using a standard factorial analysis of variance (ANOVA) to obtain estimates of the main effect of each component on *adhere* and of interactions between components. The main effects provide an assessment of the average performance of each component. The interactions provide an assessment of the extent to which components affect each other's performance. The ANOVA results form the basis for selecting components and component levels that will make up the optimized intervention. It is up to the investigator to use the information provided by the ANOVA to make decisions about which components and component levels will form the optimized intervention. This book follows an approach to decision-making based on the one outlined by Collins et al. (2014). The approach is described in Chap. 7.

In the approach to decision-making used in this book, Dr. B would first divide the components into a *screened-in set* and a *screened-out set*. Components are selected for the screened-in set because the experimental evidence suggests they have demonstrated an effect on *adhere* that is large enough to be considered important and is in the desired direction. A component may be selected for inclusion in the screened-in set because in the experiment it demonstrates an effect as an individual component or operates synergistically with one or more other components to enhance their effects. A component is selected for the screened-out set because the results of the experiment suggest it has only a weak effect on *adhere*, has a strong and positive main effect but performs poorly when combined with other effective components, or has an iatrogenic effect (i.e., the lower level performs better than the higher level).

Once the components have been sorted into the screened-in and screened-out sets, the investigator can decide which component levels will make up the optimized intervention. Why is this phrased in terms of selecting component levels rather than components? It may seem more natural to think in terms of selecting components for inclusion in the intervention, but in a sense one is always ultimately selecting component levels. To see why, compare the mindfulness meditation and behavioral skills training components. For the mindfulness meditation component, the two levels of the corresponding experimental factor are no and yes. Choosing the higher level, yes, means this component will be included in the intervention, and choosing the lower level, no, means it will not be included. For the behavioral skills training component, the two levels are workbook only and behavioral skills training plus the workbook. Here choosing the higher level means the intervention will provide both in-person behavioral skills training and a workbook, but choosing the lower level does not mean that behavioral skills training will be omitted from the intervention. Instead, choosing the lower level means that behavioral skills training will consist of the workbook only. This is why this book refers to selection of components and component levels for an intervention, rather than just selection of components. Later in this chapter and in Chap. 2 the concept of a component is discussed in more detail.

Now let us review how the optimized intervention is built by selecting component levels from the screened-in and screened-out sets. The components in the screened-out set will all be set to the lower level in the intervention. The lower level may represent the absence of the component, or it may represent the inclusion of the component at a low level, depending on what levels were examined in the experiment.

Depending on the optimization criterion and the observed results, the optimized intervention may or may not include the higher level of all of the components in the screened-in set. If not for the need to arrive at an intervention that costs no more than $500 to implement, it would be appropriate for Dr. B simply to construct an optimized intervention that includes the higher level of all of the components in the screened-in set. This straightforward approach to optimization is suitable for situations in which cost is not an explicit consideration, but there is still a desire to avoid devoting resources—including participant time and energy—to inactive or counterproductive components.

In Dr. B's case, the optimization criterion calls for identifying the combination of component levels from the components in the screened-in set that is expected to produce the best outcome on *adhere* that can be obtained without exceeding the $500 limit. Dr. B can identify this set by (a) using the results of the ANOVA to arrive at the predicted outcome for each combination of component levels and (b) using the data that were collected on cost to determine which of these combinations can be implemented for no more than $500 per person. (Exactly how to do this will be discussed in Chap. 7.) The combination of components and component levels that produces the best expected outcome on *adhere* without exceeding the $500 cost limit comprises the optimized intervention.

Once the optimized intervention has been identified, it may then be evaluated in an RCT.

1.3 Definition of Optimization of an Intervention

Up to this point, optimization of interventions has been discussed without much specificity, but now a definition of optimization is needed. In this book optimization will be defined as follows:

> Optimization of an intervention is the process of identifying an intervention that provides the best expected outcome obtainable within key constraints imposed by the need for efficiency, economy, and/or scalability.

Let us examine this definition closely.

1.3.1 Optimization Is a Process: The Continual Optimization Principle

The continual optimization principle is one of the fundamental principles of MOST. To understand this principle, consider the development of consumer products, such as automobiles, appliances, and the like. When development has reached a point at

1.3 Definition of Optimization of an Intervention

which the product is ready to market, the engineer does not declare "mission accomplished!" and walk away. Instead, work soon starts on next year's model, which will be better in a measurable, incremental way. For example, it may improve gas mileage by 3 miles per gallon, or include a new safety feature, or be more ergonomic.

The MOST framework provides an opportunity for intervention scientists to engage in the same sort of ongoing improvement. According to the continual optimization principle, optimization is a process of moving toward an ever-better intervention. Although in discussing an intervention that has been developed using MOST it may be correct to say it has been optimized, this does not mean the intervention cannot be improved further; saying an intervention has been optimized merely means it has been through one or more rounds of optimization. In fact, in theory an intervention could be optimized any number of times, with each round of optimization making it better and better by improving its effectiveness, increasing its efficiency or economy, or adapting and updating it in response to different or changing circumstances.

Suppose Dr. B has completed one round of MOST, that is, has optimized the ART adherence intervention and evaluated it in an RCT, and now has an effective intervention that can be implemented for $500 or less. According to the continual optimization principle, while the current optimized intervention is being implemented, Dr. B can use the MOST framework to begin thinking about what subsequent improvements can be made. Dr. B may wish to improve the effectiveness of the previous version of the intervention while staying within the $500 limit, perhaps by testing some new components that have been suggested by recent scientific literature. Or, the goal may be an intervention that comes as close as possible to the previous version but can be implemented for only $400; an intervention that is as effective as the previous version, but in addition to costing no more than $500, takes less than 240 min to complete; or some other quantifiable improvement.

How is Dr. B to proceed to develop the next, incrementally better version of the ART adherence intervention? Because MOST involves investigating the performance of individual intervention components and how components may affect each other's performance, it is straightforward for Dr. B to build directly on previous work. This previous work revealed which components worked well and which did not. This helps to illuminate the way forward. For example, suppose the peer mentoring component had an unacceptably small effect on *adhere* and was not included in the screened-in set, yet for theoretical reasons, peer mentoring remains an important part of the intervention in the eyes of Dr. B. In this case one next step would be to develop a revised peer mentoring component and test it, perhaps along with several other candidate additions to the intervention, in an experiment that enables assessment of the individual effects of each component and interactions between them.

Imagine that in the course of continual optimization, Dr. B and other scientists conduct more and more optimization trials and share the information gained. A body of knowledge will gradually accumulate about what intervention strategies work, under what circumstances, and for whom, in the HIV field and other fields. The knowledge gained by one scientist can be built on by others, so that every new intervention is a measureable improvement on its predecessor. Scientists working in one domain will see what strategies work in others, providing valuable starting points for intervention development in new areas.

This is not intended to imply that it is always necessary to build directly on previous work, although frequently that is arguably the most efficient way to proceed. There are times in engineering and science when a paradigm shift comes along and a wholly new approach is proposed that is qualitatively different from what has gone before. In this case, even though all of the previously tested components may be discarded, it still makes sense to establish that the new approach is a measureable improvement over the old one along one or more specific dimensions.

1.3.2 Four Desiderata: Effectiveness, Efficiency, Economy, and Scalability

In this the desired characteristics of an intervention are reviewed. In the classical approach, the emphasis is primarily on effectiveness. By contrast, in MOST four desiderata for interventions are emphasized: effectiveness, efficiency, economy, and scalability. How much emphasis is placed on each relative to the others depends on the situation.

Effectiveness. Effectiveness is a critically important concept that can be manifested in several different ways. In this book the effectiveness of an intervention or a component of an intervention is defined as the degree to which the intervention or component produces an outcome in the desired direction. It is frequently important to make a distinction between effectiveness and efficacy, as suggested by Flay (1986), who defined these terms as follows:

> Efficacy trials provide tests of whether a technology, treatment, procedure, or program does more good than harm when delivered under optimum conditions... Effectiveness trials provide tests of whether a technology, treatment, procedure, intervention, or program does more good than harm when delivered under real-world conditions. (p. 451)

This is a helpful distinction. This book takes the position that both efficacy and effectiveness are important, but establishing effectiveness is the ultimate goal. MOST can be used in both efficacy and effectiveness trials and can even be used to reduce the loss of intervention potency that is often observed between efficacy and effectiveness trials of the same intervention (e.g., Caldwell et al., 2012). To avoid the tedium of having to repeat "efficacy or effectiveness," this book will simply refer to effectiveness when what is said applies to both and highlight the distinction between efficacy and effectiveness when it is important to do so.

Efficiency, economy, and scalability. Efficiency is the degree to which the intervention produces a good outcome while avoiding wasting money, time, or any other valuable resource. Economy is the degree to which the intervention produces a good outcome without exceeding budgetary constraints (where a budget may be placed not only on money per se but time or any other resource) on implementation and the degree to which it offers a high degree of effectiveness in exchange for the resources required to implement it. Efficiency and economy are closely related but distinct. It is possible for an intervention to be efficient, that is, to be made up primarily of strongly performing components, but not to be economical because it is too expensive to be widely implemented. An intervention may be economical in the sense that it is affordable, but not efficient in the sense that it includes inactive or poorly performing components.

Scalability is defined in this book as the degree to which the intervention can be implemented widely in real-world settings exactly in the form in which it was evaluated, without the need for ad hoc adjustments. In other words, if an intervention is scalable, those tasked with implementing the intervention after its evaluation will not need to take measures to reduce its length, complexity, staff or participant burden, etc. To be scalable an intervention often has to be efficient and economical; sometimes it may need to be simple and straightforward as well. One goal of MOST is to produce interventions that are immediately scalable, that is, that can be implemented as is as soon as their effectiveness has been scientifically demonstrated in an RCT.

As will be discussed in Chap. 7, once an optimization trial has been completed, the information gathered can be used to develop different interventions that are scalable under different circumstances. In the example, Dr. B has established that to be scalable, the intervention must cost no more than $500 per participant. Suppose a health maintenance organization (HMO) that serves a resource-poor community has determined that to be scalable for them, an intervention like the one Dr. B is developing can cost no more than $400 per participant. The results of Dr. B's optimization trial could be used to identify the intervention with the best expected outcome that can be obtained for implementation costs of no more than $400 per person, provided that the results of the optimization trial can reasonably be assumed to generalize to the HMO's patient population.

1.3.3 Trade-Offs Among the Desiderata

MOST demands clarity about the relative importance of the four desiderata because, as will be discussed in detail in Chaps. 2 and 7, this clarity is necessary in order to proceed with optimization. In a public health crisis, say one involving a highly contagious and lethal disease, effectiveness might dwarf all other considerations, at least in the short term until the situation begins to be brought under control. However, in most cases, some or all of the remaining desiderata will be important too.

All else being equal, if there were unlimited resources for implementation—an intervention could cost any amount of money, take any amount of staff and participant time, and consume any amount of other resources—in most cases the result would be a more effective intervention. However, once considerations of efficiency, economy, and scalability are introduced, constraints are imposed that are likely to reduce effectiveness to some extent. Any constraints on the amount of money, time, or other resources available to implement an intervention may mean that an effective component must be eliminated to enable implementation within the allocated budget. This means there is a fundamental tension between effectiveness on the one hand and efficiency, economy, and scalability on the other. From one perspective, optimization is the process of establishing clear trade-offs and thereby resolving this tension.

The definition of optimization refers to the best expected outcome *obtainable within key constraints*. Because of the necessity for trade-offs, this is not the same as the best outcome possible in an absolute sense. Let us return for a moment to the hypothetical example. To arrive at the optimized ART adherence intervention, Dr. B starts with the screened-in set and then looks for the subset of components and component levels that is expected to produce the best outcome on *adhere* that can be obtained for an implementation cost of $500 or less. Depending on the cost associated with each component, to arrive at an intervention that will cost $500 or less, Dr. B may have to reject components that have demonstrated an effect on *adhere*. For example, suppose based on the results of the optimization trial, motivational interviewing is selected for the screened-in set. Further suppose this component is expensive to implement, because of the costs associated with training the interviewers and monitoring their performance and paying them to conduct a lengthy session with each participant. It may be that because of the high cost of motivational interviewing, the combination of components and component levels producing the best expected outcome for no more than $500 will omit this component. Including motivational interviewing would produce a more effective intervention but one that would exceed the implementation budget. As this example illustrates, an intervention that meets necessary standards of, in this instance, economy may not produce the best outcome in an absolute sense.

The quest for the ideal of absolute effectiveness without regard for practical considerations may be quixotic if the resulting intervention is so inefficient, expensive, or complex that it is never implemented widely in its intended form—or is never implemented widely at all. The reality is that nearly every intervention must operate within constraints. Optimization explicitly recognizes both the ideal and this reality and works within the reality to come as close as possible to the ideal.

1.4 The Resource Management Principle

Two fundamental principles underlie MOST. One is the continual optimization principle, discussed above. The other is the resource management principle.

The resource management principle concerns how the investigator uses whatever resources are available for conducting experimentation during the optimization phase. According to the resource management principle, an investigator using

MOST must strive to make the best and most efficient use of available resources when obtaining scientific information. Money is the primary resource because it typically can be exchanged for other resources, but time, personnel, equipment, space, experimental subjects, or anything else needed to obtain the information necessary for intervention optimization can also be considered resources.

Note that the resource management principle specifies *available* resources, meaning resources that the investigator has or can reasonably expect to obtain by, for example, writing a successful grant proposal. Of course, having more resources is always better in any scientific endeavor, because it means the investigator can obtain more and better information. But MOST can be applicable even where resources are limited. Considering the resource management principle can help the investigator take a realistic and strategic look at how best to use limited resources to move intervention science forward. This will be discussed at length in Chap. 6. As the reader of this book will see, although MOST does not necessarily require an increase in resources, in many cases it requires a realignment of resources; in other words, the investigator working within the MOST framework uses resources somewhat differently than an investigator using the classical approach.

1.5 Some Differences Between the Classical and MOST Perspectives

By now it may be evident to the reader that there are some fundamental differences in perspective and priorities between the classical approach and MOST. These are summarized in a conceptual way in Table 1.1. Later in this chapter, there will be a more specific discussion of the three phases of MOST.

The first difference between the approaches is their objectives. The objective of the classical approach is to develop an intervention that demonstrates a statistically and clinically significant effect in an RCT. The objective of MOST is to build an intervention that meets specific predetermined standards and demonstrates a statistically and clinically significant effect in an RCT. As will be discussed further, the predetermined standards must be clearly operationalized. These standards may pertain to any or all of the desiderata: effectiveness, efficiency, economy, and scalability.

An investigator using the classical approach proceeds differently from an investigator who uses MOST. As discussed above, the investigator who uses the classical approach identifies a set of intervention components and after pilot testing immediately assembles the components into an intervention and conducts an RCT to evaluate the intervention as a package. The investigator using MOST does not go directly to an RCT after pilot testing of the intervention components and instead conducts an optimization trial aimed at gathering the information needed to optimize the intervention. Only after the intervention has been optimized does the investigator consider evaluating it in an RCT. How to conduct the research necessary for optimization is a major focus of this book and the companion volume.

Table 1.1 Some differences in perspective between the classical approach and MOST

	Classical approach	MOST
Objective	To develop an intervention that demonstrates a statistically and clinically significant effect in an RCT	To build an intervention that meets specific predetermined standards of effectiveness, efficiency, cost-effectiveness, and/or scalability and demonstrates a statistically and clinically significant effect in an RCT
Next steps after identification and pilot testing of components	Intervention is assembled and then evaluated as a package in an RCT	Optimization trial is conducted; an optimized intervention is built
Experimental designs used	Primarily the RCT	For the optimization trial, experimental designs selected based on resource management principle; for evaluation of intervention as a package, primarily the RCT
Examination of effectiveness of individual components	Conducted primarily via post hoc analyses on data from RCT	Conducted primarily via experimental manipulation of components
Examination of interactions between intervention components	RCT does not permit this	Experimental designs for optimization trial selected to enable this wherever possible
Inclusion of inert or counterproductive components or unnecessarily high component levels	Generally tolerated as long as overall effectiveness of intervention can be demonstrated	Generally not tolerated because this reduces the efficiency of the intervention
Scalability of intervention	Usually dealt with after evaluation of intervention, sometimes via ad hoc modifications	Intervention built for immediate scalability

The two approaches differ in the kinds of approaches to experimentation that are used. The classical approach relies primarily on the RCT and its variants, generally to the exclusion of other kinds of experimental designs. By contrast, MOST explicitly recognizes that it is unrealistic to expect that the RCT, as valuable as it is for evaluation, is always the experimental design of choice for every research question. Thus for use in the optimization trial, which can vary considerably across applications in different settings and content areas, MOST calls for selecting from among a broad array of approaches to identify the one that is most appropriate and efficient for the particular research questions at hand, in other words, to select a design based on the resource management principle. Which is the most appropriate and efficient experimental design will depend on the exact goals of the optimization, the type of intervention that is to be optimized (this is discussed briefly below and in more detail in Chap. 8), the research questions at hand, and the resources available to conduct the research. After the intervention has been optimized, MOST relies on the RCT to address the more circumscribed question of whether the optimized intervention demonstrates a statistically and clinically significant effect.

Perhaps because of its heavy reliance on the RCT, the classical approach generally has focused on establishing whether the intervention as a package has a significant effect, rather than looking inside the intervention to understand whether and how individual components operate. When the effectiveness of individual components has been investigated, the emphasis has been on secondary analyses of the data from RCTs, such as "dose-response" analyses and mediation analyses. Post hoc "dose-response" analyses rely on naturally occurring variation in subject adherence to manipulate the dose. However, this naturally occurring variation is entirely due to self-selection; subjects decide whether to comply with any particular demand of an intervention. Thus, there are many inferential problems with such analyses. Mediation analyses on data from an RCT can reveal which mediators were affected by the treatment package and which, in turn, affected the outcome. However, they cannot reveal which components affected which mediators, so their use in determining which components had an effect is limited. In MOST, by contrast, investigation of the effectiveness of individual components is not coupled with the evaluation of the intervention as a package, but instead is accomplished in an optimization trial involving direct experimental manipulation of components.

Again perhaps because of the heavy reliance on the RCT, interactions between intervention components are seldom examined in the classical approach. This means it is unknown whether a particular component boosts or undermines the effect of another component or whether a set of two or more components should always be included together in an intervention or never included together. In MOST examination of interactions between components under consideration for inclusion in an intervention is a high priority, so experimental designs that enable this are employed wherever possible.

The two approaches also differ in their perspectives on considerations other than the effectiveness per se of the intervention. In the classical approach, inclusion of inert or counterproductive components, or inclusion of unnecessarily high levels of components, is certainly not desired. However, it is tolerated as long as the intervention as a package demonstrates an overall effect. Any adjustments to the intervention to remove inert or counterproductive components would be done after the RCT, probably informed by mediation analyses. By contrast, in MOST there is a low tolerance for inert or counterproductive components and unnecessarily high component levels. In fact, an important purpose of the optimization trial is often to identify such components and component levels so they can be eliminated from the intervention being developed.

As has been discussed, practical constraints frequently influence the success of translation of an intervention to its intended setting in homes, schools, communities, or health care. An intervention that is too expensive, lengthy, complex, or burdensome for participants or staff has little chance of being implemented widely or at least little chance of being implemented widely as it was designed and evaluated. In the classical approach, considerations related to scalability usually are taken most seriously after the intervention has been evaluated. At this point it is too late to do much about scalability. Any significant post-evaluation revision to the intervention to make it cheaper, shorter, simpler, or less burdensome will render it a different intervention from the one that was evaluated. The revised intervention may or may not exert a treatment effect comparable to the original evaluated version. By contrast, in MOST, key factors expected to affect scalability can be built in from the outset, with the goal of immediate scalability of the optimized intervention. For example, suppose clinic staff say that they can devote at best only 30 min from their work day to implement a particular intervention. MOST suggests that under these circumstances, it makes sense to build the intervention to achieve the best expected outcome that can be obtained without demanding any more than 30 min of clinic staff time, so that as soon as it is evaluated, it will be immediately scalable.

1.6 Definitions of Some Important Terms

This book draws on and integrates ideas from intervention science, statistics, and engineering, as well as other fields. Across different fields the same term may be used to refer to different things, and different terms may be used to refer to the same thing. In an effort to maintain clarity, for the purposes of this book and the companion volume, important terms are defined explicitly. Below definitions of several terms are provided; other definitions are provided as they become relevant in later chapters. A glossary of terms appears at the end of this book.

1.6.1 Design

In this book the word "design" will be used in three different ways: experimental design, research design, and intervention design. To avoid confusion, this book and the companion volume will specify intervention design, experimental design, or research design, unless the immediate context makes it clear.

Experimental design refers to the design of an experiment, in this context, to gather information needed to develop, optimize, or evaluate an intervention. Here the word "experiment" refers to manipulation of one or more independent variables for the purpose of empirically observing the effect on an outcome variable. Research design refers more broadly to the specific details of the procedures to be used in a study, such as selection and timing of measures or inclusion criteria for subjects. If the study includes an experiment, then the experimental design is one aspect of the research design.

Intervention design refers to the specific details of the approach taken by an intervention, including the intervention components (e.g., motivational interviewing, peer mentoring), the settings of the components (e.g., two 1-hour sessions of motivational interviewing; weekly half-hour sessions with a peer mentor for 5 weeks), any eligibility requirements for participants (e.g., must be HIV-positive and drink the equivalent of at least 14 grams of pure alcohol at least five times per week), and so on.

1.6 Definitions of Some Important Terms

Fixed vs. Adaptive Intervention Designs Intervention designs fall in two general categories. One is the fixed intervention. In a fixed intervention, all participants are offered the same set of intervention components in a uniform manner. This intervention design does not include any planned variability in the approach, content, or dosage of the intervention across participants. The intervention Dr. B is developing is fixed; all participants are to receive the same intervention components and levels.

Another category of intervention design is the adaptive intervention. The Almirall, Nahum-Shani, Wang, and Kasari (2018) chapter in the companion volume defines adaptive interventions as follows:

> An adaptive intervention is a sequence of pre-specified decision rules that can be used to guide whether, how, or when—and based on which measures—to alter an intervention or intervention component (e.g., treatment type, duration, frequency or amount) at critical decision points during the course of care.

In other words, in an adaptive intervention, the content, dose, or approach of certain aspects of the intervention are varied based on pre-specified decision rules. The decision rules determine how the intervention will be varied in response to measured tailoring variables. For example, suppose an adaptive version of Dr. B's intervention is developed, with the objective of achieving better HIV medication adherence by providing additional treatment to individuals who do not become adherent after an initial treatment. The tailoring variable is ART adherence assessed at 45 days, via a report provided by a medication event monitoring system (MEMS®) cap on the participants' pill bottles. Those who have been at least 80% adherent are considered to be adherent; all others are considered non-adherent. An adaptive intervention might involve the following set of decision rules:

> All participants initially are provided with behavioral skills training, a weekly meeting with a peer mentor for four weeks, and text messaging for the entire duration of the intervention. At 45 days, those who are considered to be adherent step down to daily text messaging only. Those who are considered non-adherent are provided with four more weeks of peer mentoring and one session of motivational interviewing, in addition to the ongoing text messaging.

The difference between fixed and adaptive intervention designs is important, because the experimental design chosen for the optimization trial may be different depending on whether the intervention to be optimized is fixed or adaptive. This will be discussed briefly below, in more detail in Chap. 8, and in still more detail in two chapters in the companion volume: Almirall, Nahum-Shani, Wang, and Kasari (2018) and Rivera, Hekler, Savage, and Downs (2018). The focus of the present volume is primarily fixed intervention designs. Everything that is covered in this book provides a necessary foundation for learning about optimization of adaptive interventions.

1.6.2 Component

Up to now the components of interventions have been discussed in conceptual terms. Let us now be a bit more specific about what is meant by an intervention component.

The definition of intervention component used in this book may seem a bit circular, but it is very practical: an intervention component is any part of an intervention that can be separated out for study. This means that if it can be separated out for experimental manipulation, and doing so will provide the answer to a research question and thereby potentially help improve the intervention, it is a component. Components are the building blocks of interventions.

A *candidate component* is a component from which a level (e.g., no or yes, off or on, low intensity or high intensity) is to be selected for inclusion in an intervention. This term may refer to any type of component.

Content components, the "active ingredients" (Michie et al., 2013, p. 82) of interventions, are perhaps the first type of component that will come to mind for most readers. As the name implies, these components make up the content of an intervention; that is, they are aimed at the behavioral or biological processes being intervened on. For example, in Dr. B's hypothetical ART adherence intervention described above, motivational interviewing, peer mentoring, text messaging, mindfulness meditation, and behavioral skills training are all content components. The components in the example are all behavioral, but content components can be pharmaceutical, medical, or surgical. For example, Piper et al. (2016) examined six components under consideration for inclusion in a smoking cessation intervention. Two of these were pharmaceutical content components: use of a nicotine patch and use of nicotine gum during the 3 weeks preceding the quit date. Investigators looking for a starting point to identify content components for behavioral and biobehavioral interventions are referred to the excellent behavior change taxonomy work of Michie and colleagues (e.g., Michie et al., 2013).

Engagement/adherence components are aimed at ensuring that participants remain engaged in the intervention for its entire duration, carefully follow the intervention's instructions, and otherwise comply with its requirements. For example, an electronically delivered intervention may incorporate a game of some kind, such as providing a brief joke every day the participant meets an intermediate goal. Involvement of community volunteers or peer leaders to provide encouragement and mentoring from someone who has "been there" may be an engagement/adherence component.

Schlam et al. (2016) examined five components under consideration for inclusion in a smoking cessation intervention, three of which concerned adherence to the recommended regimen of nicotine replacement therapy. The components were (a) counseling to promote adherence, (b) automated telephone calls to promote adherence, and (c) electronic medication monitoring with feedback and counseling. Because barriers to participation can interfere with engagement and adherence, some intervention components may focus on removing these barriers. Provision of transportation and child care to enable participation in an intervention are examples of engagement/adherence components. In some cases, the primary purpose of engagement/adherence components may be keeping participants interested in the intervention so that they continue with it.

Fidelity components are aimed at maintaining a high level of fidelity of intervention delivery. Such components are usually aimed at those who deliver the intervention or the environment in which the intervention is to be delivered, rather than the individuals who are the target of the intervention. For example, Caldwell et al. (2012) described a study examining three components hypothesized to affect the fidelity of delivery of the HealthWise intervention, an intervention developed for South African schoolchildren to prevent drug abuse and risky sex. The content of the HealthWise intervention had been evaluated in a previous study. In the Caldwell et al. study, two of the components, (a) enhanced teacher training and (b) structure, support, and supervision, were aimed at the teachers who delivered the intervention. A third component was aimed at (c) enhancing the school environment to make it more welcoming to and supportive of HealthWise.

In an adaptive intervention, there may also be components corresponding to aspects of adaptive intervention design such as decision points, tailoring variables, and decision rules (Collins, Nahum-Shani, & Almirall, 2014). This is discussed further in Chap. 8. In general, it is not necessary to categorize intervention components, and some components can arguably be labeled more than one way. The purpose of this section has been to encourage broad thinking about the types of components that can play different roles in an intervention and to assert that any of them can potentially be examined for the purpose of optimization.

1.6.3 Multicomponent Interventions

The focus of this book is on multicomponent interventions, which involve a strategy made up of more than one, usually numerous, tactics aimed at achieving the intervention's end goal. Nearly every behavioral and biobehavioral intervention is multicomponent. Even biomedical interventions that at first appear to have only a single component, such as a surgical procedure or a pharmaceutical, can be considered multicomponent when behavioral and biobehavioral considerations that may affect the success of the intervention are taken into account. For example, ART can be considered a single-component intervention to reduce HIV viral load. However, to reduce viral load to undetectable levels, ART requires steady and careful adherence. This suggests that simply providing patients with ART may not be sufficient to reduce viral load; a multicomponent intervention like the one Dr. B is developing may be a more effective way to administer ART.

1.7 The MOST Framework: The Three Phases

This section and the next provide an overview of the three phases of the MOST framework, represented in the flow chart depicted in Fig. 1.1. The remaining chapters in this book and those in the companion volume go into much more detail about MOST.

In Fig. 1.1 each rectangle represents one of the phases of MOST, namely, preparation, optimization, and evaluation. These three phases are reviewed in this section. The figure also contains a diamond representing a decision point and two arrows representing possible ways of returning to the preparation phase. These are reviewed in the next section.

1.7.1 The Preparation Phase

As Fig. 1.1 shows, MOST begins with the preparation phase (discussed in more detail in Chap. 2). The purpose of this phase is to lay the groundwork for optimization of the intervention.

Development or revision of a detailed conceptual model occurs during the preparation phase. This model provides the foundation for critical decisions concerning intervention development and research design, including experimental design selection. The starting point for development or revision of the conceptual model is a review of all existing information. This information will come primarily from empirical scientific literature and theory but may also be drawn from other sources, such as secondary analyses of data from prior experiments or clinical experience. A good conceptual model serves two purposes. One purpose is to depict the causal process that produces the behavioral or biological process to be intervened on. The second purpose is to specify where and how the components of the

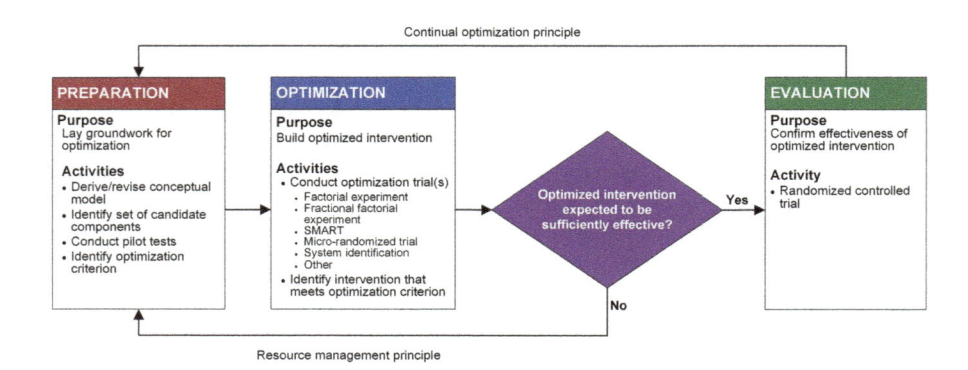

Fig. 1.1 Flow chart of the three phases of the multiphase optimization strategy (MOST)

intervention affect this causal process and how and why each component is expected to change the behavioral or biological process.

In some cases, all of the components featured in the conceptual model are selected for examination in the optimization phase, in other words, become part of the set of candidate components. In other cases, the effectiveness of some of the components may have been satisfactorily established in prior literature, and therefore it is a given that these components are to be included in the intervention. There may be no need to examine such components in the optimization phase. Or, the investigator may wish to determine whether and how the previously examined components interact with the candidate components, which would require examining them experimentally along with the others.

At this point it is necessary to determine how the components will be implemented and, usually, to pilot test the components and their implementation. After the pilot testing and any subsequent revision of components, there is one more necessary step in the preparation phase: identification of an optimization criterion. Recall the definition of optimization provided above:

> *Optimization of an intervention is the process of identifying an intervention that provides the best expected outcome obtainable within key constraints imposed by the need for efficiency, economy, and/or scalability.*

The above definition uses the terms "best expected outcome" and "constraints." The optimization criterion, which was discussed briefly earlier in this chapter, provides an operational definition of best expected outcome and also specifies the key constraints that are to be considered in optimization. In the leaf springs example, the optimization criterion is "leaf springs closest to the desired standard length that can be obtained without exceeding a manufacturing cost of $25 per leaf spring." The best outcome is "leaf springs closest to the desired standard length;" the specified constraint is on cost: the leaf springs must be produced "without exceeding a manufacturing cost of $25 per leaf spring." The ART adherence example used a conceptually similar optimization criterion, "largest expected value of *adhere* that can be obtained for an implementation cost of no more than $500 per person." There are many different types of optimization criteria; this is discussed further in Chaps. 2 and 7.

1.7.2 The Optimization Phase

The next phase of MOST is optimization. As the name implies, the purpose of this phase of MOST is to build an optimized intervention by selecting components and component levels from the set of candidates identified in the preparation phase. The decisions about selection of components and component levels are based on empirical data obtained by means of one or more carefully controlled and adequately powered optimization trials.

Choosing an Approach from the MOST Optimization Phase Toolbox There are many different approaches to experimentation that can be used in the optimization phase. Factorial experiments, fractional factorial experiments, sequential, multiple assignment, randomized trials (SMARTs), micro-randomized trials, system identification, or any other suitable approach can be considered part of what Almirall, Nahum-Shani, Wang, and Kasari (2018) in the companion volume call the MOST optimization phase toolbox. MOST does not require any particular approach to experimentation in the optimization phase, only that the approach selected from the toolbox is the best one according to the resource management principle, that is, it is the most efficient way to obtain the information needed for optimization. The choice of approach depends on the type of intervention that is to be optimized, the exact empirical information that is needed, and the resources that are available to conduct the experimentation.

This volume emphasizes optimization of fixed interventions (the difference between fixed and adaptive interventions was discussed above). Typically, although not invariably, the most efficient way to obtain the data needed to optimize a fixed intervention is via a factorial or fractional factorial experiment. Chapters 3, 4, 5, and 6 cover these experimental designs, other designs that may be useful, and how to apply the resource management principle to determine which design is the most efficient for a given application. Optimization of adaptive interventions frequently requires an experimental approach other than the traditional factorial or fractional factorial design. This is where, depending on the type of adaptive intervention being developed, SMARTs, micro-randomized trials, and system identification may merit serious consideration. Optimization of different types of adaptive interventions is introduced in Chap. 8. Optimization of particular types of adaptive interventions is discussed in more depth in the Almirall, Nahum-Shani, Wang, and Kasari (2018) and Rivera, Hekler, Savage, and Downs (2018) chapters in the companion volume. Readers who are primarily interested in adaptive interventions are strongly encouraged to read Chaps. 3, 4, 5, and 6 in the current volume, because the information contained in these chapters provides a necessary foundation for understanding optimization of adaptive interventions. In particular, SMARTs and micro-randomized trials are closely related to the factorial experiment.

Decision-Making Once the experiment has been conducted, the data are analyzed in whatever manner is appropriate given the experimental design. Then decisions about the composition of the optimized intervention are made. These decisions are based on the experimental results, along with the optimization criterion that was selected in the preparation phase. As mentioned above, Chap. 7 discusses how to identify the components and component levels that will make up the optimized intervention.

Multiple Trials Within a Single Optimization Phase Under some circumstances, development of a highly effective, efficient, economical, and scalable intervention is best facilitated by conducting a series of trials within a single optimization phase. For example, sometimes it is possible to build a strong set of components by conducting a series of optimization trials in an iterative fashion. In this approach, the experiment is used in the usual manner to determine which components are performing satisfactorily and which are performing unsatisfactorily. However, instead of building an optimized intervention after a single experiment and proceeding to the evaluation phase of MOST, an investigator using this iterative approach would revise the unsatisfactory components (or possibly replace them with new components) and then reevaluate them in a subsequent experiment. Once a set of satisfactory components has been arrived at, an optimized intervention is constructed in the usual manner, and the evaluation phase is begun. This approach has the potential to enable rapid progress in intervention development. An example of an iterative approach can be found in the Kugler, Wyrick, Tanner, Milroy, Chambers, Ma, and Collins (2018) chapter in the companion volume.

Depending on the situation, several optimization trials may be conducted. It may be expedient to use a measure of the mediator targeted by each component (see Chap. 2) as an outcome for that component, rather than the outcome of ultimate interest. Of course, MOST can help pinpoint which components are failing to have the intended effect on a mediator, but it is of limited use in determining exactly how to revise those components. Valuable leads for how to revise components may be gathered from collection of qualitative data, using sources such as focus groups and expert advisory panels.

A sequence of trials with different objectives may be conducted within a single optimization phase. Suppose an investigator who wishes to develop and optimize an adaptive intervention has the philosophy that, as a starting point, all of the intervention components to be included in the adaptive intervention should have a detectable overall effect when used in a fixed intervention. In this case the optimization phase could start with an experiment to identify which of a set of fixed components demonstrate a detectable overall effect. Then one or more subsequent experiments, perhaps SMARTs, could be conducted within the same optimization phase to build the adaptive intervention. The SMARTs could establish, for example, which of the components/component levels selected based on the first experiment represent the best initial treatment and which should be offered subsequently to those who respond to the initial treatment and to those who do not respond.

A single optimization phase of MOST can incorporate as many experiments as resources permit. To conduct several experiments within one optimization phase, it is necessary to have ready access to enough research subjects to provide sufficient power for the desired number of experiments. In addition, for each experiment the overall time frame must allow enough time for subject recruitment, conducting the experiment, data analysis, and planning of the next experiment based on the results. Investigators considering multiple experiments within a single optimization phase may be faced with the dilemma of which of two courses of action is a better use of resources within a given funding cycle: (a) completing the evaluation phase and conducting an RCT or (b) extending the optimization phase to include an additional experiment and applying for future funding to conduct the evaluation phase in a subsequent study. The former course of action has the advantage of providing a more definitive answer about the effectiveness of the optimized intervention, whereas the latter has the potential to produce an intervention that is more effective, efficient, economical, and/or scalable in the long run.

1.7.3 The Evaluation Phase

Once the optimized intervention has been identified, the investigator moves to the evaluation phase of MOST. The purpose of this phase is to confirm the effectiveness of the optimized intervention by means of an RCT. The objective of the RCT is to enable the investigator to decide whether the optimized intervention has a statistically and clinically significant treatment effect. Here effectiveness is expressed in terms of the size of the treatment/control difference. The treatment is the optimized intervention, and the control group could be current standard of care, a wait-list control, or any other suitable comparison group. If the results of the RCT indicate that the optimized intervention has a statistically and clinically significant effect, then the intervention may be released, in whatever manner is appropriate, for implementation in the intended setting.

1.8 Reasons for Returning to the Preparation Phase

1.8.1 Returning to the Preparation Phase Immediately After the Optimization Phase

The diamond in Fig. 1.1 represents a decision that must be made between the optimization and evaluation phases. The decision is whether or not the optimized intervention is expected to be sufficiently effective to justify continuing on to the evaluation phase.

It is possible for an intervention to be optimized and yet have an effect that is not expected to be large enough to be likely to achieve statistical significance in a reasonably sized RCT or to make much of a difference clinically. To understand how this can happen, let us return to the example. Suppose Dr. B identifies the set of intervention components that produces the best outcome that can be obtained without spending more than $500 per person, that is, optimizes the ART adherence intervention.

Now consider three different scenarios.

In the first scenario, the optimized intervention comprises a set of components that has many very potent members with strong individual and combined effects. In this happy scenario, it is likely that the optimized intervention will produce a statistically and clinically significant treatment effect when examined in an adequately powered RCT, and so the investigator would be justified in moving on to the evaluation phase of MOST.

In the second scenario, none of the components under consideration have very strong effects. It is possible to identify the best outcome that can be obtained without exceeding $500, but this best outcome is not likely to be much better than would be expected from a reasonable control or comparison treatment, and the cost is expected to be roughly the same. In this case, going to the trouble and expense of an RCT makes little sense. The resource management principle suggests that the resources that would have been used in an RCT can more profitably be used to revise the conceptual model and develop and test a new set of components, with the objective of arriving at a new and more effective optimized intervention.

The third scenario is a bit more complicated. Here some of the components under consideration do have substantial effects, but the best combination of components that can be implemented within the $500 limit set by the insurer happens to be made up primarily of components with small effects. In other words, the optimized intervention, the one that produces the best outcome subject to the $500 constraint, is not very potent; there are combinations of components that do exert a potent effect, but they exceed the upper limit on cost. One possibility would be for Dr. B to ask the insurer whether the upper limit on cost can be raised, so that a more potent optimized intervention can be identified. A related possibility would be to suggest to the insurer that cost-effectiveness might be used in the optimization criterion rather than a fixed upper limit on cost. In other words, the optimized intervention would be the most cost-effective intervention, irrespective of absolute cost. These alternatives would be workable only if the insurers were willing to accept the idea of an intervention that will cost more than $500.

If the idea of a costlier intervention is acceptable, there is no need to redo the optimization trial. A new optimized intervention can be identified using the new criterion and the results of the experiment that was done to examine the existing components. Assuming the new optimized intervention looks more promising, it can be evaluated via an RCT. However, if the optimization criterion cannot be changed, it will be necessary to return to the preparation phase.

Whenever the results of the optimization phase suggest that the optimized intervention is likely to have an effect that will not achieve statistical or clinical significance, the resource management principle and common sense both suggest that it would be a poor use of resources to continue to the evaluation phase and subject this intervention to an RCT. Instead, it would be better to take the resources that would have been spent on an RCT and use them to return to the preparation phase to begin a new cycle of MOST focused on identifying some new, more effective components and, ultimately, arriving at a more effective intervention.

This option is represented in Fig. 1.1 by the arrow leading from the diamond immediately after the optimization phase back to the preparation phase. The investigator who follows this arrow is starting a new cycle of MOST but is not starting from square one. Whatever has been learned in the current cycle of MOST can be built upon and used to illuminate the way forward. Any components that performed well can be retained and do not necessarily have to be retested (although replicating their effects and seeing how they interact with any new components are probably good ideas); any components that performed poorly can be revised. Even if all of the components previously tested turned out to be poor performers, this at least has demonstrated that none of those components work.

Mediation analyses may be helpful in planning a strategy for building a more potent intervention in the next round of MOST (see the Smith, Coffman, & Zhu (2018) chapter in the companion volume). Depending on the experimental design used in the optimization phase, it may be possible to fit models that enable examination of mediation of individual components. If a particular component fails to have an effect on a mediator but the mediator has an effect on the outcome, it may be helpful to try to revise the component or try a different strategy to affect the mediator. If the component affects the mediator but the mediator does not affect the outcome, then it may be necessary to rethink the conceptual model.

1.8.2 Returning to the Preparation Phase After the Evaluation Phase

In Fig. 1.1 there is an arrow leading from the evaluation phase back to the preparation phase, indicating that at the conclusion of the evaluation phase, there is always a return to the preparation phase. This return will occur after one of two possible outcomes: the optimized intervention either has or has not been determined to have a statistically and clinically significant effect.

If, based on the results of the optimization phase of MOST, a careful decision was made about whether or not to conduct the RCT, a null result is less likely than it would have been if the classical approach was taken. This is because in MOST, as discussed above, the decision about whether or not to go ahead with the RCT is informed by empirical evidence about the effectiveness of the components, which provides a sense of the expected potency of the intervention. This evidence is obtained not from pilot studies with low statistical power, but from one or more carefully controlled and adequately powered optimization trials. By contrast, in the classical approach, a set of components is typically assembled into an intervention a priori, informed by little, if any, empirical evidence about the likely effectiveness of the intervention except what has been obtained from pilot studies.

However, a null result on the RCT is always a possibility. This may occur because the statistical conclusion drawn from the RCT is the result of a Type II error; in other words, the intervention is effective in the population, but by chance a sample has been drawn that does not reflect this. Type II errors are always possible, particularly when the RCT is underpowered. Another possibility is that one or more of the conclusions upon which the optimization was based were the result of Type I errors; in other words, the investigator concluded that one or more components were effective when in reality they were ineffective. In this case the effect size of the optimized intervention in reality may have been smaller than would have been expected based on the results observed during the optimization phase. There are many other possibilities for why an intervention might fail to show a significant effect in an RCT. These include the usual issues that may affect RCTs, such as poor implementation of the intervention or compensatory behavior in the control subjects. A careful analysis of such issues should be completed as part of the preparation phase for the subsequent cycle of MOST.

If the results of the RCT indicate that the optimized intervention has a statistically and clinically significant effect, the return to the preparation phase is consistent with the continual optimization principle, discussed above. This principle states that once an intervention has been optimized and evaluated—in other words, a cycle of MOST has been completed—subsequent work can begin to improve the intervention further. Interventions could even be assigned consecutive version numbers, and release notes could be provided, in much the way successive versions of software are released!

1.9 The Distinction Between Optimization and Evaluation

The optimization of an intervention and the evaluation of the resulting optimized intervention are related but distinct concepts. In this section some of the differences between optimization and evaluation are reviewed.

Optimization and evaluation have different objectives. The objective of optimization is to arrive at an intervention that meets the optimization criterion that was identified in the preparation phase. By contrast, the objective of evaluation is to determine whether an intervention has a statistically and clinically significant effect. An intervention may have demonstrated a statistically significant effect in an RCT and not have been optimized; in fact, at this writing this is true of most evidence-based interventions. Conversely, as discussed above, an intervention may have been optimized, and yet its anticipated effect may be small. A large sample size may be required to achieve adequate power for detection of such an effect in an RCT; moreover, a small effect may not be clinically meaningful.

Optimization and evaluation require different approaches to research. Optimization requires going "under the hood" to assess the performance of *individual components* of the intervention. This information enables the selection of the components and component levels that best meet the optimization criterion. In general, the standard two-arm RCT is not the most efficient experimental design for addressing the research questions that are posed in the optimization phase, although it is the most appropriate experimental design when it comes time to assess the performance of the intervention *as a package*. Much of the remainder of this book and the companion volume are devoted to research methods for the optimization of interventions.

1.10 A Different Way of Thinking About Intervention Research: MOST-Induced Dilemmas

In some ways, MOST calls for a different way of thinking about intervention research, on the part of investigators, funders, and the field in general. When an investigator is thinking about intervention research from a MOST perspective and the investigator's mentors, employers, and funders are not, this can lead to MOST-induced dilemmas.

For example, consider Dr. B, who is building the intervention to improve adherence to ART that has been discussed in this chapter. Suppose Dr. B has 5 years of funding, with the usual expectation of development of an intervention followed by its evaluation in an RCT. It has been made clear to Dr. B by several mentors that the greatest professional rewards come from an RCT that shows a statistically and clinically meaningful effect.

Recall that based on extended discussions with insurers, Dr. B has established that they are unwilling to pay more than $500 per person for implementation. Therefore, Dr. B has selected an optimization criterion that states the optimized intervention will be the one that produces the best expected outcome for $500 or less. Suppose an optimization trial reveals several effective components, but not many that are both effective and inexpensive. Thus because of the $500 limit, the optimized intervention is expected to have an effect that is so small it would be unlikely to be deemed statistically and clinically significant in an RCT. The results of the optimization trial suggest that if, instead of using the optimization criterion demanded by the insurers, cost was disregarded and all of the components were included at the higher level, the resulting intervention would be likely to demonstrate a statistically and clinically significant effect in an RCT. The insurers are not interested in this intervention; they are adamant that it is too expensive to be scalable, and the $500 limit is firm.

Thus, Dr. B is faced with a dilemma. The resource management principle of MOST, as depicted in Fig. 1.1, would suggest that available resources should not be spent on an RCT. Instead, they would be better spent on a return to the preparation phase and a search for additional candidate components that will contribute to intervention effectiveness without adding too much cost. But this means that in the current 5-year funding period at least, there will be no RCT. This is professionally disappointing to Dr. B and may be disappointing to the organization funding the research as well. The other course of action is to ignore issues of scalability and build and evaluate the more costly intervention. This is consistent with the classical approach and is a promising path to a successful RCT and all the professional rewards that go along with it. However, because Dr. B is familiar with MOST, there is the nagging concern that in the long run, it is best to avoid devoting resources to evaluation of an intervention unlikely ever to be widely implemented and therefore unlikely to make much of an impact on public health.

From one perspective, the root of this dilemma is a disconnect between advancing public health on the one hand and the reward structure for academic intervention scientists on the other. If publication of an RCT that reveals an intervention to have a statistically and clinically significant effect is a major career maker, to the exclusion of other kinds of publications, then it is not surprising Dr. B would seriously consider conducting an RCT on an intervention known not to be scalable. After all, at this time there is no expectation that a report of an RCT will make any argument that the intervention evaluated is efficient, economical, or scalable—only that it is effective.

By publishing the results of the optimization trial, Dr. B will increase the knowledge base about which strategies work and which do not. This is an important contribution and one that arguably should not be disappointing to funders or anyone else. Nevertheless, today such a publication may not be as highly valued, for example, by a promotion and tenure committee, as a report of an RCT.

Now suppose Dr. B wants to use the remaining funds in the grant to return to the preparation and optimization phases, in particular, to conduct another optimization trial. This leads to another dilemma: how to approach the funding agency to ask permission to replace the RCT with another optimization trial. Intervention science based on testing strategies to see whether they are effective, and then moving ahead if they are or going back to the drawing board if they are not, is intuitively appealing and consistent with how most of us have been taught to conduct scientific inquiry. Yet this approach requires a level of flexibility in resource management that at this writing is foreign to many funding agencies. Most grant proposals are expected to provide a plan for 5 years of research, with no provision for making a midcourse decision about how best to use resources based on results obtained.

It could be argued that in some areas, intervention science would be moved forward faster if investigators were encouraged to submit a research strategy to funders that contained a clearly specified if-then branching. One example of such branching might be

> *If at the conclusion of the optimization phase the empirical evidence suggests that the optimized intervention package is expected to have an effect size of at least d = 0.3, then its performance will be compared to a suitable control group in an RCT. Otherwise, we will return to the preparation phase, work on improving or replacing any poorly performing components, and build a new optimized intervention.*

If necessary, separate budgets could be provided for each branch. Such an approach to funding research could ultimately lead to better interventions.

1.11 What's Next

This chapter has provided a conceptual overview of MOST, without going into much detail. The remaining chapters in this book are intended to complete the picture and provide the reader with the necessary background to implement MOST in his or her own research. In the next chapter, Chap. 2, the first phase of MOST, preparation, is described. In the preparation phase, considerable attention is paid to the articulation of a conceptual model of the behavioral, biobehavioral, or biomedical process of interest and how the intervention under development is to intervene on this process. Chapter 2 also discusses the optimization criterion and its role in MOST.

References

Almirall, D., Nahum-Shani, I., Wang, L., & Kasari, C. (2018). Experimental designs for research on adaptive interventions: Singly and sequentially randomized trials. In L. M. Collins & K. C. Kugler (Eds.), *Optimization of behavioral, biobehavioral, and biomedical interventions: Advanced topics* (forthcoming). New York, NY: Springer.

Caldwell, L. L., Smith, E. A., Collins, L. M., Graham, J. W., Lai, M., Wegner, L., ... Jacobs, J. (2012). Translational research in South Africa: Evaluating implementation quality using a factorial design. *Child and Youth Care Forum, 41*, 119–136.

Collins, L., Trail, J., Kugler, K., Baker, T., Piper, M., & Mermelstein, R. (2014). Evaluating individual intervention components: Making decisions based on the results of a factorial screening experiment. *Translational Behavioral Medicine, 4*, 238–251.

Collins, L. M., & Kugler, K. C. (Eds.) (2018). *Optimization of multicomponent behavioral, biobehavioral, and biomedical interventions: Advanced topics*. New York, NY: Springer.

Collins, L. M., Kugler, K. C., & Gwadz, M. V. (2016). Optimization of multicomponent behavioral and biobehavioral interventions for the prevention and treatment of HIV/AIDS. *AIDS and Behavior, 20*, 197–214.

Collins, L. M., Nahum-Shani, I., & Almirall, D. (2014). Optimization of behavioral dynamic treatment regimens based on the sequential, multiple assignment, randomized trial (SMART). *Clinical Trials, 11*, 426–434.

Flay, B. R. (1986). Efficacy and effectiveness trials (and other phases of research) in the development of health promotion programs. *Preventive Medicine, 15*, 451–474.

Kugler, K. C., Wyrick, D. L., Tanner, A. E., Milroy, J. J, Chambers, B. D., Ma, A., ... Collins, L. M. (2018). Using the multiphase optimization strategy (MOST) to develop an optimized online STI preventive intervention aimed at college students: Description of conceptual model and protocol. In L. M. Collins & K. C. Kugler (Eds.), Optimization of multicomponent behavioral, biobehavioral, and biomedical interventions: Advanced topics (forthcoming). New York, NY: Springer.

Leon, A. C., Davis, L. L., & Kraemer, H. C. (2011). The role and interpretation of pilot studies in clinical research. *Journal of Psychiatric Research, 45*, 626–629.

Michie, S., Richardson, M., Johnston, M., Abraham, C., Francis, J., Hardeman, W., ... Wood, C. E. (2013). The behavior change technique taxonomy (v1) of 93 hierarchically clustered techniques: Building an international consensus for the reporting of behavior change interventions. *Annals of Behavioral Medicine, 46*, 81–95.

Pignatiello, J. J., Jr., & Ramberg, J. S. (1985). Discussion of "off-line quality control, parameter design, and the Taguchi method" by Kackar, R.N. *Journal of Quality Technology, 17*, 198–206.

Piper, M. E., Fiore, M. C., Smith, S. S., Fraser, D., Bolt, D. M., Collins, L. M., ... Baker, T. B. (2016). Identifying effective intervention components for smoking cessation: A factorial screening experiment. *Addiction, 111*, 129–141.

Rivera, D. E., Hekler, E. B., Savage, J. S., & Downs, D. S. (2018). Intensively adaptive interventions using control systems engineering: Two illustrative examples. In L. M. Collins & K. C. Kugler (Eds.), *Optimization of multicomponent behavioral, biobehavioral, and biomedical interventions: Advanced topics* (forthcoming). New York, NY: Springer.

Schlam, T. R., Fiore, M. C., Smith, S. S., Fraser, S., Bolt, D. M., Collins, L. M., ... Baker, T. B. (2016). Comparative effectiveness of intervention components for producing long-term abstinence from smoking: A factorial screening experiment. *Addiction, 111*, 142–155.

Smith, R. A., Coffman, D. L., & Zhu, X. (2018). Investigating an intervention's causal story: Mediation analysis using a factorial experiment and multiple mediators. In L. M. Collins & K. C. Kugler (Eds.), *Optimization of multicomponent behavioral, biobehavioral, and biomedical interventions: Advanced topics* (forthcoming). New York, NY: Springer.

West, S. G., Aiken, L. S., & Todd, M. (1993). Probing the effects of individual components in multiple component prevention programs. *American Journal of Community Psychology, 21*, 571–605.

Wu, C. J., & Hamada, M. S. (2011). *Experiments: Planning, analysis, and optimization*. New York, NY: Wiley.

Chapter 2
The Preparation Phase of MOST

Abstract This chapter reviews the first of the three phases of the multiphase optimization strategy (MOST), namely, preparation. In the preparation phase, the investigator develops a conceptual model of the behavioral, biobehavioral, or biomedical process of interest and how the intervention under development is to intervene on this process. As will be discussed, this conceptual model is even more detailed than a standard logic model. The conceptual model will guide every decision that is made throughout the subsequent phases of MOST. Based on the conceptual model, the investigator identifies a set of components that are candidates for inclusion in the optimized intervention and, if necessary, pilot tests the components. The investigator also specifies the criterion that will define the optimized intervention, that is, the optimization criterion. A clear and well-specified conceptual model, a set of candidate components that follow directly from this model, and an appropriate optimization criterion are all critical aspects of MOST. It is assumed that readers are familiar with the material in Chap. 1.

Contents

2.1	The Conceptual Model: The "Engine" that Drives the Intervention	36
	2.1.1 Specifying the Conceptual Model	37
	2.1.2 Targeted Populations and Participant Heterogeneity	38
2.2	Using a Figure to Represent a Conceptual Model	38
	2.2.1 A Hypothetical Example	39
	2.2.2 Tracing the Causal Chain	42
2.3	Using the Conceptual Model to Select Outcome Variables for the Optimization and Evaluation Phases of MOST	43
	2.3.1 When the Outcome of Primary Interest Is Far in the Future	44
2.4	The Living Conceptual Model	45
2.5	Correspondence Between Components and Target Mediators	46
	2.5.1 The Granularity of Components	49
2.6	Meta-Analysis and the Conceptual Model	51
2.7	Including Moderation (Effect Modification) in a Figure Representing a Conceptual Model	51
2.8	Why Bother With a Conceptual Model?	53
2.9	The Role of Pilot Testing in MOST	54

© Springer International Publishing AG 2018
L. M. Collins, *Optimization of Behavioral, Biobehavioral, and Biomedical Interventions*, Statistics for Social and Behavioral Sciences,
https://doi.org/10.1007/978-3-319-72206-1_2

2.10 The Optimization Criterion 55
    2.10.1 Reasons for Identifying the Optimization Criterion in the Preparation Phase Rather Than the Optimization Phase 56
    2.10.2 The Role of the Screened-In Set in Optimization 56
    2.10.3 All Active Components 57
    2.10.4 Optimization Criteria Involving Specific Constraints 58
    2.10.5 Overview of Incorporating Constraints Into the Optimization Criterion 58
    2.10.6 Other Kinds of Constraints 60
    2.10.7 Constraints on Multiple Resources 61
2.11 Identifying Key Constraints 62
2.12 What's Next 62
References 63

2.1 The Conceptual Model: The "Engine" that Drives the Intervention

A thoughtful, well-specified, clear, and detailed conceptual model is an essential starting point for MOST. The conceptual model is referred to throughout the process of developing and optimizing an intervention. Its role in the preparation phase is to guide the choice of candidate intervention components and primary outcome variables. In the optimization phase, the conceptual model will inform the critical decision about what experimental design makes the best use of available resources.

    A conceptual model is very similar to a logic model, and, in fact, a logic model may serve as an excellent starting point for development of a conceptual model. According to the W.K. Kellogg Foundation (2004), "[a] program logic model links outcomes (both short- and long-term) with program activities/processes and the theoretical assumptions/principles of the program" (p. iii). The conceptual model takes this a step further by looking "under the hood" to describe the "engine" that drives the intervention. This means the conceptual model details the mechanisms by which the intervention is expected to effect change in the primary outcome(s). The conceptual model links each individual intervention component with short- and long-term outcomes, to express which specific components target which specific outcomes. The conceptual model also specifies whether and where a component is expected to exert its effects in whole or in part by enhancing the effectiveness of another component (this will be illustrated below).

    The conceptual model expresses all of what is known or hypothesized about how the intervention under development is to intervene on the behavioral, biobehavioral, or biomedical process. The strongest conceptual models are explicitly theory-based and directly informed by peer-reviewed empirical literature. If different theories apply to different aspects of a model, a conceptual model may be informed by more than one theory. Where there is currently little theory or peer-reviewed empirical literature to inform a part of a conceptual model, the model may be informed by other sources, such as unpublished literature, conference presentations, secondary data analyses, and clinical experience. A conceptual model should be as comprehensive as possible, even if not every aspect of it is to be investigated in a particular study. In this chapter some recommendations for specifying a conceptual model are offered.

2.1.1 Specifying the Conceptual Model

Conceptual models vary in specificity, depending on factors such as the availability of relevant theory to inform the model and the consistency of the empirical literature in the area. The more specific a conceptual model is, the more helpful it will be in optimization of an intervention. To be most helpful, the conceptual model should offer a high level of specificity in three areas.

First, the conceptual model should clearly describe the causal process to be intervened upon. This description should include key outcome variables, each variable that is a hypothesized causal influence on these outcome variables, and any mediators of these causal influences.

Second, the conceptual model should express the investigator's approach to intervening on this causal process. Most interventions operate not by directly changing the key outcome variables, but instead by changing the causal influences on those outcome variables. These causal influences then become mediators of the effect of the intervention on the outcomes. Put another way, causal influences on the outcomes are also target mediators for the intervention components. The conceptual model should specify the target mediators for each intervention component; in other words, the conceptual model should express the investigator's strategy for changing each causal influence via one or more intervention components. This is one area in which the conceptual model required by MOST is more detailed than a typical logic model. In a logic model, which components target which mediators is not always specified, because the emphasis is on the treatment package as a whole rather than on individual components.

Third, a conceptual model should detail any anticipated moderation of the effects of one component by another (also known as treatment interactions), for example, when the presence of one component is hypothesized to enhance the effect of another. This is another area in which additional detail is required in a conceptual model to be used in MOST as compared to a logic model to be used in the classical approach; in the classical approach, a logic model typically does not specify moderation among intervention components. Moderation is represented in statistical modeling by interactions, which are discussed further in Chaps. 3 and 4.

There are two additional types of information that investigators may wish to include in a conceptual model. First, there may be causal factors that are not targeted by the intervention. Because the emphasis of the conceptual model is on explaining how the intervention works, it is not necessary to include these (at least for the purpose of MOST), but the investigator may wish to do so for the sake of completeness, or to highlight them as a direction for future improvement of the intervention.

Second, variables such as characteristics of the individual (e.g., gender, depression) or the environment (e.g., neighborhood socioeconomic status) may moderate the effects of intervention components. This is known as effect modification in the epidemiology literature. Even if untargeted causal factors and effect modification are included in the conceptual model, investigators may wish to be cautious about including this information in a figure representing the model, because if a figure is too complex, the main points may not be conveyed effectively.

2.1.2 Targeted Populations and Participant Heterogeneity

Sometimes an intervention is aimed at a specific subgroup of people. For example, the hypothetical intervention introduced in Chap. 1 is aimed at HIV-positive individuals who drink heavily. In this case, the conceptual model should clearly specify that the intervention is aimed only at a particular subgroup and is not intended to be suitable for a wider population of participants.

Other times it is hypothesized that different approaches to intervention will benefit different subgroups of people (e.g., males and females) or the same population at different stages of development (e.g., pre-pubertal and adolescent) or at different points in a time-varying process (e.g., different points during recovery from a chronic relapsing disorder such as addiction to nicotine). Dealing with this kind of heterogeneity to ensure good outcomes for all participants, which is a significant scientific challenge, is one of the objectives of adaptive interventions. Adaptive interventions vary, sometimes repeatedly and even rapidly over time, in response to characteristics of the individual or group. Adaptive interventions are discussed in Chap. 8 in this volume and in two chapters in the companion volume (Collins & Kugler, 2018): Almirall, Nahum-Shani, Wang, and Kasari (2018); and Rivera, Hekler, Savage, and Downs (2018).

It is insufficient to approach the issue of participant heterogeneity simply by including a variety of components in an intervention in the hope of offering something for everyone, without specifying how each component is hypothesized to respond to this heterogeneity. Such an approach is not systematic enough to produce effective, efficient, economical, and scalable interventions.

2.2 Using a Figure to Represent a Conceptual Model

A figure can be an indispensable tool for conveying the essence of a conceptual model, particularly to readers of a journal article or reviewers of a grant proposal. Above, the conceptual model was described metaphorically as the engine that drives the intervention. The emphasis in a figure representing a conceptual model should be on providing a clear depiction of the essential parts of this engine. In other words, someone viewing the figure must be able to see at a glance what components target each causal factor, that is, how each component of the intervention is intended to intervene on the behavioral, biobehavioral, or biomedical process. For this reason it is usually impossible for a figure to represent every detail of the various types of information discussed above. Expressing the entire conceptual model in a figure may be counterproductive if too much detail draws the viewer's attention away from the critical parts of the intervention strategy. Thus, the figure should be seen as a supplement to a complete narrative description.

At this writing, there are no generally agreed-upon conventions for a figure representing the conceptual model underlying development of an intervention. This book follows conventions for these figures that are similar to those followed by most of the behavioral and biobehavioral sciences. The arrows have roughly the same meaning as those in diagrams that represent structural equation models, that is, they represent non-zero regression coefficients. However, the figures used in this book to depict the conceptual model are not structural equation modeling figures and do not include all the paths that would be included in such a figure. Because the main purpose of the figures in this book is to provide a clear and succinct representation of the investigator's strategy for effecting change on the primary outcome variable (s) via the intervention under development, the figures in this book include only paths corresponding to causal relations and moderation of causal relations.

2.2.1 A Hypothetical Example

Collins, Kugler, and Gwadz (2016), in an article introducing MOST to the HIV/AIDS research community, described a conceptual model for a hypothetical intervention aimed at reducing HIV viral load among HIV-positive individuals who drink heavily. These individuals are particularly vulnerable for two reasons. First, the effectiveness of antiretroviral therapy (ART) depends on strict adherence to its protocol (Glass et al., 2010; Grant et al., 2010), and it has been shown that the use of alcohol and other substances is related to inadequate adherence (Azar, Springer, Meyer, & Altice, 2010; Braithwaite & Bryant, 2010; Hendershot, Stoner, Pantalone, & Simoni, 2009). Second, in addition to its negative effects on ART adherence, alcohol use is associated with increased viral load (Wu, Metzger, Lynch, & Douglas, 2011) and poor health outcomes (Azar et al., 2010). Thus even moderate alcohol use is contraindicated for individuals on ART (Shuper et al., 2010).

Suppose the scientist introduced in Chap. 1, Dr. B, develops a conceptual model based on the Collins et al. model. (This is a hypothetical model that has been informed by the scientific literature but is not intended to be an accurate representation of current thinking in the HIV area.) Dr. B's model, which is depicted in Fig. 2.1, provides a detailed description of how each component in the hypothetical intervention is designed to intervene on the process by which social, cognitive, and behavioral causal factors affect HIV viral load.

As Fig. 2.1 illustrates, the model specifies that alcohol use leads directly to an increase in HIV viral load and that ART adherence leads to a decrease in HIV viral

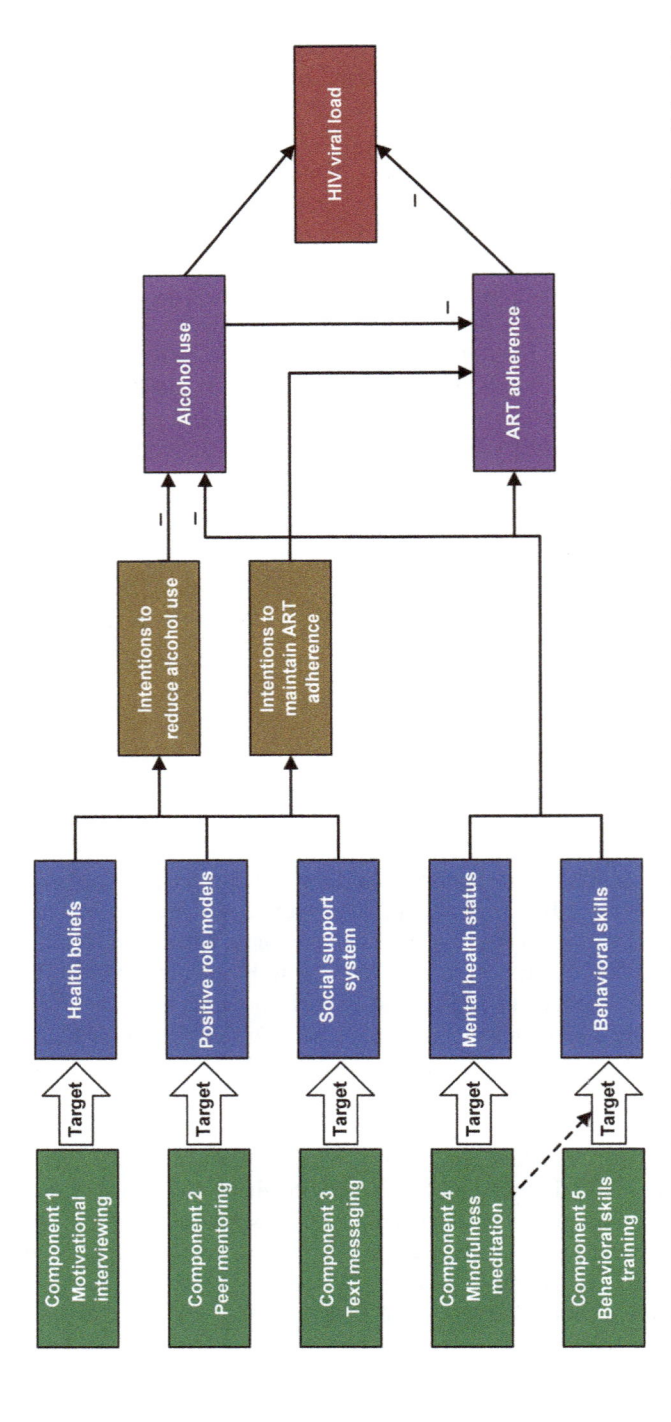

Fig. 2.1 Conceptual model for a hypothetical behavioral intervention to reduce viral load in HIV-positive individuals who are heavy drinkers (Adapted from model in Collins et al. (2016))

load (note the minus sign indicating a negative relation). In addition, alcohol use leads indirectly to an increase in HIV viral load by reducing ART adherence; a reduction in ART adherence would, in turn, lead to an increase in viral load. Thus, the objective of the intervention is to reduce alcohol use and increase ART adherence.

Showing at a glance which components target which mediators is an essential feature of any figure depicting a conceptual model. Figure 2.1 indicates that there are five intervention components. Although the components are intended ultimately to affect alcohol use, ART adherence, and HIV viral load, each is aimed at a single target mediator. This is represented by the large hollow arrows marked with the word *Target*. The target mediators in this conceptual model are five distal causal influences on alcohol use and ART adherence drawn from social-cognitive theory (Bandura, 1977, 1986), which specifies that behavior change involves both social (e.g., positive role models, social support) and individual cognitive (e.g., health beliefs, mental health status, behavioral skills) factors.

The motivational interviewing component targets health beliefs (Bandura, 1977) by engaging the participant in the process of examining alcohol use patterns in the context of HIV infection and ART adherence, considering his or her normative beliefs about alcohol toxicities and ART, examining perspectives on alcohol use and ART adherence (Parsons, Golub, Rosof, & Holder, 2007), and considering his or her intentions to use alcohol and adhere to ART.

The peer mentoring component is designed to provide contact with a positive role model (Bandura, 1977; Nyamathi, Flaskerud, Leake, Dixon, & Lu, 2001; Purcell et al., 2007). The role model will be a trained and supervised peer who is living with HIV, has experienced alcohol problems in the past, is presently managing alcohol at a non-problem level, and has been taking ART with excellent adherence for at least 12 months (Harris & Larsen, 2007).

The text messaging component is designed to provide the participant with access to a strong social support system (Brown, Littlewood, & Vanable, 2013; Langebeek et al., 2014). This component uses short message service (SMS/text messaging) technology to increase the participant's experience of social support for improving alcohol reduction and/or ART adherence intentions, reducing alcohol use behavior, improving ART adherence behavior, or other relevant attitudinal or behavioral changes. Text messaging has the potential to accomplish this more efficiently than one-on-one structured sessions (Mo & Coulson, 2008; Uhrig et al., 2012).

The mindfulness meditation component targets mental health status. It is expected that the vast majority of the sample will exhibit clinically significant mental health distress (e.g., perceived stress, anxiety, depression) as a result of the demands of coping with HIV diagnosis and ART, stigma (including sexual minority status), and, typically, low socioeconomic status (Duncan et al., 2012; Gonzalez, Batchelder, Psaros, & Safren, 2011; McIntosh & Rosselli, 2012; Safren, Reisner, Herrick, Mimiaga, & Stall, 2010). Based on research suggesting that individuals with untreated depression are at increased risk for both alcohol use (Sullivan, Goulet, Justice, & Fiellin, 2011) and non-compliance with ART (Cruess et al., 2012; Gonzalez et al., 2011; Springer, Dushaj, & Azar, 2012), the conceptual model in

Fig. 2.1 specifies that improved mental health status leads to a decrease in alcohol use and an increase in ART adherence. The mindfulness meditation component will provide training in meditation practices intended to reduce stress and improve mental health.

Finally, the behavioral skills training component targets behavioral skills for both managing alcohol use and adhering to ART. There is evidence that individuals who have developed behavioral skills (Bandura, 1977) for reducing alcohol use and managing the ART treatment regimen are less likely to use alcohol and more likely to adhere to treatment than those who have not developed behavioral skills (Norton et al., 2010).

The first three target mediators—health beliefs, contact with positive role models, and access to a strong social support system—are each hypothesized to affect intentions both to reduce alcohol use and to maintain ART adherence. Consistent with the theory of planned behavior (Ajzen, 1985), in this model intentions to reduce alcohol use and intentions to maintain ART adherence are critically important causal influences on alcohol use and ART adherence, respectively. The two remaining target mediators in Fig. 2.1, mental health status and behavioral skills, directly affect alcohol use and ART adherence without being mediated by intentions.

In figures depicting behavioral models, moderation is typically depicted as an arrow running from the moderator to another arrow, rather than to a box. In this book a dashed arrow is used to signify moderation. In Fig. 2.1, there is a dashed arrow running from mindfulness meditation to the large hollow arrow between the behavioral skills training component and behavioral skills. This depicts the hypothesis that the presence of the mindfulness meditation component will increase the effect of the behavioral skills training component on behavioral skills by enhancing the participant's receptivity to the training.

2.2.2 Tracing the Causal Chain

Conceptual models of interventions often can be considered a causal chain. Imagine a series of dominos (small oblong tiles used as game pieces) standing vertically on the short edge and set up in a line, so that if the first is knocked over, it knocks over the next one, which in turn knocks over the next one, and so on. This can be seen as a metaphor for the mechanisms by which most interventions work. In a sense, an intervention component knocks over the first domino by changing a causal variable. Then that causal variable knocks over the next domino, and so on until the primary outcome variable is reached.

When a figure representing a conceptual model includes sufficient detail, it is possible to trace the series of dominos as they are knocked over and, in other words, to trace the causal chain via which an intervention component is hypothesized to have its effect on the primary outcome. Sometimes there is more than one causal chain. For example, in Fig. 2.1, it is possible to trace the ways in which mindfulness meditation is hypothesized to reduce viral load. One causal chain is as follows:

Mindfulness meditation improves mental health, which leads to reduced alcohol use. Reduced alcohol use directly leads to a reduced HIV viral load. Reduced alcohol use also leads to better ART adherence, which in turn reduces viral load. Another causal chain is as follows: *Mindfulness meditation improves mental health, which leads to better ART adherence, which directly reduces HIV viral load.* As another example, it is possible to trace how text messaging is hypothesized to reduce HIV viral load. One causal chain is *text messaging helps to establish a strong social support system; this leads to increased intentions to reduce alcohol use, which in turn leads to reduced alcohol use. Reduced alcohol use in turn leads to a reduced HIV viral load directly and indirectly through improved ART adherence,* as described above. Another causal chain is *text messaging helps to establish a strong social support system; this leads to increased intentions to maintain ART adherence, which in turn leads to better ART adherence, which directly reduces viral load.*

In the behavioral and biobehavioral sciences, causal chains are often represented by mediation models (e.g., MacKinnon, 2008). For example, the relation between the mindfulness meditation component and HIV viral load is mediated by mental health status, alcohol use, and ART adherence. In this case mental health status would be a proximal mediator of the mindfulness meditation component, and alcohol use and ART adherence would be intermediate mediators. It would also be accurate to say that mental health status mediates the relation between the mindfulness meditation component and both alcohol use and ART adherence, that both alcohol use and ART adherence mediate the relation between mental health status and HIV viral load, and that the relation between alcohol use and HIV viral load is partially mediated by ART adherence.

It is particularly important to remember that the target mediators are also causal variables in the conceptual model. Every target of a component must have a causal influence on the key outcome(s) if the intervention is to be effective.

2.3 Using the Conceptual Model to Select Outcome Variables for the Optimization and Evaluation Phases of MOST

One use of the conceptual model is in selecting the primary outcome variable or variables to be used in the optimization and evaluation phases of MOST. Optimization and evaluation are conducted in separate experiments using separate subject samples. In the evaluation phase, in which the optimized intervention is evaluated in an RCT, the outcome variable is often the outcome of most salient public health significance. In the example, the conceptual model suggests that the outcome of most salient public health significance is HIV viral load, so it is likely that the investigators would use this as the primary outcome variable for the RCT.

It is not necessary, and not always even advisable, to use the same primary outcome variable in the optimization phase of MOST as is used in the evaluation phase. Sometimes in the optimization phase, it is more practical and expedient to use a measure of a mediating variable, ideally one proximal to the outcome, as a proxy for the outcome. In the example, there are good reasons not to use HIV viral load as the outcome when experimenting to assess the individual effects of the five intervention components. HIV viral load is expected to have a severely non-normal distribution, with many observations of zero—at least, it is hoped the intervention will be successful enough to produce some undetectable viral loads—but others ranging up to as high as one million. A clinically meaningful approach would be to use a dichotomous outcome with levels undetectable and detectable viral load. All else being equal, the power of this hypothesis test will be lower compared to a hypothesis test based on a normally distributed outcome. Because the RCT will evaluate the optimized intervention as a package, the effect size may be large enough that a reasonable sample size will provide sufficient power even with a dichotomous outcome variable. However, the effect of any single component making up an intervention is expected to be smaller than the effect of the entire intervention, so the power to detect the effects of the individual components may be unacceptably low, unless a very large sample size can be obtained or the Type I error rate is set high.

The conceptual model described above and illustrated in Fig. 2.1 provides the basis for the example threaded throughout the discussions of various aspects of the optimization phase of MOST that appear in this book. In these discussions, the primary outcome variable will be the proximal mediator ART adherence. The logic, informed by the conceptual model, is that if components that have an effect on ART adherence can be identified and assembled to form an optimized intervention, the resulting intervention will have an effect on HIV viral load.

The outcome variable will be operationalized as number of days of adherence to ART in the 30-day period following the conclusion of the intervention and abbreviated *adhere*. (This is the same outcome variable that was used in Chap. 1.) For the remainder of this book, it will be assumed that *adhere* is approximately normally distributed, although it is acknowledged that in practice this could be subject to debate.

2.3.1 When the Outcome of Primary Interest Is Far in the Future

Sometimes the outcome of primary interest may occur months or even years after participation in the intervention. For example, consider a school-based drug abuse prevention intervention delivered in sixth grade. Drug use is the primary outcome. Although the goal is to prevent drug use beginning immediately, the rates of drug use are expected to be so low in sixth grade that any treatment/control comparison would be meaningless. To make a meaningful treatment/control comparison with drug use as the outcome, the investigators must wait until drug use becomes more normative, say in tenth or eleventh grade. As another example, consider a lifestyle modification program aimed at preventing a second heart attack in individuals who have recently had a first heart attack. It would be necessary to follow the patients for several years to make a meaningful treatment/control comparison in number of heart attacks or time to next heart attack. In both of these examples, if the primary outcome were used, optimization would take many years, rendering it impractical.

When the outcome of primary interest occurs in the distant future, optimization often can be conducted relatively quickly by using measures of the target mediators specified in the conceptual model as short-term outcomes for the optimization trial, instead of using the primary outcome. Suppose the school-based drug abuse prevention intervention includes components targeting mediators such as norms about drug use, expectations about the effects of drugs, and self-efficacy for resisting offers to use drugs. The performance of each of these components could be evaluated shortly after implementation of the optimization trial by using a measure of its respective target mediator. Thus the component targeting norms about drug abuse would be evaluated using a measure of norms, the component targeting expectations about the effects of drugs would be evaluated using a measure of expectations, and the component targeting self-efficacy for resisting offers to use drugs would be evaluated using a measure of self-efficacy. Similarly, suppose the lifestyle modification program for cardiac patients includes components targeting mediators such as diet, physical activity, stress management, and positive thinking. The performance of each of these components could be evaluated using a measure of its respective target mediator.

The logic behind this approach is that if the intervention works by changing the target mediators, which in turn change the primary outcome, it makes sense to optimize based on the impact on the mediators; if an intervention is developed that has a potent effect on the mediators, and the mediators have a causal relation to the outcome, then an effect on the primary outcome will follow. If the conceptual model is valid, then this logic holds. If the conceptual model is flawed, and some or all of the mediators do not have a causal relation to the outcome, then an intervention that has a potent effect on the mediators may ultimately have little or no effect on the primary outcome. The significance of the effect of the optimized intervention on the primary outcome is tested in the evaluation phase by means of an RCT.

Employing measures of target mediators as short-term outcomes for the purpose of optimization can complicate or even rule out the use of some optimization criteria. This is discussed further in Chap. 7.

2.4 The Living Conceptual Model

There are several reasons why an investigator may wish to return to the preparation phase and take a fresh look at the conceptual model. First, recall the continual optimization principle (Chap. 1), which holds that optimization of an intervention is an ongoing process of moving toward an ever-better intervention. This opens up the possibility of optimizing an intervention repeatedly, with the objective of moving toward a better and better intervention each time. Second, the resource management principle specifies that if, at the conclusion of the optimization phase, the optimized intervention is deemed unlikely to demonstrate a statistically or clinically significant effect, the investigator returns to the preparation phase rather than expending resources on an RCT. Third, if in the evaluation phase the optimized intervention does not demonstrate a statistically and clinically significant effect, common sense suggests a return to the preparation phase without releasing the intervention for implementation.

This fresh look at the conceptual model involves updating based on any information that may have become available since the last time the model was updated. This information includes (a) the results of any optimization trials and (b) any new scientific literature or theoretical developments. In this sense, a conceptual model is a living thing; over repeated rounds of MOST, it will be updated and refined along with the intervention.

When a conceptual model is being updated, this is a good time to improve its specificity, for example, to clarify exactly which causal factor is targeted by each component. It is also a good time to conduct secondary analyses to examine why a particular component was ineffective. Did the component demonstrate an effect on the targeted causal variable, but this hypothesized causal variable did not demonstrate an effect on the outcome? In this case, the investigators may want to rethink the conceptual model, possibly replacing the mediator (although it might be worth considering whether the hypothesized causal variable is measured well). Did the component demonstrate an effect on the targeted causal variable but only for a subgroup of subjects? This means subgroup membership moderates this component's effect. In this case, the investigators may wish to include this moderator of the component's effect in the conceptual model and either seek a more robust component that works across the board or consider developing an adaptive intervention that would provide different versions of the component, or entirely different components, to the different subgroups. Did the hypothesized causal variable demonstrate an effect on the outcome, but the component failed to demonstrate an effect on the causal variable? Here it may be necessary to find a different approach so that the component can be replaced by a more effective one.

2.5 Correspondence Between Components and Target Mediators

Figure 2.1 shows that each of the five intervention components directly targets one and only one of the five hypothesized causal influences, that is, each component is aimed at a single target mediator. Although it is not always possible, it is often desirable to develop an intervention strategy in which each component has a single target mediator. This strategy can simplify decision-making in the optimization phase and later can help illuminate the way forward for improvement of the

2.5 Correspondence Between Components and Target Mediators

intervention. This is not to rule out the possibility that a component will have incidental effects on other causal variables. This possibility is acknowledged and can be explored in data analysis but is not depicted in Fig. 2.1 because it is not a part of the intervention strategy; the intervention is expected to be effective whether or not such incidental effects occur.

To see why it is desirable to have each component target a single causal variable, consider the peer mentoring component and the text messaging component. Both are hypothesized to increase ART adherence, but they target different causal variables. The peer mentoring component is designed to operate by increasing contact with positive role models, whereas the text messaging component is designed to operate by increasing perceived social support. Suppose an experiment has been conducted that enables estimation of individual component effects and interactions between components (the design of such experiments will be discussed later in this book). Further suppose statistical analysis of the data from the experiment shows that peer mentoring has a desired effect on *adhere*; text messaging has no effect on *adhere*; and there is no interaction between the two components. These results provide unambiguous information for making decisions about selection of components into the screened-in set (recall from Chap. 1 that the screened-in set includes the components that have demonstrated an effect on the outcome large enough to be considered important and in the desired direction). The results suggest that peer mentoring should be included in the screened-in set, and text messaging should not be included in the screened-in set. (Even if there had been an interaction, the information may have been unambiguous; see Chaps. 4 and 7.)

Now suppose a mediation analysis shows that the effect of peer mentoring on *adhere* is mediated by the establishment of positive role models, so the conclusion is that the peer mentoring component is operating as hypothesized. Additional analyses show that a high level of perceived social support is associated with a larger value of *adhere*, but text messaging has no effect on a subject's perceived social support. If the text messaging component was poorly implemented in some way, that would be one possible explanation for these results. Assuming that the component was properly implemented, these results send a clear message about one step that could be taken to improve the intervention the next time it is to be optimized. They suggest it would be advisable either to revise the text messaging component so that it does a better job of increasing social support or else replace it with a different component aimed at increasing social support.

Now let us turn back the clock and imagine that the investigator starts the preparation phase by developing a slightly different conceptual model. In this model, depicted in Fig. 2.2, everything starting from the target mediators moving to the far right is the same as in Fig. 2.1. The only difference is that in the model in Fig. 2.2, peer mentoring and text messaging have been combined into a single component called social support. In the new social support component, both peer mentoring and text messaging are provided. The combined component targets two causal variables: positive role models and social support system.

Suppose an experiment has been conducted to examine these four components, and statistical analysis of the data shows that the social support component has a

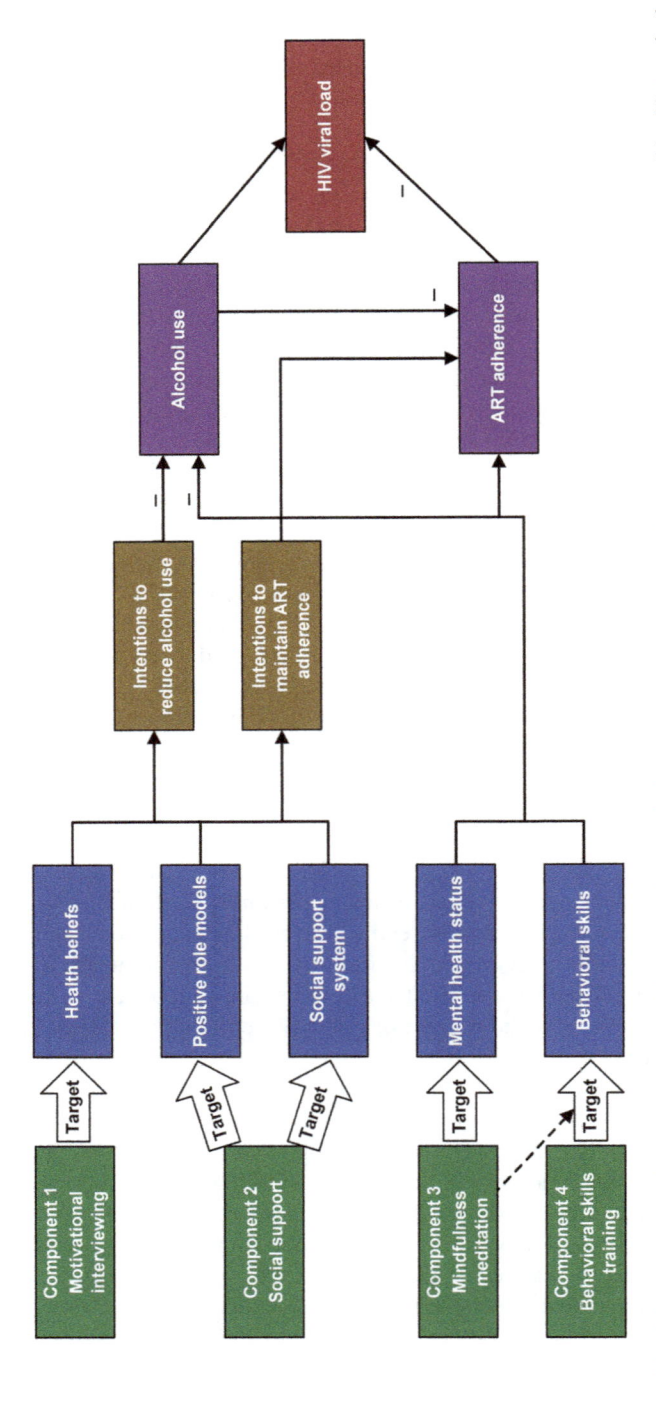

Fig. 2.2 Variation on the conceptual model depicted in Fig. 2.1. In this model, peer mentoring and text message support components are combined into a single component called social support

detectable but small effect on *adhere*. The investigator conducts further analyses to help determine what changes in the social support component could be made in the next cycle of MOST to help to increase its strength. The analyses indicate that the effect of the social support component is mediated by positive role models but not by the social support system variable. The component does help establish a positive role model, but it does not lead to an increased perception of social support, even though a high level of perceived social support is associated with a lower viral load. In other words, the combined social support component targeted two mediators, but it successfully changed only one of them.

These results give less information concerning how to improve the intervention in the next cycle of MOST. Should the social support component be left as is, because it had an effect on positive role models, or should it be revised and reevaluated in an attempt to do a better job on perceived social support? If it is to be revised, what revisions should be made? The text messages were intended to increase perceived social support, so it makes sense to consider revising them. However, because in the combined component the text messaging was always provided along with the peer mentor, it is impossible to determine whether the observed effect on establishment of a positive role model was due to the peer mentor alone, text messaging alone, or a combination of the two. Thus any change made to text messaging risks changing the effect on establishment of a positive role model.

Whenever a component targets more than one causal variable, these kinds of questions can arise. For this reason, it is recommended to try to set up the conceptual model so that each component targets a single causal variable (again, remembering that there may be incidental effects on other causal variables). If it is unavoidable to have a component target more than one causal variable, it is helpful to be aware of the kinds of inferential difficulties that may arise.

Although having a single component target more than one causal variable can lead to the kinds of ambiguities described above, having more than one component target a single causal variable typically does not create the same kinds of problems. In fact, this can even be desirable. In situations where a particular causal variable plays a critical role in a process, it may be helpful to include several different components aimed at it, if this can be accomplished without creating redundancy. If an appropriate experimental design is selected for the optimization phase (see Chap. 6), it will be possible to investigate which components are successful at affecting the causal variable.

2.5.1 *The Granularity of Components*

Interventions and intervention components can be conceptualized in many ways, and in general there is not a right way or wrong way to identify components. In Chap. 1 the term component was defined as any part of an intervention that can be separated out for study. In other words, if it would be useful to find out whether it works, and it fits into a well-articulated conceptual model, it can be considered a component. Any investigator who is struggling to express an intervention as a set of components is encouraged to put aside the idea of components temporarily and concentrate on identifying the rest of the conceptual model. Once the causal variables affecting the outcome have been articulated, this will provide a firm basis for identifying a set of intervention components.

How intervention components are conceptualized partly determines which experimental design will emerge as most appropriate in the optimization phase. Consider the granularity of components. Often a component can conceivably be conceptualized in a range of granularity, from relatively micro to relatively macro. For example, it would be possible to identify the text messaging component in Fig. 2.1 at a more micro level of granularity by breaking it down into different categories of messages, such as, say, instrumental support ("Don't forget to pick up your free bus pass!") and encouragement ("You did great taking your meds this week!"). The instrumental support and encouragement components, in turn, could be broken down into even more micro-level individual messages. What is the appropriate level of granularity?

Just as there are not right and wrong ways to identify components, there are not right and wrong levels of granularity. In choosing a level of granularity, there are several considerations to keep in mind. First, for the reasons discussed above, wherever possible it is a good idea to select a level of granularity that is micro enough so that no component targets more than one mediator variable. Second, as components become more fine-grained, there are likely to be more of them. Third, as components become more fine-grained, it is likely that the size of each individual component's effect on its target mediator, and by extension on the outcome, will decrease. For example, compare two different ways of conceptualizing text messaging, one at a relatively macro level of granularity, and one at a relatively micro level. The macro level would be a single component made up of 12 text messages, delivered in a rotation so that 4 of them are sent out per week. The corresponding micro-level conceptualization would consist of 12 components, each made up of a single text message delivered once every 3 weeks. It is probable that each of these individual micro-level components would have a smaller effect than the macro-level component.

Unfortunately, and perhaps rather frustratingly, the first consideration listed in the preceding paragraph in some ways stands in opposition to the second and third. A finer level of granularity provides additional information that is useful in optimization of the intervention; on the other hand, obtaining additional information usually demands expending additional resources. Because a finer level of granularity may mean there are more candidate components, the number of factors to be included in the optimization trial may increase, which means there will be additional expense associated with managing more experimental conditions (this is discussed in Chap. 6). Even with a highly efficient experimental design, the number of components that can be examined in a single optimization trial is limited. This is fundamentally situation-specific, but a typical range is from four to seven components. Although an increase in the number of experimental conditions in an optimization trial does not in and of itself imply that an increase in the number of experimental subjects will be needed (see Chap. 3), detection of smaller effects will require an increase in sample size, all else being equal. For example, an optimization trial to examine the effect of individual text messages would probably require a larger sample size to maintain adequate power than a trial aimed at examining the effects of a set of more macro-level components.

Thus the resource management principle should be considered when choosing a level of granularity. Given finite resources, breaking one component into smaller components and examining the smaller components in an optimization trial will, in most cases, use resources that could be devoted instead to examination of other components. Only the investigator can determine which information is most scientifically important and therefore most deserving of having resources devoted to obtaining it.

2.6 Meta-Analysis and the Conceptual Model

Meta-analyses are sometimes conducted to gain a sense of which components are likely to be useful to include in an intervention and which are not (e.g., van Ryzin, Roseth, Fosco, Lee, & Chen, 2016). Meta-analyses of this type can be extremely helpful for informing the development of a conceptual model and for hypothesis generation. A strong conceptual model informed by meta-analysis is an excellent starting point for selection of candidate intervention components.

However, meta-analyses do not take the place of an optimization trial. At this writing, meta-analyses aimed at understanding which components are essential in an intervention must be based on data from RCTs. These RCTs have combined components in different ways, but nevertheless, every component is confounded with other components. Perhaps in the future, it will be possible to conduct meta-analyses of all the optimization trials that have examined particular components. This would be likely to provide a more definitive assessment of component effectiveness than meta-analyses based on RCTs.

2.7 Including Moderation (Effect Modification) in a Figure Representing a Conceptual Model

Above it was mentioned that the hypothetical conceptual model discussed in this chapter includes modification of one component's effect by another component. This is depicted in Fig. 2.1. It was also mentioned that an investigator may wish to include other kinds of moderation, specifically effect modification, in a conceptual model. In this section an example of inclusion of effect modification in a figure is provided. Figure 2.3 is identical to Fig. 2.1, except for the addition of several instances of effect modification.

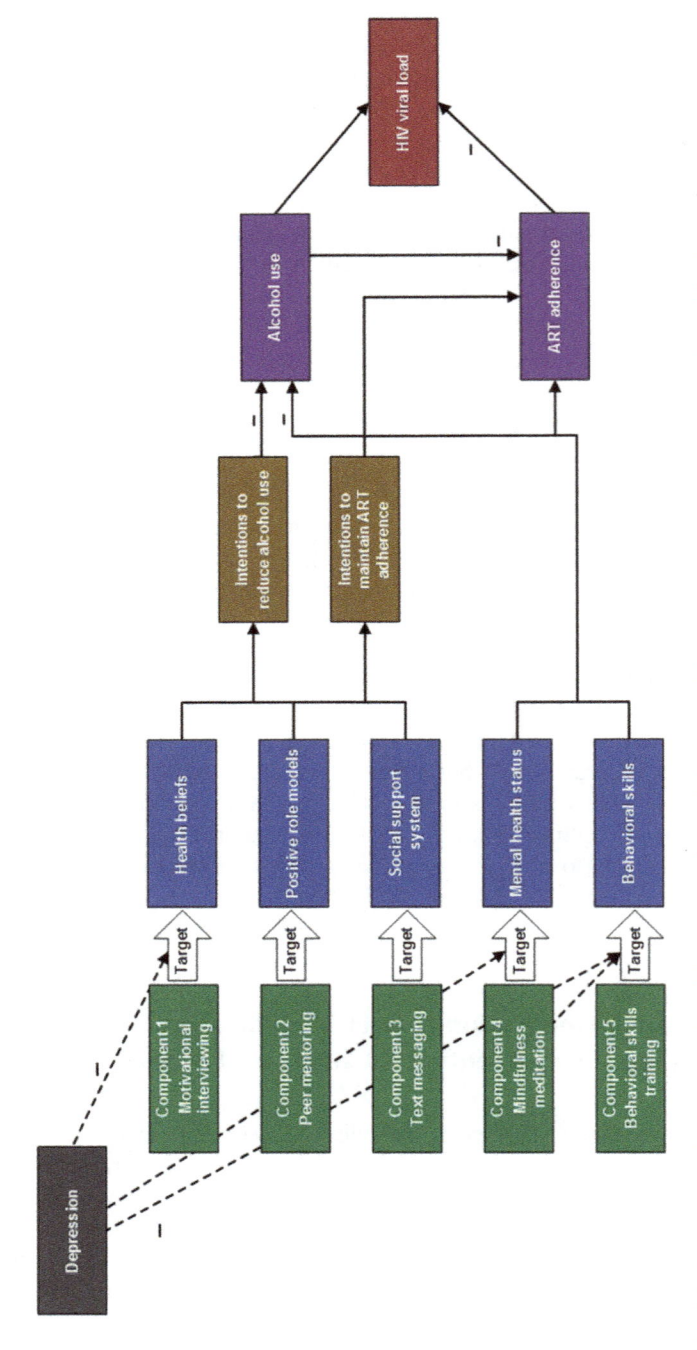

Fig. 2.3 Variation on the conceptual model depicted in Fig. 2.1. In this model, certain component effects are moderated by depression

Suppose Dr. B hypothesizes that individuals with depression will be less receptive to motivational interviewing and behavioral skills training, and therefore these components will be less effective for depressed individuals. However, Dr. B hypothesizes that mindfulness meditation will be more effective with depressed individuals, who will find it particularly helpful in coping with stress. This effect modification is depicted in Fig. 2.3 by a shaded box representing the moderator, depression, and dashed arrows indicating which effects are moderated. Where the moderator's impact is to reduce the modified effect, there is a minus sign next to the arrow.

Dr. B also plans to conduct exploratory analyses to examine other potential moderators, such as gender, age, race/ethnicity, sexual orientation, use of substances other than alcohol, and comorbidities other than depression. It would be possible to include all of these potential relations in the figure, but it is evident that this would add many arrows to the figure and could result in excessive complexity. Because Dr. B has not formulated specific hypotheses about these potential moderators, it seems reasonable to omit them from the figure but to discuss them in the narrative description of the conceptual model.

How important is it to include effect modification in a figure depicting a conceptual model? From one perspective, effect modification is not part of the engine driving an intervention; the intervention is expected to be effective on average. This would argue for omitting effect modification from the figure. However, from a different perspective, an understanding of effect modification can be an important way to illuminate the way forward to a better intervention. In the example, if the impact of depression on the effectiveness of each intervention component could be understood, this might suggest strategies for creating an intervention that is equally effective for depressed and non-depressed individuals. This would argue for including some effect modification in a figure, particularly effect modification considered critical with respect to the continual optimization principle.

The moderation depicted in Fig. 2.3 would be expressed statistically as two-way interactions (Chap. 4), which are interactions involving two predictor variables. In theory, interactions can involve any number of predictor variables, although in practice interpretation becomes increasingly difficult as more variables are involved. The convention of representing an interaction in a figure by using an arrow running from one of the predictor variables to another arrow works well for two-way interactions but is less useful for representing three-way and higher-order interactions. If a conceptual model includes some of these more complex instances of moderation, it is probably clearer to note and describe them in the narrative description of the conceptual model than to attempt to represent them in a figure.

2.8 Why Bother With a Conceptual Model?

Development of a well-specified conceptual model as described above is demanding and time-consuming. It is not unusual for a research team to find that it takes some effort to arrive at consensus about many of the details and that this may require hours of discussion and numerous revisions of the model. Why bother? A different approach would be for the investigators simply to generate a list of components without taking the trouble to articulate a conceptual model and investigate as many of these components as possible in the optimization trial.

The conceptual model is essential in MOST for several reasons. First, as will be discussed in the remainder of this book, the conceptual model is needed to guide subsequent decision-making in MOST, in particular during the choice of experimental design. Second, if two or more of the components represent alternative methods of changing the same mediator, there could be an undesirable degree of redundancy within some experimental conditions. Third, if the components are radically different in viewpoint or philosophy, they could work at cross purposes, or subjects could become confused or frustrated. Fourth, with such an approach, some mediators critical for effecting change on the outcome may not be targeted by any components, which could result in a less effective optimized intervention. Relying on a well-specified conceptual model when selecting components to examine helps the investigator avoid problems like these and provides valuable guidance throughout the MOST process.

2.9 The Role of Pilot Testing in MOST

Once a set of candidate intervention components has been identified, the next step is to determine the content of each component and how the components will be administered. For example, consider the components listed in Fig. 2.1. It will be necessary to determine what technique will be used for conducting the motivational interviews and how the staff who conduct the interviews will be trained, what the content of the text messages will be and how frequently they will be delivered, and so on. Development of the content and procedural aspects of intervention components varies widely from field to field and is outside the scope of this book. Once the content and procedural aspects of each component have been decided upon, it is often desirable to pilot test the components before moving on to the optimization phase of MOST.

The pilot study, which is conducted in the preparation phase of MOST, and the optimization trial, which is conducted in the optimization phase of MOST, have different purposes. (The optimization trial will be discussed at length in subsequent chapters.) They are not interchangeable. Before an optimization trial is done, it is a good idea to establish whether each level of each candidate intervention component is feasible to implement, understood by and acceptable to participants, safe, not unduly demanding of participants or staff, no more expensive or time-consuming than expected, and in general ready for inclusion in an intervention. Pilot studies are a way of accomplishing this.

There is wide agreement that pilot studies play a major role in science, but subdisciplines disagree about the definition of pilot study and its purpose. This book will use Leon, Davis, and Kraemer's (2011) helpful assessment of what is, and is not, the purpose of a pilot study:

> *The purpose of conducting a pilot study is to examine the feasibility of an approach that is intended to be used in a larger scale study... A pilot study can be used to evaluate the feasibility of recruitment, randomization, retention, assessment procedures, new methods, and implementation of the novel intervention. A pilot study is not a hypothesis testing study... a pilot study does not provide a meaningful effect size estimate...* (p. 626)

In other words, pilot studies are meant to be an early check on the approaches used in a research endeavor, with the idea that any needed modifications will be made before the actual experiment begins. Typically, resources are not devoted to maintaining a high level of statistical power when a pilot test is planned and executed. This is appropriate for an experiment with the objectives described by Leon et al. It follows that pilot studies are not powered for hypothesis testing.

Because they are not powered for hypothesis testing, pilot studies do not provide a basis for formal assessment of the effectiveness of a component. Thus experiments that are pilot studies by the Leon et al. definition cannot serve as optimization trials. Conversely, if an experiment is powered sufficiently to support optimization of an intervention, it is not a pilot study according to this definition. Decisions about which components comprise an optimized intervention in MOST are never made based on the results of a pilot study. However, a pilot study could serve a prescreening function if, based on its results, a component was deemed unready for consideration for inclusion in the intervention and, therefore, was not included in the set of candidate components examined in the optimization trial.

2.10 The Optimization Criterion

This section provides an overview of the optimization criterion, for two reasons. First, selection of the optimization criterion is a part of the preparation phase. Second, having an overview of the role of the optimization criterion will be helpful in understanding the material in Chaps. 3, 4, 5, and 6 and in seeing the relevance of this material to optimization of interventions. The optimization criterion will be discussed in more detail in Chap. 7, particularly its role in selection of the components and component levels that constitute the optimized intervention.

Recall that in Chap. 1 optimization of an intervention was defined as follows:

> *Optimization of an intervention is the process of identifying an intervention that provides the best expected outcome obtainable within key constraints imposed by the need for efficiency, economy, and/or scalability.*

This definition suggests that in order to start with a set of candidate intervention components and component levels and select the subset that makes up the optimized intervention, it is necessary to specify what is meant by (a) the best expected outcome and (b) key constraints. This specification is expressed in the optimization criterion. The expected outcome is expressed in terms of the primary outcome variable that has been identified for the optimization phase. The constraints may be expressed in units of money, time, or any other resource. For example, the optimization criterion "Best expected outcome on *adhere* for an implementation cost of no more than $500 per participant" states that the preferred combination of components/component levels is the one that is expected to provide the most days adherent to ART in the 30 days following the conclusion of participation, and the key constraint is that implementation cost cannot exceed $500.

Both the primary outcome variable and the constraints should be clearly operationalized. Note that the definition refers to key constraints. Although there may be many constraints, it is prudent to identify one or two critically important constraints and base the optimization on them. Otherwise, the optimization process may become overly complex.

2.10.1 Reasons for Identifying the Optimization Criterion in the Preparation Phase Rather Than the Optimization Phase

The reader may wonder why the optimization criterion is identified in the preparation phase, rather than the optimization phase. There are two reasons.

First, although usually the investigator will have a rough idea of the cost of each component in advance, to include money, time, or any other resource as a key constraint in the optimization criterion, it is a good idea to have accurate data on how much of the resource is demanded by each component and each combination of components. Data on cost can be collected during the experimentation conducted in the optimization phase. One reason for clearly specifying the optimization criterion during the preparation phase is to ensure that the research design includes plans for collecting whatever data are needed for optimization.

Second, if the optimization criterion involves an upper limit on cost or time, the investigator must avoid including any components that alone exceed or nearly exceed the upper limit. For example, if Dr. B is using the optimization criterion "Best expected outcome on *adhere* for an implementation cost of no more than $500 per participant," it would be a waste of resources to include a component that cost more than $500, because that component would be ineligible for selection into the optimized intervention.

2.10.2 The Role of the Screened-In Set in Optimization

In thinking about how the optimization criterion will be used, it is helpful to look ahead to the optimization phase of MOST. Suppose Dr. B has conducted an optimization trial to examine the effectiveness of each of the five candidate components under consideration. In this optimization trial, Dr. B compared two levels of each component: included in the intervention (yes) and not included in the intervention (no). Dr. B examined the results to determine which effects were large enough to be important in decision-making and which were not. Based on the important effects, Dr. B has divided the candidate components into the screened-in set and the screened-out set.

Recall from Chap. 1 that a component is selected for the screened-in set because it has demonstrated an empirically detectable and sufficiently large effect in the desired direction, in this case on the outcome *adhere*, or because it operates synergistically with one or more other components to enhance their effects. Suppose the screened-in set includes (note that these are hypothetical results) motivational interviewing, text messaging, mindfulness meditation, and behavioral skills training. The remaining component, peer mentoring, has been placed in the screened-out set. For now, do not be concerned about how these components were selected for inclusion in the screened-in set. Chap. 7 discusses this process in detail. The objective here is to provide a conceptual understanding of optimization and the role of the optimization criterion.

Once the components have been divided into the screened-in set and the screened-out set, any screened-out components, in this case peer mentoring, are set to the lower level. Then the screened-in set is considered further in relation to the optimization criterion, to select the levels of these components that will be included in the optimized intervention.

2.10.3 All Active Components

Sometimes no specific key constraints on time, money, or other resources have been identified, but nevertheless there is a desire to develop an intervention that is efficient in the sense that it does not waste resources on inactive or weak components. In this case a simple optimization criterion can be used, which will be called the all active components criterion. This criterion can be expressed as "best expected outcome irrespective of cost," or in the example, "best expected outcome on *adhere* irrespective of cost."

The all active components criterion is satisfied in a very straightforward manner, by comprising the optimized intervention out of the entire screened-in set, with each component set to the level associated with the better outcome. In the example, the optimized intervention would include motivational interviewing, text messaging, mindfulness meditation, and behavioral skills training.

Where implementation cost is not much of an issue, using the all active components optimization criterion can be a straightforward and reasonable strategy. For example, the financial costs associated with internet-delivered interventions are mostly in up-front development; implementation and dissemination are typically not very costly and also do not vary much according to the length or complexity of the intervention.

2.10.4 Optimization Criteria Involving Specific Constraints

Why not simply use the all active components criterion and seek an efficient and highly effective intervention irrespective of cost? Upon reflection it becomes evident that in most situations, even internet-delivered interventions, there may be costs of a kind associated with each component. For example, with internet-delivered interventions length may be a kind of cost; a participant may be more likely to abandon an overly long intervention without finishing it, as compared to a briefer one.

If identification of a limited number of sensible key constraints is thought through carefully during the preparation phase, the constraints can be included in the optimization criterion. For example, if the literature suggests that participants are most likely to complete an internet-delivered intervention if it takes no longer than, say, 45 min to complete, the optimization criterion could include this as a constraint. In this case the optimization criterion might be stated as "most effective that can be completed in 45 min or less." This approach can help produce an immediately scalable intervention and also help reduce the difference between efficacy and effectiveness.

The optimization criterion provides a basis for making the decisions necessary to arrive at an optimized intervention. Depending on the optimization criterion, all, some, or even (rarely) none of the components in the screened-in set may be included in the intervention. Thus, different optimization criteria may call for the inclusion of different components from the screened-in set in the intervention. This means that different investigators starting with the same empirical results but different optimization criteria could, and often would, arrive at different optimized interventions. This is not a problem; on the contrary, it is appropriate, because different optimization criteria make different statements about the objectives of the optimization, and in a sense offer different visions for the optimized intervention. Below several possible optimization criteria are reviewed briefly, to provide a sense of their role. These are merely examples out of many possibilities; different optimization criteria are suited to different applications of MOST. Details of how to use an optimization criterion to arrive at an optimized intervention are provided in Chap. 7 in the present volume and the chapter by Dziak (2018) in the companion volume.

2.10.5 Overview of Incorporating Constraints Into the Optimization Criterion

Here a brief overview of the use of optimization criteria involving constraints is provided, to convey a conceptual feel for the role of the optimization criterion. A more in-depth presentation can be found in Chap. 7.

Recall from Chap. 1 that Dr. B has contacted several government-funded HIV clinics to discuss how much they would be willing to pay to implement an intervention aimed at HIV-positive heavy drinkers. The conclusion was that if the

2.10 The Optimization Criterion

intervention effectively improves adherence to ART, and thereby reduces viral load, the clinics would be willing to pay up to $500 per participant. Dr. B will use the optimization criterion "best expected outcome on *adhere* that can be obtained for up to $500 per person." If the intervention that meets the all active components criterion costs more than $500, then Dr. B needs to decide which components to select (or set to the higher level) and which to jettison (or set to the lower level) to arrive at the best intervention that can be delivered within this budget constraint.

Suppose, as discussed above, Dr. B has conducted an optimization trial with *adhere* as the outcome variable, identified the important effects, and based on these effects selected which components go into the screened-in and screened-out sets. Motivational interviewing, text messaging, mindfulness meditation, and behavioral skills training go in the screened-in set. Because, as described above, each of the components has 2 possible levels, there are 16 different combinations of the levels of the components in the screened-in set. These are listed in Table 2.1. The combinations range from all four components set to the lower level, to all set to the higher level. Dr. B must now decide which of these combinations of component levels will, along with the no level of peer mentoring, form the optimized intervention.

A parsimonious prediction model that includes only the important effects can be used to arrive at a predicted *adhere* for each of the combinations of component levels listed in Table 2.1. Do not worry about the details of how this is accomplished; this is discussed in later chapters in this book, particularly Chap. 7. Hypothetical predicted outcomes are shown in Table 2.1. As the table shows, the best predicted outcome is associated with Combination 16, which corresponds to the all active components

Table 2.1 All possible combinations of levels of components in the screened-in set and predicted outcomes

Combination number	Motivational interviewing	Text messaging	Mindfulness meditation	Behavioral skills training	Predicted *adhere*
1	No[a]	No	No	Low	4
2	No	No	No	High	8
3	No	No	Yes	Low	10
4	No	No	Yes	High	14
5	No	Yes	No	Low	12
6	No	Yes	No	High	16
7	No	Yes	Yes	Low	18
8	No	Yes	Yes	High	22
9	Yes	No	No	Low	8
10	Yes	No	No	High	12
11	Yes	No	Yes	Low	14
12	Yes	No	Yes	High	18
13	Yes	Yes	No	Low	16
14	Yes	Yes	No	High	20
15	Yes	Yes	Yes	Low	22
16	Yes	Yes	Yes	High	26

[a]No means not included in intervention; yes means included in intervention

optimization criterion. However, as will be demonstrated below, Combination 16 may or may not be selected if a different optimization criterion is used.

Dr. B can take the following approach to identifying the optimized intervention. Dr. B determines the expected implementation cost of each combination of components. Sometimes this can be accomplished simply by noting the individual cost of providing each component. However, sometimes a combination of components may cost more or less than the simple sum of the individual costs; where this occurs, it must be factored into any cost computations. Once the expected cost of each combination of components is determined, the combinations that exceed the cost limit can be ruled out immediately. Then, of the remaining combinations, the one with the best predicted outcome is selected. (Examples of this procedure are discussed in Chap. 7.)

Now suppose instead of money, Dr. B has identified participant burden, operationalized as participant time spent on the intervention, as a key constraint, and wants to develop a very brief intervention that will take no more than 3.5 h of participant time. In this case the optimization criterion can be expressed as "Best expected outcome on *adhere* obtainable while demanding no more than 3.5 h of participant time." Dr. B can take an approach to identifying the optimized intervention that is essentially the same as the one taken when the optimization criterion involved an upper limit on money. To do this, it is necessary to determine the anticipated amount of time that will be required for each combination of components. Any combinations that require more than 3.5 h of participant time are then ruled out immediately. Of the remaining combinations, the one with the best expected outcome is selected.

One alternative to a firm upper limit on cost is cost-effectiveness. Here the optimization criterion could be stated something like "Best expected outcome on *adhere* per dollar (or other unit of currency) spent." The most cost-effective combination of components may be neither the least costly nor the most effective, and is likely to be different from the most effective combination that does not exceed a defined upper limit on cost. Making a decision based on cost-effectiveness can be complex; this approach is discussed in the Dziak (2018) chapter in the companion volume.

2.10.6 Other Kinds of Constraints

An optimization criterion can be based on key constraints on nearly any kind of resource, very broadly defined, as long as the constraint can be operationalized. For example, complexity of the intervention is often a consideration that affects scalability; if an intervention is highly complex, it can be difficult to ensure that staff are trained properly, and even properly trained staff can find it difficult to administer an intervention with fidelity in hectic environments. Complexity potentially can be included in the optimization criterion, but it may be a challenge to operationalize it. One possibility would be to limit complexity by putting an upper limit on the number of components that can be included in the intervention, so that the optimization criterion would be, for example, "best expected outcome on *adhere* obtainable with no more than three components." Another example of a constraint that may be a concern, particularly in cancer treatment, is cognitive burden. Cancer patients often experience chemotherapy-induced cognitive dysfunction, or "chemo brain," (e.g., Boykoff, Moieni, & Subramanian, 2009) making it imperative to keep cognitive demands as low as possible if good compliance is to be achieved. Provided that level of cognitive burden can be operationalized, it can form the basis of an optimization criterion.

2.10.7 *Constraints on Multiple Resources*

A single optimization criterion can include constraints on more than one resource. In this case it is necessary to specify how the constraints relate to each other. For example, suppose Dr. B decides to include constraints on both money and time in the optimization criterion. The optimization criterion must specify whether the constraint is on time AND money, or on time OR money. The following optimization criteria could result in different optimized interventions: "best expected outcome on *adhere* obtainable for no more than $500 per person AND while demanding no more than 10 h of participant time" and "best expected outcome on *adhere* obtainable for no more than $500 per person OR while demanding no more than 10 h of participant time." The former optimization criterion expresses the idea that to be acceptable the intervention must both cost $500 or less and require no more than 10 h of participant time. The latter expresses the idea that either an intervention costing $500 or less per person or an intervention requiring 10 h or less of participant time is acceptable. In other words, the intervention can either be more expensive than $500 or take more than 10 h, but not both. To identify the intervention meeting the latter optimization criterion, it will be necessary to identify (a) the intervention with the best expected outcome on *adhere* obtainable for no more than $500 per person and (b) the intervention with the best expected outcome on *adhere* obtainable while demanding no more than 10 h of participant time. Then the intervention that satisfies the optimization criterion is the one out of these two that has the better expected outcome on *adhere*.

It is also possible to combine cost-effectiveness with another constraint. For example, an alternative to "Best expected outcome on *adhere* per dollar spent" might be "Best expected outcome on *adhere* per dollar spent, obtainable for no more than $500 per person." This states that the desired intervention is the most cost-effective one that does not exceed an implementation cost of $500 per person. This intervention may or may not be the most cost-effective out of the alternatives that can be formed from the components in the screened-in set, and it may or may not be the most effective that can be obtained for $500 per person. Another alternative that could be considered is "Best expected outcome on *adhere* per hour of participant time expended, while demanding no more than 10 h of participant time."

2.11 Identifying Key Constraints

When specifying an optimization criterion, it is necessary to identify the key constraints. If data from a related previous study are available, it may be possible to examine them for suggestions about key constraints. For example, suppose a previous version of an intervention involved 12 sessions, but few participants attended more than 6 of these sessions. It may be worth considering limiting the intervention to a maximum of six sessions. However, there are constraints that may not arise until an intervention is implemented in a community setting, as opposed to the controlled settings in which most interventions are developed and evaluated. Examples are constraints on available staff time or on the expense or complexity of the intervention. Constraints of this type often become most salient after the intervention has gone to scale.

One approach to identifying key constraints is to use the RE-AIM framework and begin with the application in mind, as suggested by Klesges, Estabrooks, Dzewaltowski, Bull, and Glasgow (2005). Klesges et al. recommended "formative work with intervention settings and agents," noting such work is "helpful in... addressing potential staff and setting barriers to participation... and identifying unanticipated costs..." (p. 69). In other words, it may be helpful during the preparation phase to visit environments in which the planned intervention eventually is to be delivered and talk with administrators who will be making decisions about whether or not to adopt the intervention, personnel who would be expected to deliver or administer the intervention, and typical participants. Such individuals can be interviewed about what they see as the most important practical limits. Administrators may indicate that their budget would not allow consideration of an intervention that costs more than a particular amount of money. Intervention delivery personnel may say they would have to take measures to shorten any intervention that requires more than a certain amount of time to deliver, or simplify an intervention that exceeds a particular level of complexity. Participants may indicate that they can see themselves attending a 1 h session once a week for 5 weeks and would not have the time or inclination to do more. Information of this type can be very helpful in developing the optimization criterion.

2.12 What's Next

This chapter has reviewed the preparation phase of MOST. Chaps. 3, 4, and 5 discuss factorial and fractional factorial experiments, which can be a highly efficient approach to experimentation in the optimization phase. Chapter 6 discusses how to select the experimental design that is consistent with the resource management principle in a given situation.

References

Ajzen, I. (1985). From intentions to actions: A theory of planned behavior. In J. Kuhl & J. Beckman (Eds.), *Action control* (pp. 11–39). Berlin, Germany: Springer Heidelberg.

Almirall, D., Nahum-Shani, I., Wang, L., & Kasari, C. (2018). Experimental designs for research on adaptive interventions: Singly and sequentially randomized trials. In L. M. Collins & K. C. Kugler (Eds.), *Optimization of multicomponent behavioral, biobehavioral, and biomedical interventions: Advanced topics* (forthcoming). New York, NY: Springer.

Azar, M. M., Springer, S. A., Meyer, J. P., & Altice, F. L. (2010). A systematic review of the impact of alcohol use disorders on HIV treatment outcomes, adherence to antiretroviral therapy and health care utilization. *Drug and Alcohol Dependence, 112*(3), 178–193.

Bandura, A. (1986). *Social foundations of thought and action*. Englewood Cliffs, NJ: Prentice Hall.

Bandura, A. (1977). *Social learning theory*. Englewood Cliffs, NJ: Prentice Hall.

Boykoff, N., Moieni, M., & Subramanian, S. K. (2009). Confronting chemobrain: An in-depth look at survivors' reports of impact on work, social networks, and health care response. *Journal of Cancer Survivorship, 3*(4), 223–232.

Braithwaite, R. S., & Bryant, K. J. (2010). Influence of alcohol consumption on adherence to and toxicity of antiretroviral therapy and survival. *Alcohol Research & Health, 33*(3), 280.

Brown, J. L., Littlewood, R. A., & Vanable, P. A. (2013). Social-cognitive correlates of antiretroviral therapy adherence among HIV-infected individuals receiving infectious disease care in a medium-sized northeastern US city. *AIDS Care, 25*(9), 1149–1158.

Collins, L. M., & Kugler, K. C. (Eds.). (2018). *Optimization of multicomponent behavioral, biobehavioral, and biomedical interventions: Advanced topics*. New York, NY: Springer.

Collins, L. M., Kugler, K. C., & Gwadz, M. V. (2016). Optimization of multicomponent behavioral and biobehavioral interventions for the prevention and treatment of HIV/AIDS. *AIDS and Behavior, 20*, 197–214.

Cruess, D. G., Kalichman, S. C., Amaral, C., Swetzes, C., Cherry, C., & Kalichman, M. O. (2012). Benefits of adherence to psychotropic medications on depressive symptoms and antiretroviral medication adherence among men and women living with HIV/AIDS. *Annals of Behavioral Medicine, 43*(2), 189–197.

Duncan, L. G., Moskowitz, J. T., Neilands, T. B., Dilworth, S. E., Hecht, F. M., & Johnson, M. O. (2012). Mindfulness-based stress reduction for HIV treatment side effects: A randomized, waitlist controlled trial. *Journal of Pain and Symptom Management, 43*(2), 161–171.

Dziak, J. (2018). Optimizing the cost-effectiveness of a multicomponent intervention using data from a factorial experiment: Considerations, open questions, and tradeoffs among multiple outcomes. In L. M. Collins & K. C. Kugler (Eds.), *Optimization of multicomponent behavioral, biobehavioral, and biomedical interventions: Advanced topics* (forthcoming). New York, NY: Springer.

Glass, T. R., Battegay, M., Cavassini, M., De Geest, S., Furrer, H., Vernazza, P. L., . . . Swiss HIV Cohort Study. (2010). Longitudinal analysis of patterns and predictors of changes in self-reported adherence to antiretroviral therapy: Swiss HIV cohort study. *Journal of Acquired Immune Deficiency Syndromes, 54*(2), 197–203.

Gonzalez, J. S., Batchelder, A. W., Psaros, C., & Safren, S. A. (2011). Depression and HIV/AIDS treatment nonadherence: A review and meta-analysis. *Journal of Acquired Immune Deficiency Syndromes (1999), 58*(2), 181–187.

Grant, R. M., Lama, J. R., Anderson, P. L., McMahan, V., Liu, A. Y., Vargas, L., . . . Glidden, D. V. (2010). Preexposure chemoprophylaxis for HIV prevention in men who have sex with men. *New England Journal of Medicine, 363*(27), 2587–2599.

Harris, G. E., & Larsen, D. (2007). HIV peer counseling and the development of hope: Perspectives from peer counselors and peer counseling recipients. *AIDS Patient Care and STDs, 21*(11), 843–860.

Hendershot, C. S., Stoner, S. A., Pantalone, D. W., & Simoni, J. M. (2009). Alcohol use and antiretroviral adherence: Review and meta-analysis. *Journal of Acquired Immune Deficiency Syndromes, 52*(2), 180.

Klesges, L. M., Estabrooks, P. A., Dzewaltowski, D. A., Bull, S. S., & Glasgow, R. E. (2005). Beginning with the application in mind: Designing and planning health behavior change interventions to enhance dissemination. *Annals of Behavioral Medicine, 29*(2), 66–75.

Langebeek, N., Gisolf, E. H., Reiss, P., Vervoort, S. C., Thóra, B., Richter, C., ... Nieuwkerk, P. T. (2014). Predictors and correlates of adherence to combination antiretroviral therapy (cART) for chronic HIV infection: A meta-analysis. *BMC Medicine, 12*(1), 142.

Leon, A. C., Davis, L. L., & Kraemer, H. C. (2011). The role and interpretation of pilot studies in clinical research. *Journal of Psychiatric Research, 45*(5), 626–629.

MacKinnon, D. P. (2008). *Introduction to statistical mediation analysis.* New York: Lawrence Erlbaum Associates.

McIntosh, R. C., & Rosselli, M. (2012). Stress and coping in women living with HIV: A meta-analytic review. *AIDS and Behavior, 16*(8), 2144–2159.

Mo, P. K., & Coulson, N. S. (2008). Exploring the communication of social support within virtual communities: A content analysis of messages posted to an online HIV/AIDS support group. *Cyberpsychology & Behavior, 11*(3), 371–374.

Norton, W. E., Amico, K. R., Fisher, W. A., Shuper, P. A., Ferrer, R. A., Cornman, D. H., ... Fisher, J. D. (2010). Information–motivation–behavioral skills barriers associated with intentional versus unintentional ARV non-adherence behavior among HIV+ patients in clinical care. *AIDS Care, 22*(8), 979–987.

Nyamathi, A., Flaskerud, J. H., Leake, B., Dixon, E. L., & Lu, A. (2001). Evaluating the impact of peer, nurse case-managed, and standard HIV risk-reduction programs on psychosocial and health-promoting behavioral outcomes among homeless women. *Research in Nursing & Health, 24*(5), 410–422.

Parsons, J. T., Golub, S. A., Rosof, E., & Holder, C. (2007). Motivational interviewing and cognitive-behavioral intervention to improve HIV medication adherence among hazardous drinkers: A randomized controlled trial. *Journal of Acquired Immune Deficiency Syndrome, 46*(4), 443.

Purcell, D. W., Latka, M. H., Metsch, L. R., Latkin, C. A., Gómez, C. A., Mizuno, Y., ... INSPIRE Study Team. (2007). Results from a randomized controlled trial of a peer-mentoring intervention to reduce HIV transmission and increase access to care and adherence to HIV medications among HIV-seropositive injection drug users. *Journal of Acquired Immune Deficiency Syndromes, 46,* S35–S47.

Rivera, D. E., Hekler, E. B., Savage, J. S., & Downs, D. S. (2018). Intensively adaptive interventions using control systems engineering: Two illustrative examples. In L. M. Collins & K. C. Kugler (Eds.), *Optimization of multicomponent behavioral, biobehavioral, and biomedical interventions: Advanced topics* (forthcoming). New York, NY: Springer.

Safren, S. A., Reisner, S. L., Herrick, A., Mimiaga, M. J., & Stall, R. (2010). Mental health and HIV risk in men who have sex with men. *Journal of Acquired Immune Deficiency Syndromes (1999), 55*(Suppl 2), S74.

Shuper, P. A., Neuman, M., Kanteres, F., Baliunas, D., Joharchi, N., & Rehm, J. (2010). Causal considerations on alcohol and HIV/AIDS—A systematic review. *Alcohol and Alcoholism, 45*(2), 159–166.

Springer, S. A., Dushaj, A., & Azar, M. M. (2012). The impact of DSM-IV mental disorders on adherence to combination antiretroviral therapy among adult persons living with HIV/AIDS: A systematic review. *AIDS and Behavior, 16*(8), 2119–2143.

Sullivan, L. E., Goulet, J. L., Justice, A. C., & Fiellin, D. A. (2011). Alcohol consumption and depressive symptoms over time: A longitudinal study of patients with and without HIV infection. *Drug and Alcohol Dependence, 117*(2), 158–163.

Uhrig, J. D., Lewis, M. A., Bann, C. M., Harris, J. L., Furberg, R. D., Coomes, C. M., & Kuhns, L. M. (2012). Addressing HIV knowledge, risk reduction, social support, and patient

References

involvement using SMS: Results of a proof-of-concept study. *Journal of Health Communication, 17*(sup1), 128–145.

Van Ryzin, M. J., Roseth, C. J., Fosco, G. M., Lee, Y. K., & Chen, I. C. (2016). A component-centered meta-analysis of family-based prevention programs for adolescent substance use. *Clinical Psychology Review, 45,* 72–80.

W. K. Kellogg Foundation. (2004*). Using logic models to bring together planning, evaluation, and action: Logic model development guide.* Battle Creek, MI W.K. Kellogg Foundation. https://www.wkkf.org/resource-directory/resource/2006/02/wk-kellogg-foundation-logic-model-development-guide, downloaded 6/29/2017.

Wu, E. S., Metzger, D. S., Lynch, K. G., & Douglas, S. D. (2011). Association between alcohol use and HIV viral load. *Journal of Acquired Immune Deficiency Syndromes (1999), 56*(5), e129.

Chapter 3
Introduction to the Factorial Optimization Trial

Abstract In the multiphase optimization strategy (MOST), an intervention is optimized before it is evaluated in an RCT. The optimization is based on empirical evidence gathered in an optimization trial. This chapter reviews the factorial experiment, which is often a highly efficient way of conducting an optimization trial. The factorial experiment can be used to estimate the effectiveness of each candidate component and also to estimate the extent to which the effect of a component depends on the levels of one or more other components. This chapter also reviews the conclusion-priority and decision-priority perspectives on evaluating empirical data and when it is appropriate to take each perspective. Readers should be familiar with the material in Chaps. 1 and 2.

Contents

3.1	Introduction	68
3.2	The Basics of the Factorial Experiment	70
	3.2.1 Intervention Components vs. Factors	71
3.3	Scientific Information Provided by the Factorial Experiment: Main Effects and Interactions	72
	3.3.1 Main Effects	73
	3.3.2 Two-Way Interactions	74
	3.3.3 Three-Way and Higher-Order Interactions	75
	3.3.4 The Importance of Interactions in Building an Intervention	76
3.4	The Balance Property and the Efficiency of Factorial Experiments	76
	3.4.1 Criteria that Define Balance	76
	3.4.2 Balance and Efficiency	77
3.5	The Concept of Experimental Control	79
	3.5.1 The Concept of Control in the RCT and the Factorial Experiment	79
	3.5.2 Different Control for Different Factors in a Factorial Experiment	80
3.6	Including a Constant Component in a Factorial Experiment	81
3.7	Translating Intervention Components Into Factors and Levels of Factors	83
	3.7.1 Ensuring the Factors Can Be Fully Crossed	83
	3.7.2 Ensuring the Factors Can Be Manipulated Independently	84
3.8	When Subjects are Clustered	85
3.9	A Very Brief Review of Statistical Power	87
3.10	Conclusion-Priority and Decision-Priority Perspectives on Research	88

© Springer International Publishing AG 2018
L. M. Collins, *Optimization of Behavioral, Biobehavioral, and Biomedical Interventions*, Statistics for Social and Behavioral Sciences,
https://doi.org/10.1007/978-3-319-72206-1_3

    3.10.1 Conclusion-Priority Perspective vs. Decision-Priority Perspective 88
    3.10.2 When to Take Each Perspective 93
3.11 The General Linear Model (GLM) Approach to Classical Factorial Analysis
    of Variance (ANOVA) 94
    3.11.1 Effect Coding 94
    3.11.2 Dummy Coding 97
    3.11.3 Why Effect Coding Is Preferred for Analysis of Data from Factorial
           Optimization Trials 97
3.12 Powering a Factorial Experiment: Main Effects 99
    3.12.1 Power and Factorial Experiments 99
    3.12.2 Identifying Effect Sizes for Power Analysis From the Conclusion-Priority
           and Decision-Priority Perspectives 100
    3.12.3 Comparison of Two Putatively Active Levels of a Component 100
    3.12.4 Demonstration of Power Analysis for a Factorial Experiment 101
    3.12.5 The Negligible Impact on Power of Adding a Factor
           to a Factorial Experiment 102
    3.12.6 The Large Impact on Power of Adding a Level to a Factor: Why 2^k
           Factorial Optimization Trials are Recommended 105
    3.12.7 Recommendations Based on the Resource Management Principle 107
    3.12.8 Small N Situations 109
3.13 The Coefficient Correction: Why Is There a 2 in the Denominator
    of the Two-Way Interaction? 109
3.14 Summary: The Efficiency of the Factorial Experiment and the Efficiency
    of the RCT 111
3.15 What's Next? 112
References 113

3.1 Introduction

The previous chapters have mentioned that optimization of an intervention is based on scientific information about the performance of the components that are under consideration. Many behavioral, biobehavioral, and biomedical scientists have been trained primarily in the randomized controlled trial (RCT), and so when they are designing an experiment, they naturally turn to this familiar approach and its relatives. The standard two- or three-arm RCT is essential for obtaining the kind of information that is needed in the evaluation phase of MOST, where the research question concerns the performance of the intervention as a package. However, as has been stated previously, the information that is necessary for optimization generally cannot be obtained via this experimental design. In the optimization phase, the research questions concern the performance of the individual components that are under consideration for inclusion in the optimized intervention and whether the performance of one component varies depending on the presence or setting of one or more other components. The RCT is not an efficient way to collect information on the former, and it offers no information about the latter.

3.1 Introduction

Different experimental designs provide different types of information, and so a variety of approaches to experimentation can be considered for the optimization trial, as was discussed briefly in Chap. 1. Because the optimization trial determines the type, amount, and quality of information that will form the basis for optimization of the intervention, selecting the design of this experiment is one of the critical decisions made by an investigator using MOST.

For many applications of MOST, some variant of the factorial experiment will be used to collect the necessary information in an efficient and economical manner. Factorial experiments are routinely used in engineering, pharmacy, agriculture, and related fields, and they could play a much larger role in behavioral, biobehavioral, and biomedical research than they do today. Factorial designs, particularly those with multiple factors, can be highly efficient for optimization trials. When a factorial experiment is used for an optimization trial, this is sometimes called a screening experiment (e.g., McClure et al., 2012; Piper et al., 2016) if a primary objective is to screen out poorly performing candidate components. The term screening experiment is also frequently used in engineering to refer to factorial experiments.

Although the logical underpinnings of the factorial experiment are straightforward, they are very different from those of the RCT and therefore can initially be counterintuitive for scientists coming from an RCT background. However, many scientists find that once they become familiar with the ideas behind the factorial experiment, they appreciate its elegance and efficiency. This chapter will contrast the factorial experiment to the RCT as a way of helping readers who have been trained primarily in the RCT to gain an understanding of the factorial experiment.

When an investigator is considering what experimental design to use in the optimization phase, it is essential to compare a variety of approaches, including a standard factorial design as well as other appropriate designs. Chapter 5 discusses a more complex but potentially more economical special case of the factorial experiment, the fractional factorial. Chapter 6 discusses how to compare the resource demands of different experimental designs. The final decision about which experimental design to use in the optimization phase of MOST will be made based on the resource management principle; the approach selected will be the one that makes the best use of available resources while providing the necessary scientific information.

This chapter provides an elementary introduction to factorial experiments. It is assumed the reader has some familiarity with multivariate statistics, particularly multivariate regression. This book does not provide a comprehensive treatment of factorial experimental designs and analysis of the resulting data, because there are already many excellent books on this topic. Readers are referred to Kirk (2013) for a treatment of factorial experiments from a social and behavioral science perspective and Montgomery (2009) or Wu and Hamada (2009) for a treatment from an engineering perspective.

3.2 The Basics of the Factorial Experiment

To assist in the discussion of factorial experiments, let us return to the hypothetical example presented in Chap. 1, in which Dr. B is developing an intervention to increase adherence to the ART treatment regimen in HIV-positive individuals who drink heavily. For simplicity, in this chapter, only three components will be considered: (a) motivational interviewing, (b) peer mentoring, and (c) text messaging. The primary outcome variable is number of days of adherence to ART in the 30-day period following the conclusion of the intervention. This will be abbreviated *adhere*.

In a factorial experiment, two or more independent variables, which are called factors, are manipulated in a systematic manner so as to achieve a critical property called balance, which will be discussed below. Suppose Dr. B wishes to conduct a randomized factorial experiment to examine the effects of the three components in the hypothetical example. In the experiment there would be a factor corresponding to each candidate component. This book will use italicized uppercase abbreviations to represent experimental factors. In the hypothetical example here, *MI* will refer to the factor corresponding to the motivational interviewing component; *PEER* will refer to the factor corresponding to the peer mentoring component; and *TEXT* will refer to the factor corresponding to the text messaging component.

For now, let us assume that for each of these factors there are two possible levels: yes, which means the corresponding component is present, and no, which means the component is absent. Because there are three factors, and each factor has two possible settings, this would be a $2 \times 2 \times 2$, or 2^3, factorial experiment; thus there are $2^3 = 8$ experimental conditions. The conditions in this factorial experiment are listed in Table 3.1. Each experimental condition corresponds to a combination of levels of the three factors. For example, a subject randomly assigned to experimental condition 3 would be provided peer mentoring only and not provided with motivational interviewing or text messaging. An experimental condition in a factorial experiment is often referred to as a cell.

This book emphasizes experiments in which all the factors have two levels. These are known as 2^k designs. In general, 2^k designs are much more efficient for optimization trials than those in which one or more factors have more than two levels, for reasons that will be explained later in this chapter.

Table 3.1 Experimental conditions in 2^3 factorial design

Experimental condition	*MI*	*PEER*	*TEXT*
1	No[a]	No	No
2	No	No	Yes
3	No	Yes	No
4	No	Yes	Yes
5	Yes	No	No
6	Yes	No	Yes
7	Yes	Yes	No
8	Yes	Yes	Yes

[a]No means not included in intervention; yes means included in intervention

3.2 The Basics of the Factorial Experiment

3.2.1 Intervention Components vs. Factors

In discussing the use of factorial experiments in the optimization phase of MOST, it is necessary to be clear at all times about the distinction between an intervention component and an experimental factor. In this book they often will closely resemble each other, but they are different. A component is a candidate for inclusion in an intervention, whereas a factor is an independent variable in a factorial experiment. In this book, a factor is manipulated to obtain information about the performance of a component.

Suppose Dr. B is considering whether to include motivational interviewing as a component in an intervention. To gather the information needed to make this decision, Dr. B will conduct an experiment. In this experiment, there will be a factor corresponding to the motivational interviewing component. This factor will be manipulated in a principled way; in other words, in the experiment, the factor will take on more than one level so that the outcomes produced by different levels can be compared. The levels of the factor should correspond to possible levels or values that would be reasonable for the component to take on in the intervention. Dr. B is considering whether to include or not to include the motivational interviewing component, therefore, the corresponding factor should take on the levels no and yes. If instead Dr. B was sure that motivational interviewing would be included in the intervention, and wanted to decide whether to include a minimal or intense version, the levels of the corresponding factor would be minimal and intense.

Based on the results of the factorial experiment, a decision is made about which level of the component to select. Ultimately, Dr. B needs to select one and only one level for each component to include in the intervention. Suppose the levels of the factor being compared are no and yes. If the experiment suggests that there is a difference between these levels, then consideration may be given to setting the component to the yes level, in other words, including it in the intervention. If the experiment suggests that there is not a detectable difference between these levels, then the component will be set to the no level and not included. Now suppose the levels of the factor being compared in the experiment are minimal and intense. If the experiment suggests that the intense level outperforms the minimal level, consideration may be given to setting the component to the intense level. By contrast, if the experiment suggests that there is no detectable difference between the minimal and intense levels, the component will be set to minimal.

This suggests that the levels of a factor can be conceptualized in different ways. Consider several possibilities for the levels of the factor *TEXT*. (i) The levels might correspond to the presence or absence of text messaging; for example, the levels of *TEXT* could be no (do not provide text messaging) and yes (e.g., provide text messaging five times per day). (ii) The levels might correspond to particular doses or intensities that are being considered, for example, four per week and one per day. (iii) The levels might represent different treatment alternatives, say text messages and telephone calls.

Usually a component could conceivably be included in an intervention set to any of many different levels (including the special case of a level representing the absence of the component), but only a small number of these possibilities, frequently only two, will be examined in a particular experiment. Whenever an independent variable is manipulated in an experiment, the goal is to address a scientific question. The levels of a component that will be examined in an experiment, that is, the levels that will make up a factor, must be selected so as to enable the experiment to address this scientific question appropriately. The three different ways of conceptualizing the factor listed above address three different scientific questions.

If the levels selected for the factor are no and yes, the scientific question is whether *adhere* is greater when subjects are provided with text messages (say, five times per day) as compared to when they are not provided with any text messages. When the levels represent different doses, the scientific question is whether text messages four times per week or once per day produce the greater *adhere*. With this framing of the factor, text messaging is not compared to a no-treatment control; the results would help to determine whether the more intense dose of text messages is better than the less intense dose but would not help to determine whether text messaging is better than no text messaging. Manipulating the factor in this way could be appropriate if previous research had established the efficacy of text messages, and now the question concerns the appropriate dose. When the levels are text message and phone call, the scientific question is whether *adhere* is greater when subjects receive text messages or phone calls. With this framing of the factor, there would not be a way to determine whether either text messages or phone calls are better than a control consisting of not receiving either one. Manipulating the factor in this way could be appropriate if (a) previous research had established the efficacy of both text messages and phone calls, and (b) it had been decided a priori that one *and only one* of these methods of providing support would be included in the intervention.

In summary, ultimately the investigator must use the results of the optimization trial to choose the level of each component that will be included in the intervention. If for a particular factor the levels are no and yes, the results can be used to help the investigator decide whether to include the component (by selecting the yes level) or not to include the component (by selecting the no level). If the levels represent alternative intensities of the component, for example, a low dose and a high dose, there has been an implicit a priori decision that the component is going to be included in the intervention; the question at hand concerns not whether to include the component but what its intensity should be. This book will often refer to selecting components and component levels, to emphasize that both kinds of decisions, that is, whether to include a component at all and what its intensity should be, may be needed in the process of optimization.

3.3 Scientific Information Provided by the Factorial Experiment: Main Effects and Interactions

On the surface, the factorial experiment looks like a variant of the RCT. In fact, it would be possible (but not recommended) to treat the data from a factorial experiment like the one in Table 3.1 as if it had come from an eight-arm RCT, by analyzing the data using a one-way ANOVA that directly compared the means of the

3.3 Scientific Information Provided by the Factorial Experiment: Main... 73

Table 3.2 Notation for a factorial design with two levels per factor

Experimental condition (cell)	MI	PEER	TEXT	Mean
1	−	−	−	μ_{---}
2	−	−	+	μ_{--+}
3	−	+	−	μ_{-+-}
4	−	+	+	μ_{-++}
5	+	−	−	μ_{+--}
6	+	−	+	μ_{+-+}
7	+	+	−	μ_{++-}
8	+	+	+	μ_{+++}

experimental conditions on the *adhere* outcome variable. Despite this superficial resemblance to a multi-arm RCT, the factorial experiment's purpose is not to address questions about comparisons between its individual experimental conditions. In fact, as will be explained below, using a factorial experiment for this purpose would overlook its signature strengths.

In the classical approach to ANOVA, introduced by Fisher (e.g., Fisher, 1925, 1971), data from a factorial experiment are analyzed to examine two types of effects: main effects and interactions. Before defining these effects, it is necessary to introduce some notation. Table 3.2 shows the conditions in the factorial experiment depicted in Table 3.1, expressed in a slightly different way. In Table 3.2 the higher level, yes, is represented by a plus sign +, and the lower level, no, is represented by a minus sign −. (Which level is assigned + and which is assigned − is largely arbitrary.) The Greek letter μ will be used to represent the mean of each experimental condition on the outcome variable, with subscripts to refer to the settings of the independent variables in a specific condition. For example, μ_{+--} represents the mean of *adhere* in experimental condition 5, in other words, the mean for cell 5, in which *MI* is set to yes and *PEER* and *TEXT* are set to no.

3.3.1 Main Effects

When a factor has two settings, the classical definition of the main effect is the difference between the mean response at one setting of the factor and the mean response at the other setting, collapsing over the levels of all remaining factors (e.g., Montgomery, 2009). In the example, the main effect of *MI* is defined as

$$ME_{MI} = \bar{\mu}_{+..} - \bar{\mu}_{-..}$$

(e.g., Kirk, 2013, p. 360). Here the notation $\bar{\mu}$ denotes a mean of means, that is, a mean taken across several experimental conditions, and the dot subscript means "averaged over." $\bar{\mu}_{+..}$ and $\bar{\mu}_{-..}$ represent the mean of *adhere* for the higher (yes) and lower (no) settings, respectively, of the first factor, *MI*, averaging over all the levels of the second and third factors, *PEER* and *TEXT*. In the hypothetical example, $\bar{\mu}_{+..}$

represents the mean of *adhere* for cells 5–8 in Table 3.2, and $\bar{\mu}_{-..}$ represents the mean of *adhere* for cells 1–4. Thus, in the factorial experiment described above, the main effect of *MI* is the difference between the mean of *adhere* when motivational interviewing is provided as part of the intervention and the mean of *adhere* when it is not provided.

Similarly,

$$ME_{PEER} = \bar{\mu}_{.+.} - \bar{\mu}_{.-.};$$

that is, the main effect of *PEER* is the mean of cells 3, 4, 7, and 8 in Table 3.2 minus the mean of cells 1, 2, 5, and 6; or the difference between the mean of *adhere* when peer mentoring is provided as part of the intervention and when it is not. Finally, the main effect of *TEXT* is the mean of cells 2, 4, 6, and 8 minus the mean of cells 1, 3, 5, and 7,

$$ME_{TEXT} = \bar{\mu}_{..+} - \bar{\mu}_{..-};$$

in other words, the difference between the mean of *adhere* when text messaging is provided as part of the intervention and when it is not.

If a factor demonstrates a large main effect, this does not guarantee that it will have the same effect at any particular combination of settings of the other factors. As discussed briefly in the next section and at length in Chap. 4, if a factor interacts with one or more other factors in the experiment, its effect may vary depending on the levels of the other factors.

3.3.2 Two-Way Interactions

Two factors interact if the effect of one is different depending on the level of the other. A two-way interaction is half of the difference in the effect of a particular factor across the level of a second factor, averaging over all other factors (Montgomery, 2009; p. 217). In the example, the *MI*×*PEER* interaction is defined as

$$INT_{MI \times PEER} = \frac{(\bar{\mu}_{++.} - \bar{\mu}_{-+.}) - (\bar{\mu}_{+-.} - \bar{\mu}_{--.})}{2}. \tag{3.1}$$

(e.g., Montgomery, 2009, p. 217). The numerator of the *MI*×*PEER* interaction is the difference between the effect of *MI* on *adhere* when *PEER* is set to yes and when *PEER* is set to no.

$$INT_{MI \times PEER} = \frac{\overbrace{(\bar{\mu}_{++.} - \bar{\mu}_{-+.})}^{PEER=yes} - \overbrace{(\bar{\mu}_{+-.} - \bar{\mu}_{--.})}^{PEER=no}}{2}$$

In other words, Eq. (3.1) represents the difference between the effect of *MI* on *adhere* when the intervention also includes peer mentoring and when the intervention does not include peer mentoring, averaging over the two possible levels of *TEXT*. If the effect of *MI* on *adhere* is different when peer mentoring is provided—for example, if the effect of MI on *adhere* increases when peer mentoring is also provided—then *MI* and *PEER* interact. If the effect of *MI* on *adhere* is the same irrespective of whether or not peer mentoring is provided, this difference is zero—*MI* and *PEER* do not interact. It is equivalent to state this in terms of the effect of *PEER* when *MI* is set to yes or no.

In this book, the denominator of the expression for the interaction, in this case 2, will be called the coefficient correction. An explanation of why the coefficient correction is necessary appears later in this chapter.

3.3.3 *Three-Way and Higher-Order Interactions*

Three factors interact when a two-way interaction is different depending on the level of a third factor. In the example, the $MI \times PEER \times TEXT$ interaction is defined as

$$INT_{MI \times PEER \times TEXT} = \frac{[(\mu_{+++} - \mu_{-++}) - (\mu_{+-+} - \mu_{--+})] - [(\mu_{++-} - \mu_{-+-}) - (\mu_{+--} - \mu_{---})]}{4}.$$

(3.2)

The numerator of the $MI \times PEER \times TEXT$ interaction is the difference between the $MI \times PEER$ interaction when *TEXT* is set to yes and when *TEXT* is set to no. In other words, Eq. (3.2) represents the difference between the $MI \times PEER$ interaction when the intervention includes text messaging and when the intervention does not include text messaging.

$$INT_{MI \times PEER \times TEXT} = \frac{\overbrace{[(\mu_{+++} - \mu_{-++}) - (\mu_{+-+} - \mu_{--+})]}^{TEXT=yes} - \overbrace{[(\mu_{++-} - \mu_{-+-}) - (\mu_{+--} - \mu_{---})]}^{TEXT=no}}{4}$$

If the $MI \times PEER$ interaction effect is different when text messaging is provided as compared to when it is not provided, there is an $MI \times PEER \times TEXT$ interaction. There are no bars over the μs in the above expression because this particular example involves only three factors, so the means are not averaged over settings of any additional factors. If there were additional factors in the experiment, there would be bars over the μs, indicating that the means were averaged over those factors.

Note that for the three-way interaction the denominator, or coefficient correction, is increased to 4; this will be explained later in this chapter.

Four-way interactions, five-way interactions, and so on are defined using straightforward extensions of the definitions above. The coefficient correction increases as the order of the interaction increases. Interactions can involve up to all the factors in a factorial experiment. The order of an interaction is the number of factors involved in the interaction, with interactions involving more factors referred to as higher-order interactions.

3.3.4 The Importance of Interactions in Building an Intervention

As will be discussed at length in Chap. 7, the approach to decision-making presented in this book is based first on main effects, followed by careful consideration of interactions. The critical role played by interactions in decision-making in MOST and the importance of estimating them is one reason why this book emphasizes factorial experiments. Interactions need to be taken into account when selecting components and component levels because some components may enhance or reduce the effects of others. For example, in a meta-analysis of data primarily from RCTs and quasi-experimental evaluations, Michie, Abraham, Whittington, McAteer, and Gupta (2009) found that healthy eating and physical activity interventions that included self-monitoring along with at least one other component were the most effective. This suggests that self-monitoring components may interact with other components to enhance their effects. A factorial experiment would be a good way to test this hypothesis. Interactions are discussed at length in Chap. 4.

3.4 The Balance Property and the Efficiency of Factorial Experiments

3.4.1 Criteria that Define Balance

The ideal factorial experiment is balanced. A factorial experiment is balanced if it meets two criteria. The first criterion is that the correct array of experimental conditions has been included in the design. To explain, let us return to the hypothetical factorial design depicted in Table 3.1. It is evident that *MI* is set to yes in half of the conditions and set to no in half of the conditions. Similarly, it is evident that *PEER* is set to yes in half of the conditions and set to no in half of the conditions. In fact, examination of the experimental conditions that make up any balanced factorial experiment will show that each level of each factor appears the same number of

3.4 The Balance Property and the Efficiency of Factorial Experiments

times. The second criterion to achieve balance is that every experimental condition has the same number of subjects.

Assume for the moment that the second criterion, that of equal cell sizes, is met. Then if an experimental design includes all possible combinations of the levels of the factors, it will be balanced. This is sufficient to achieve balance but not necessary. A factorial design may be balanced even if it does not include all of the possible combinations of levels of factors, as long as each level of each factor appears the same number of times. Fractional factorial designs are balanced factorial designs that do not include all possible combinations of levels of the factors. These designs, which can be very efficient under some circumstances, are discussed in Chap. 5.

3.4.2 Balance and Efficiency

There are two reasons why balance is important in factorial experiments. First, balance ensures that the estimates of the regression coefficients are uncorrelated, assuming the data are analyzed using effect coding, which produces estimates of main effects and interactions that are equivalent to the definitions presented above. The difference between effect coding and dummy coding is discussed briefly later in this chapter and at length in the Kugler, Dziak, and Trail (2018) chapter in the companion volume (Collins & Kugler, 2018). Of course, in practice most factorial experiments, particularly those conducted in field settings (i.e., most factorial experiments of interest to readers of this book), have some variability in cell sizes and therefore are unbalanced to an extent. Fortunately, minor statistical adjustments can deal with variability in cell sizes and will produce uncorrelated or nearly uncorrelated effect estimates, unless the imbalance is extreme. (Discarding data to achieve balance is almost never a good idea.) Second, balanced factorial experiments make extremely efficient use of experimental subjects for estimation of main effects and interactions. To see why, let us contrast estimation of main effects and interactions based on a factorial experiment to estimation of effects based on an RCT.

RCTs are motivated by scientific questions that revolve around direct comparison of the outcome in one experimental condition to the outcome in a control condition or a different experimental condition. Therefore, when data from an RCT are analyzed, effects are estimated by directly comparing the means of individual experimental conditions. By contrast, in factorial ANOVA, individual cell means generally are NOT directly compared. Instead, main effects and interactions are estimated based on comparison of aggregate means obtained by combining cells in different ways. *This is a fundamental conceptual difference between the RCT and the factorial experiment.*

Table 3.3 provides the same illustration of the experimental conditions in the example as was provided in Table 3.1, for the purpose of illustrating how the main effects would be computed. Consider the top panel of Table 3.3, which illustrates the main effect of *MI*. Following Eq. (3.1) above, this main effect would be estimated by

Table 3.3 Illustration of computation of three main effects

	Main effect of *MI*: mean of experimental conditions 5–8 compared to mean of experimental conditions 1–4		
Experimental condition (cell)	MI	PEER	TEXT
1	No[a]	No	No
2	No	No	Yes
3	No	Yes	No
4	No	Yes	Yes
5	Yes	No	No
6	Yes	No	Yes
7	Yes	Yes	No
8	Yes	Yes	Yes

	Main effect of *PEER*: mean of experimental conditions 3, 4, 7, 8 compared to mean of experimental conditions 1, 2, 5, 6		
Experimental condition (cell)	MI	PEER	TEXT
1	No	No	No
2	No	No	Yes
3	No	Yes	No
4	No	Yes	Yes
5	Yes	No	No
6	Yes	No	Yes
7	Yes	Yes	No
8	Yes	Yes	Yes

	Main effect of *TEXT*: mean of experimental conditions 2, 4, 6, 8 compared to mean of experimental conditions 1, 3, 5, 7		
Experimental condition (cell)	MI	PEER	TEXT
1	No	No	No
2	No	No	Yes
3	No	Yes	No
4	No	Yes	Yes
5	Yes	No	No
6	Yes	No	Yes
7	Yes	Yes	No
8	Yes	Yes	Yes

[a]No means not included in intervention; yes means included in intervention

subtracting the mean of the no level of *MI* (i.e., the mean of experimental conditions 1–4 (light shading)), from the mean of the yes level of *MI* (i.e., the mean of experimental conditions 5–8 (dark shading)). Now consider the main effect of *PEER*, illustrated in the middle portion of Table 3.3. This would be estimated by subtracting the mean of the no level of *PEER* (i.e., the mean of experimental conditions 1, 2, 5, and 6 (light shading)), from the mean of the yes level of *PEER* (i.e., the mean of conditions 3, 4, 7, and 8 (dark shading)). Finally, consider the main effect of *TEXT*, illustrated in the bottom portion of Table 3.3. This would be estimated by subtracting the mean of the no level of *TEXT* (i.e., the mean of the odd-numbered experimental conditions (light shading)), from the mean of the yes level of *TEXT* (i.e., the mean of the even-numbered experimental conditions (dark shading)).

In the RCT, each subject is assigned to a particular treatment group or the control group, and a subject's data contributes only to comparisons involving that treatment group's mean. For example, consider an RCT involving three experimental conditions: Treatment A, Treatment B, and a control. If a subject is randomly assigned to Treatment A, that subject's data can contribute to the comparison of Treatment A to the control, and to the comparison of Treatment A to Treatment B, but not to the comparison of Treatment B to the control. This approach is very efficient for direct comparison of individual experimental condition means.

By contrast, as Table 3.3 illustrates, *in a 2^k factorial ANOVA based on effect coding, all main effect estimates are based on the entire sample of subjects.* (This is true of interactions also, but it is easier to illustrate with main effects.) This is possible because when the main effect of each factor is estimated, in a sense the subjects are re-sorted into different levels of each factor. In a factorial design, the same subject may be a treatment subject for one factor and a control subject for another factor. For example, a subject in experimental condition 4 is in the no level for purposes of estimation of the main effect of *MI* and in the yes level for purposes of estimation of the main effects of *PEER* and *TEXT*. Collins, Dziak, and Li (2009) called this "recycling" of subjects. This approach is very efficient for estimation of main effects and interactions. Later in this chapter, the implications for statistical power in factorial experiments will be discussed.

3.5 The Concept of Experimental Control

3.5.1 The Concept of Control in the RCT and the Factorial Experiment

The RCT and related designs use a different concept of control group than factorial experiments. In RCT-type designs, there are one or more treatment conditions and a single control group. Usually each treatment condition is compared to the control group.

By contrast, in factorial experiments, each factor has its own control. Let us refer again to Table 3.3. As was discussed previously, the main effect of *MI* is found by comparing the mean of Conditions 5–8 to the mean of Conditions 1–4. Thus, the aggregate of Conditions 5–8 is the treatment group for the *MI* factor, and the aggregate of Conditions 1–4 is the control group for the *MI* factor. Similarly, the aggregate of Conditions 3, 4, 7, and 8 is the treatment group for the *PEER* factor, and the aggregate of Conditions 1, 2, 5, and 6 is the control group for this factor.

In an RCT a subject is either a treatment subject or a control subject. As mentioned above, in a factorial experiment, the same subject may be a treatment subject for one factor and a control subject for another. For example, consider a subject who is assigned to Condition 3. This individual is a treatment subject for *PEER* and a control subject for *MI* and *TEXT*.

In most (but not all; see Chap. 5) factorial experiments, there is a condition in which all of the factors are set to the lowest level. Superficially, such a condition closely resembles the control condition in an RCT. However, the role of this condition in a factorial experiment is not the same as that of a control condition in an RCT. To see why, consider Condition 1 in the example. All three factors are at the no level in Condition 1, so it may be tempting to refer to this as "the control condition." But as was demonstrated above, conducting an effect-coded ANOVA to estimate the main effects and interactions based on data from this factorial experiment NEVER calls for direct comparison of the mean of Condition 1 alone to any other condition mean. Thus Condition 1 is not "the control condition" for the experiment. In general, it is best to refrain from referring to any single condition in a factorial experiment as "the control condition."

3.5.2 *Different Control for Different Factors in a Factorial Experiment*

It follows from the discussion in the above section that in a factorial experiment, experimental control is approached on a factor-by-factor basis. In other words, when designing a carefully controlled factorial experiment, the investigator proceeds by ensuring that there is appropriate control for each factor, rather than identifying a single control group.

A no level, in which nothing is provided, may be the most appropriate control for a particular factor in an experiment. For a different factor, it may be more appropriate for ethical reasons to provide the current standard of care. For yet another factor in the experiment, it may be scientifically important to provide some kind of placebo, such as a placebo pill or an attention placebo. Various kinds of control can readily be incorporated into a factorial experiment, resulting in several different kinds of controls in a single factorial experiment.

In the hypothetical example, suppose an attention placebo is to be provided to subjects in the control level of the *PEER* factor. Further suppose that subjects in the control level of the *TEXT* factor will receive as many text messages as the subjects in the yes condition, but instead of pertaining to ART adherence, these texts will pertain to healthy eating. Table 3.4 lists the conditions in this experimental design, illustrating how each factor has its own control and how the controls may be different across factors.

Table 3.4 Hypothetical experiment with different controls for different factors

Experimental condition (cell)	MI	PEER	TEXT
1	No[a]	Placebo	Healthy eating
2	No	Placebo	ART
3	No	Yes	Healthy eating
4	No	Yes	ART
5	Yes	Placebo	Healthy eating
6	Yes	Placebo	ART
7	Yes	Yes	Healthy eating
8	Yes	Yes	ART

[a]No means not included in intervention; yes means included in intervention

3.6 Including a Constant Component in a Factorial Experiment

Sometimes a particular component at a particular setting, or a set of components, is to be provided to all subjects. One reason may be that there is specific intervention content that must be provided to all subjects for ethical or logistical reasons. In the hypothetical example, the investigators may decide that it is necessary for anyone who is HIV+ to be provided with some basic information about the physiological effects of alcohol and the importance of ART adherence. Another reason may be that prior research has established the effectiveness of several intervention components. Because empirical research has established that these components work, it would be unethical to withhold them in an experiment, and so it is a given that they are to be included, even if other components are to be examined experimentally. (This is not meant to imply that such components cannot be investigated in an experiment. If it were desired to test new, updated versions of components that previously have been shown to be effective, the corresponding factor levels could be current and new.) In this book the term constant component will be used to refer to a component or set of components that is provided to all subjects in a factorial optimization trial rather than experimentally manipulated.

Because the constant component is not manipulated, there is no factor corresponding to this component in the experiment. Nevertheless, it is a good idea to note the presence of a constant component when reporting on the experimental design, because the constant component is in a sense a part of the experiment. Table 3.5 illustrates one way of doing this. This table shows the design in Table 3.1 with the addition of a constant component providing basic information.

The constant component is not allowed to vary in a factorial experiment, so it is not possible to estimate whether it has a main effect or interacts with any of the factors in the experiment. However, this does not imply it has no effect! In fact, it is precisely because the constant component can have an effect that it is advisable to include mention of it when describing the experimental design, even though the constant component does not correspond to a manipulated factor. If the constant

Table 3.5 Hypothetical experiment in Table 3.1 with constant component added

Experimental condition	Basic information	MI	PEER	TEXT
1	Yes[a]	No	No	No
2	Yes	No	No	Yes
3	Yes	No	Yes	No
4	Yes	No	Yes	Yes
5	Yes	Yes	No	No
6	Yes	Yes	No	Yes
7	Yes	Yes	Yes	No
8	Yes	Yes	Yes	Yes

[a]Yes means included in intervention; no means not included in intervention

component exerts a main effect only, it will simply increase or decrease the grand mean and will have no impact on the estimates of the effects of the experimental factors. In this case, any observed main effects or interactions involving the factors in the experiment would be generalizable to an experiment that does not include the constant component. However, if the constant component interacts with any of the experimental factors, then some effects would be different depending on whether or not the constant component is present. For example, suppose a main effect for *MI* is observed. Further suppose the basic information about the physiological effects of alcohol and the importance of ART adherence that is included in the constant component is necessary for motivational interviewing to be effective, because without this information subjects do not attend sufficiently to *MI*. In this case, the constant component interacts with *MI*; *MI* has an effect when the constant component is present and would have no (or a smaller) effect if the constant component were absent.

Any effects estimated when a constant component is included are conditioned on the presence of that constant component. Where interactions exist involving the constant component and experimental factors, the effects observed in the experiment that includes the constant component may be different than they would have been in an experiment without a constant component. It follows that the performance of the corresponding components may be different in an intervention package that includes the constant component as compared to one that does not include it. However, the experimental design does not permit empirical investigation of this possibility, because by definition a constant component is not manipulated. For this reason, in general it is best if the constant component is limited to content that is already identified for inclusion in the optimized intervention.

Another situation in which a constant component may be considered *with caution* is when there are insufficient resources to examine all of the components that are candidates for inclusion in the optimized intervention. This situation may arise when an existing intervention that is made up of many components is to be optimized. In this case the investigators may wish to select a subset of components to be examined experimentally and to provide the remaining components to research subjects in a constant component. For example, suppose Dr. B is interested in optimizing an existing ART adherence intervention that had previously been developed and evaluated using the classical approach. Resources are available to investigate three components in a factorial experiment, but this intervention has more than three components. Dr. B may then select, say, motivational interviewing, peer mentoring, and text messaging to examine in the experiment, and provide the remaining components to all subjects in a combined constant component. In this design it will not be possible to investigate whether the constant component interacts with any of the experimental factors, so this can be a risky approach unless the investigator is confident that such interactions are likely to be small.

3.7 Translating Intervention Components Into Factors and Levels of Factors

When designing a factorial optimization trial, it is helpful to keep two important considerations in mind.

3.7.1 Ensuring the Factors Can Be Fully Crossed

The first consideration is that it must be possible to fully cross all the factors. This means that every combination of levels of the factors, that is, every experimental condition, must be logically implementable. (This is highly desirable even if a fractional factorial design is planned in which only a subset of experimental conditions will be run; see Chap. 5.) Examination of any of the tables in this chapter listing the experimental conditions in a factorial experiment shows that, in every case, all of the conditions are made up of logical combinations of factors.

As an example of a situation in which two factors cannot be crossed, consider the following scenario. Suppose Dr. B wishes to examine whether better outcomes can be obtained by including an online component to follow up on and reinforce what is covered in the motivational interviewing sessions. Dr. B therefore plans to add a fourth factor to the experiment, *ONLINE*, with levels no and yes. However, this is not possible, because *ONLINE* cannot be crossed with *MI*. The combination of *MI* set to no and *ONLINE* set to yes does not make sense; how can subjects be presented with an online component to follow up on a motivational interviewing session they did not have?

If the effectiveness of the inclusion of an online follow-up to motivational interviewing is to be examined in the experiment, either the experimental design or the research question, or both, will have to be revised. This can be accomplished in several ways. (a) The online component can be reconceptualized so that it is a stand-alone component that can be delivered irrespective of whether *MI* is set to no or yes. For example, the online component could contain material intended to be helpful whether or not the subject has been assigned to receive motivational interviewing. There would then be an additional factor, *ONLINE*, with levels no and yes. This factor could be crossed with *MI* and the other two factors. As explained later in this chapter, this approach will not require many additional subjects as long as the expected effect size of *ONLINE* is no smaller than the effect size used to power the experiment. However, adding this two-level factor will double the number of experimental conditions. (b) The *MI* factor can be changed so that it has three levels: no, motivational interviewing only, and motivational interviewing plus online follow-up. However, as will be explained below, changing *MI* from a two-level factor to a three-level factor will increase the number of experimental conditions from 8 to 12 and will also increase the sample size requirements by at least 50%. (c) The *MI* factor can be reframed so that it is still two levels, but now the two levels are no and yes plus online follow-up. If the experimental results suggest that this difference is large enough to be considered important, a subsequent experiment may be considered to determine whether the online follow-up is necessary, or the effect is due primarily to motivational interviewing alone. If the experimental results suggest that there is little or no effect of providing motivational interviewing plus the online component, then obviously there would be no need to disentangle the in-person and online parts of *MI*.

In deciding among these alternatives, it is important to weigh the resource demands against the information gained. Alternative (a) requires a negligible amount of additional subjects, at least under some conditions, but requires doubling the number of experimental conditions. Alternative (b) requires increasing the number of experimental conditions by 50% and also increasing the number of subjects by at least 50%. Whether (a) or (b) is more resource-intensive depends on which is more expensive—obtaining additional subjects or mounting more experimental conditions; this will be discussed further in Chap. 6. Option (c) is the least resource-intensive in the short run, because it does not require any increase in subjects or experimental conditions, but it does not address the question of whether providing an online follow-up improves the effectiveness of the intervention. This option will at least provide an answer to one important practical question: Does the combination of motivational interviewing and an online follow-up work? However, if this option is selected, it will not be possible to disentangle the effects of motivational interviewing and the online follow-up without additional experimentation.

3.7.2 Ensuring the Factors Can Be Manipulated Independently

The second consideration is to ensure that the factors can be manipulated completely independently, so that each level of each factor is implemented in as uniform a manner as possible across experimental conditions. In other words, in a factorial experiment, each level of each factor always should be implemented in exactly the same way, no matter what are the levels of the other factors in a given experimental condition. For example, an experimental subject who is in Condition 4 in Table 3.1 should experience peer mentoring in the same way as a subject who is in Condition 7 experiences peer mentoring. A violation of this principle would occur if the peer mentors delivered a different type or intensity of peer support to subjects in Condition 4 because these subjects do not receive motivational interviewing. Where possible, it is a good idea to keep staff who are delivering each component blind to the experimental condition a subject has been assigned.

In some cases, careful thought and planning may be required to achieve this aim without introducing excessive redundancy or overlap into the experiment. For example, suppose information about the effects of alcohol on HIV+ individuals and the importance of ART adherence for general health is considered essential for both motivational interviewing and peer mentoring. If exactly the same protocol is to be followed in all conditions, subjects in the conditions in which both *MI* and *PEER* are set to the yes level will receive this background information twice. A reasonably small amount of redundancy is probably tolerable in most experiments. However, it can be a problem if it reaches a point at which subjects find it irritating. They may stop paying attention, reduce compliance with the demands of the intervention, or even drop out of the study.

Including a constant component with information, materials, procedures, etc. to be provided to all subjects can be one way to avoid redundancy. In the example above, the experimenters may determine that every subject, even those in Condition 1, who are not assigned any components, should be provided with basic information about the effects of alcohol and the importance of ART adherence. Of course, anything that is not to be provided to all subjects cannot be included in a constant component.

3.8 When Subjects are Clustered

There are many situations in behavioral, biobehavioral, and biomedical research where experimental subjects are clustered because they are sampled from within clinics, schools, and so forth (Murray, 1998). Usually the individuals within a unit are to some degree more similar to each other than they are to an individual randomly sampled from a different unit. This relation among the observations is expressed in the intra-class correlation. Under many circumstances, when subjects are clustered, the data must be analyzed using multilevel models, also known as hierarchical linear models (e.g., Goldstein, 2011; Raudenbush & Bryk, 2002).

Suppose the hypothetical experiment discussed in this chapter will be conducted in a sample of HIV clinics. Here the clinics are clusters, and Dr. B must consider whether to conduct within-cluster or between-cluster randomization (or some hybrid of the two; see the Nahum-Shani and Dziak (2018) chapter in the companion volume).

Within-cluster randomization means that the individual is the unit of random assignment, and the entire experiment will be conducted within each clinic. This means that an individual in Clinic A has the same probability of being assigned to a particular experimental condition as an individual in Clinic B or any other clinic. Another way to think of this is that knowing which clinic the individual is in provides no information about what experimental condition the individual has been assigned.

Between-cluster randomization means that entire clinics (or, in other studies, schools, classrooms, therapy groups, and so on) are the unit of random assignment. Between-cluster randomization is usually necessary when the experiment is wholly or in part delivered in intact groups (e.g., classrooms) or when there is likely to be contamination across experimental conditions due to contact between subjects. If Dr. B uses between-cluster randomization, in any particular clinic, only one experimental condition will be implemented. When between-cluster randomization is used, if an individual's cluster membership is known, then the individual's assigned experimental condition is known, because everyone in a given cluster is assigned to the same condition.

There are two big disadvantages to between-cluster randomization. First, when between-cluster randomization is planned, usually more subjects are needed to maintain the same level of power, as compared to an experiment in which within-cluster randomization is used (all else being equal and assuming the data are to be analyzed using an appropriate multilevel model). This is explained in detail in Murray (1998). Second, aside from the overall sample size, it is necessary to have enough clusters so that at least one cluster can be assigned to each experimental condition. For example, suppose $N = 360$ has been determined to be sufficient to provide a desired level of power in a particular 2^4 factorial experiment (how to power a factorial experiment is discussed below), but the subjects are clustered in 10 units of 36 subjects each. With only 10 clusters and 16 experimental conditions, there will not be enough clusters to assign to experimental conditions. Even though the subject sample size is large enough to provide sufficient power, more clusters will have to be recruited to conduct the experiment. For these reasons, within-cluster randomization is preferred whenever possible. However, in most educational settings, many clinic settings, and many situations where one or more components involve group-delivered treatments, between-cluster randomization is necessary.

Factorial experiments are often feasible even if a cluster structure is present. For example, Schlam et al. (2016) implemented a 2^5 factorial experiment to examine five candidate components related to smoking cessation treatment. The experiment was conducted in 11 primary care clinics, and within-clinic randomization was used. An example of between-cluster randomization can be found in Caldwell et al. (2012), who assigned 56 schools to experimental conditions in a 2^3 factorial experiment. Their objective was to examine three components hypothesized to affect fidelity of delivery of Healthwise, a school-based drug abuse and HIV prevention program. For a thorough treatment of statistical considerations related to factorial experiments in which there is a cluster structure, see Nahum-Shani and Dziak (2018) in the companion volume. Also, see Chap. 5 for an introduction to how fractional factorial designs may be helpful when between-cluster randomization is required.

Investigators may wish to note that assuming the per-cluster sample size is large enough to maintain an adequate level of statistical power, one cluster per experimental condition is usually workable. One cluster per condition can be risky if there is a chance than an entire cluster may drop out of the experiment, leaving an empty condition. However, even if this happened, it could limit data analysis options, but the results would still be useful.

3.9 A Very Brief Review of Statistical Power

Much of the remainder of this chapter provides a very brief review on statistical power in factorial experiments and how to manage research resources when planning an optimization trial that will use a factorial design. For more about statistical power analysis, the reader is referred to the classic book by Cohen (1988). In this chapter power for detection of main effects is discussed; power for detection of interactions is discussed in Chap. 4.

In classical statistical hypothesis testing, two types of errors are identified. A Type I error is defined as rejecting the null hypothesis—in other words, concluding that an effect has been detected—when in fact the null hypothesis is true. The Type I error rate is represented by the Greek letter α. A Type II error is defined as failing to reject the null hypothesis—in other words, failing to detect an effect—when in fact the null hypothesis is false. The Type II error rate is represented by the Greek letter β. Statistical power is defined as $1-\beta$, which is the probability of concluding that the null hypothesis is false when in fact it is false, or, put another way, the probability of detecting an effect given that the effect is truly there.

In general, in experiments in which individual units are randomly assigned to experimental conditions, statistical power is influenced by three factors: the chosen α, the size of the effect to be detected, and the sample size N. Given a fixed effect size and N, and all else being equal, a smaller α means there will be less statistical power, and a larger α means there will be greater statistical power. Given a fixed α and N, and all else being equal, smaller effects are associated with less statistical power, and larger effects are associated with greater statistical power. Given a fixed α and effect size, and all else being equal, a smaller N will produce less statistical power, and a larger N will produce greater statistical power.

In general, power analysis is conducted in the planning stages of research to determine what sample size is needed to achieve a particular desired level of statistical power to be able to detect an effect of a given size using a particular α. Traditionally $\alpha = .05$ is used for the Type I error rate. Cohen's (1988, p. 56) recommendation of .8 is widely considered an acceptable minimum level of statistical power, which means that $\beta = .20$ is used for the Type II error rate. As will be discussed below, although these traditional rules of thumb are appropriate for the evaluation phase of MOST, there are good reasons to reconsider them for the research conducted during the optimization phase.

The next section compares and contrasts two different perspectives on research: the conclusion-priority perspective and the decision-priority perspective. Which of these perspectives is taken by an investigator is likely to influence the investigator's approach to resource management and powering a planned experiment. As will be discussed, both perspectives are important but are relevant to different phases of MOST. For this next section, do not be concerned about the details of how to power a factorial experiment; these details are reviewed later in this chapter. Instead, it is best to approach the next section conceptually.

3.10 Conclusion-Priority and Decision-Priority Perspectives on Research

When considering how to best to manage resources when planning an optimization trial, it is helpful to draw a distinction between the conclusion-priority perspective and the decision-priority perspective. This distinction is important, because the optimization phase of MOST typically calls for the decision-priority perspective, whereas the evaluation phase always calls for the conclusion-priority perspective.

The conclusion-priority perspective is the one most scientists have been trained to take. From this perspective, the top priority is to draw a scientific conclusion that will stand up when critically examined by other scientists. By contrast, the decision-priority perspective is consistent with how engineers often go about their work. Here, although scientific conclusions are of interest, the top priority is to make a practical decision about which components and component levels are to be selected for the screened-in set. Depending on the objective at a given time, the same experimental results may be approached from a conclusion-priority or decision-priority perspective. The distinction between the two perspectives is subtle.

Suppose Dr. B has sufficient resources to conduct a 2^3 factorial experiment like the one depicted in Table 3.1, with $N = 96$ subjects. This sample size is sufficient to power the experiment at .8 to detect a particular main effect size ES (how to express effect sizes is discussed later in this chapter) using a Type I error rate of $\alpha = .05$. The data have been analyzed using factorial ANOVA, and the results have been obtained. To illustrate the difference between the conclusion-priority and decision-priority perspectives, let us consider how taking each perspective might lead Dr. B to take a different approach to evaluate the empirical results about the main effect of one of the factors, say *PEER*. As will be seen, ultimately the differences in these perspectives can affect how an investigator chooses to manage resources when planning an experiment.

3.10.1 Conclusion-Priority Perspective vs. Decision-Priority Perspective

Suppose Dr. B is evaluating the estimated main effect of *PEER* with the primary objective of drawing a publishable scientific conclusion about whether peer mentoring demonstrates a detectable effect on ART adherence. This result, along

3.10 Conclusion-Priority and Decision-Priority Perspectives on Research

with the other results of the experiment, will serve as the basis for a manuscript that will be submitted to a peer-reviewed scientific journal. Dr. B knows that the review process is, in a sense, "science court" in which each scientific conclusion is put "on trial" before a jury of scientific peers. Any conclusion drawn about the effectiveness of peer mentoring must be defensible in this trial if the manuscript is to be published.

Dr. B's conclusions are most likely to be defensible in the eyes of scientific peers if they are based on generally agreed-upon conventions for drawing these conclusions. One such convention is that a Type I error rate no larger than .05 is tolerated. Another convention is that .8 is an acceptable level for statistical power. If statistical power is .8, the Type II error rate is $.1-.8 = .2$. This means it is acceptable, even common, for the Type II error rate to be four times greater than the Type I error rate. In other words, from the conclusion-priority perspective, one would much rather make a Type II error, that is, overlook the effect of peer mentoring, than a Type I error, that is, mistakenly conclude that peer mentoring has an effect when in reality no effect is present.

Dr. B will adhere strictly to $\alpha = .05$ in drawing scientific conclusions about the effectiveness of peer mentoring. If the main effect of *PEER* is significant at $p \leq .05$, according to this convention, the evidence suggests the null hypothesis is false, and therefore Dr. B will conclude that peer mentoring demonstrates an effect on ART adherence. Otherwise, the null hypothesis cannot be rejected. If the null hypothesis is not rejected, it is not known whether peer mentoring genuinely has no effect, or a Type II error has been made; it will be left to future research to make this determination. Based on currently available evidence, the conclusion is essentially "we do not know whether or not peer mentoring has an effect on ART adherence."

By contrast, now suppose Dr. B is evaluating the results of the same experiment, but with a different primary objective. Although Dr. B would like to make scientific conclusions, the primary objective is to make decisions about what components are to be included in the screened-in set. Let us focus on the decision about whether the peer mentoring component should be included. (In this discussion, for purposes of illustration a greatly simplified version of this decision-making process, based only on main effects, is reviewed. See Chap. 7 for a more complete description of a suggested process for decision-making that relies on both main effects and interactions.) Here concluding "We do not know" and awaiting the results of future research is not an option. A decision must be made, and made now, about whether or not to include peer mentoring in the screened-in set. This decision is to be based on the information provided by the experiment, that is, the observed main effect of *PEER*.

The decision must be made even if the available information is imperfect. Of course, the available information is usually imperfect to some degree, and there is always the possibility of an incorrect decision when inferences about a population are being made based on a sample. There are two ways Dr. B can make an incorrect decision about peer mentoring based on the observed results. One way is to decide to include peer mentoring in the screened-in set when it does not in fact exert a positive effect on ART adherence. The other way is to send peer mentoring to the screened-out set when it does in fact exert a positive effect on ART adherence. In hypothesis-testing terms, the former is a Type I error and the latter is a Type II error.

Hypothesis testing as master and as servant When the conclusion-priority perspective is taken, hypothesis testing is, in a sense, the master. There are strict rules about how hypothesis testing is to be conducted and about how scientific conclusions are to be drawn, if the results are to hold up in "science court." The established conventions for the appropriate Type I error rate, usually $\alpha = .05$, and level of statistical power, usually in the .8 range, are seldom deviated from when the conclusion-priority perspective is taken. For example, imagine a particular scientifically important, but small, effect has been hypothesized. To conduct a conclusive investigation of this effect would require a study expected to achieve statistical power of at least .8 using $\alpha = .05$. Suppose given the small expected effect size, this study would require a sample size in the several thousands. The scientific community would not draw any conclusions until such a study has been conducted. If the resources to conduct this study never become available, then the hypothesis is never formally tested, and no conclusions are ever drawn.

By contrast, when the decision-priority perspective is taken, the investigator uses the information provided by the experiment in any reasonable manner to make sound decisions. Here hypothesis testing is more a servant than a master. Its main purpose is to serve as an aid to decision-making and as a useful organizing framework for resource management. In fact, as is discussed further in Chap. 7, it is not necessary to use classical hypothesis testing in decision-making in MOST. At the investigator's option, decision-making may involve classical hypothesis testing, or it may be based on examination of effect sizes expressed in any metric, or a Bayesian approach could be used. (Bayesian approaches are not covered in this book. To the author's knowledge, at this writing, they have not yet been used in MOST, although there is tremendous potential there.)

The decision-priority perspective and classical hypothesis testing For now, imagine Dr. B is using classical hypothesis testing to decide which components to include in the screened-in set. When taking the decision-priority perspective, Dr. B is responsible for establishing the Type I and Type II error rates that are most consistent with the resource management principle, that is, that are most helpful in decision-making and make the best use of available resources. For example, perhaps Dr. B has determined that incorrectly sending a component to the screened-out set and incorrectly including a component in the screened-in set are equally undesirable. Therefore, Dr. B would like to keep the Type I and Type II error rates about the same.

One obvious option would be to increase the N in the experiment, thereby increasing power and decreasing the Type II error rate. However, it was said above that there are resources available to collect data on $N = 96$ subjects only; additional subjects cannot be obtained. Thus, a way of increasing power without increasing N is needed. An alternative option is to conduct hypothesis testing using an increased Type I error rate. Recall that the Type I error rate is one determinant of statistical power. All else being equal, a larger Type I error rate produces greater power and a smaller Type II error rate.

As explained above, in this example, $N = 96$ produces power of .8 to detect an effect of size *ES*, using $\alpha = .05$. Suppose raising the Type I error rate to $\alpha = .10$ raises power to .90 and, correspondingly, reduces the Type II error rate to .10. In other words, Dr. B establishes that the peer mentoring component will be included in the screened-in set if *PEER* has a main effect significant at $p \leq .10$. If the *p*-value associated with the main effect of *PEER* is greater than .10, Dr. B will conclude that whatever the true effect of peer mentoring may be, it is too small to be useful in the intervention being developed. Note that, just as in the conclusion-priority perspective, no conclusion will be drawn about whether or not the null hypothesis with respect to the main effect of *PEER* is true. Dr. B has a 10 percent chance of mistakenly concluding that *PEER* has a main effect when in fact it does not and therefore mistakenly placing the peer mentoring component in the screened-in set. Dr. B also has a 10 percent chance of mistakenly concluding that *PEER* does not have a main effect when in fact it does and therefore mistakenly placing the peer mentoring component in the screened-out set. Consistent with the continual optimization principle (see Chap. 1), any decisions will be reconsidered in the future as more and better evidence becomes available.

Imagine a skeptical colleague, Dr. S, expresses disapproval of two aspects of Dr. B's approach. First, Dr. S is shocked at Dr. B's choice of an α greater than .05, arguing that this runs counter to a century of scientific hypothesis-testing tradition. Dr. B explains this choice as a strategic tradeoff guided by the resource management principle. The tradeoff is to increase the Type I error rate so as to decrease the Type II error rate correspondingly and increase statistical power without having to increase the sample size. The only way to reduce the Type II error rate without increasing the Type I error rate would be to include more subjects in the experiment. Maintaining both the Type I and Type II error rates at .05 would have required increasing the number of subjects from $N = 96$ to about $N = 152$. Dr. B acknowledges that this would have been the preferred approach had the level of resources permitted it, but in this case funds were not available to increase the sample size. In Dr. B's view, raising the α upon which the decisions will be based is a cost-effective way of reducing the Type II error rate without increasing the sample size and is justified based on the resource management principle of MOST, even though the reduction in the Type II error rate is accompanied by an increased risk of a Type I error.

Second, Dr. S points out that in this 2^3 factorial experiment, there are hypothesis tests associated with three main effects, three two-way interactions, and one three-way interaction. Dr. S argues that a Bonferroni or similar correction for multiple hypothesis tests should be applied. Dr. B reminds Dr. S that when effect coding (discussed below) is used, all of the effect estimates are uncorrelated when the experiment is balanced and nearly uncorrelated when there is some variability in cell size. This means that the hypothesis tests are independent. However, the main point is that building an optimized intervention, not hypothesis testing, is the primary focus here. Later, once the optimized intervention has been built based on the empirical evidence provided by the experiment, its effectiveness will be examined in an RCT in the evaluation phase of MOST. For this RCT, the conclusion-priority perspective will be taken, and all of the hypothesis-testing conventions will be followed.

This is one hypothetical example. In any particular application, an investigator taking the decision-priority perspective may or may not wish the Type I and Type II error rates to be equal, may wish to use a Type I error rate less than .05, may wish to use a Bonferroni or similar adjustment, and so on. The point is that the investigator taking the decision-priority perspective must think carefully about how best to use resources to obtain the information needed for decision-making and how best to use the information in decision-making once it has been obtained. This means not feeling constrained by the conventions that appropriately are followed when the conclusion-priority perspective is taken. If it makes the best use of available resources and is most helpful in decision-making, it may be prudent for the investigator to use unconventional Type I and Type II error rates.

Basing decisions on effect sizes Suppose instead of focusing on hypothesis testing per se, Dr. B decides a priori that the peer mentoring component will be included in the screened-in set if the *PEER* factor demonstrates an observed main effect size, ES_{PEER}, larger than some minimum effect size cutoff, ES_{MIN}. In other words, peer mentoring will be selected for the screened-in set if $ES_{PEER} > ES_{MIN}$. Even though in this case the focus is not on hypothesis testing per se, the concepts of Type I and Type II errors and error rates remain relevant. Whatever approach to decision-making is taken, the investigator wants to select only effective components (i.e., does not want to make any Type I errors) while not overlooking any effective components (i.e., not making any Type II errors).

Within a single factorial experiment, establishing that peer mentoring will be selected for the screened-in set if $ES_{PEER} > ES_{MIN}$ is equivalent to selecting a Type I error rate. To see why, imagine Dr. B has decided to use a Type I error rate of $p < .10$. The value of *ES* that would be significant at exactly $p < .10$ can easily be identified and that can be used as ES_{MIN}. Conversely, for any ES_{MIN} that may be selected, there is a corresponding Type I error rate. Of course, if more than one optimization trial were conducted with different sample sizes, the same value of ES_{MIN} would be associated with different Type I error rates.

The Type II error rate determines the precision with which ES_{PEER} is estimated. This can perhaps be seen most clearly in the size of the standard errors associated with regression weights. The regression weight corresponding to the main effect of *PEER*, b_{PEER}, is one way of expressing ES_{PEER}. All else being equal, a larger N will produce a smaller Type II error rate, which means smaller standard errors and, therefore, a smaller confidence interval about b_{PEER}. In other words, when the standard error is smaller, b_{PEER} is estimated more precisely. This extends to any way of expressing ES_{PEER} the investigator may select; all else being equal, a larger sample size means the estimate of the effect size will be more precise, and the more precise the estimate, the firmer basis the estimate provides for decision-making.

Thus, whether or not classical hypothesis testing per se is to be used in decision-making, it is a good idea to pay attention to Type I and Type II error rates, because these error rates can be useful in resource management. The resource management principle of MOST dictates that the investigator is responsible for ensuring that the Type I and Type II error rates are as low as possible given the available resources and that the relative sizes of the Type I and Type II error rates are appropriate for the given situation. Due to resource limitations, it may not be possible to keep the Type I and Type II error rates as low as might be wished or as low as would be acceptable if the conclusion-priority perspective was taken. Of course, both rates need to be low enough so that the investigator can be reasonably confident in the decisions made.

3.10.2 When to Take Each Perspective

The optimization phase of MOST requires decision-making about which components and component levels will make up the screened-in set and, ultimately, the optimized intervention. Therefore, in the optimization phase of MOST, the decision-priority perspective is generally more appropriate than the conclusion-priority perspective. By contrast, for the evaluation phase of MOST, which requires a scientific test of the effectiveness of the optimized intervention, the resource management principle suggests that the conclusion-priority perspective is needed to gather the kind of information that will hold up to scientific scrutiny. This is one reason why Dr. B feels comfortable with taking the decision-priority perspective during the optimization phase—in the evaluation phase, the effectiveness (or lack thereof) of the optimized intervention will be established by means of an RCT using the conclusion-priority approach with a strict $\alpha = .05$.

The same experimental results from an optimization trial can be interpreted from the decision-priority perspective when assembling the screened-in and screened-out sets *and* from the conclusion-priority perspective when writing up results for publication. In a particular situation, the decision-priority perspective may suggest that a larger α should be used than would be acceptable if the conclusion-priority perspective was taken. This opens up the possibility that the same investigator, based on the same results, will report in a published article that it cannot be concluded that a particular component is effective and will nevertheless include that component in an optimized intervention. In the example, suppose the results showed that the main effect of *PEER* was significant at $p = .09$. For purposes of publication, Dr. B would write that it cannot be concluded that the peer mentoring component was effective—yet at the same time would have decided to include the peer mentoring component in the screened-in set. This is not evidence of inconsistency, but rather an example of two rational but different approaches to decision-making using the same evidence for different purposes. From the conclusion-priority perspective, there was not enough evidence to draw a scientific conclusion about the effectiveness of peer mentoring. From the decision-priority perspective, although the evidence may not be strong enough to support a scientific conclusion, it is strong enough to justify including peer mentoring in the screened-in set. These are different decisions based on the same evidence but made using criteria appropriate for their respective purposes.

3.11 The General Linear Model (GLM) Approach to Classical Factorial Analysis of Variance (ANOVA)

Sections 3.11, 3.12, and 3.13 contain somewhat more technical material that may be skipped without loss of continuity.

3.11.1 Effect Coding

The meaning of any parameter estimates based on data from a factorial experiment (or any other experiment) depends on how the data are modeled. In factorial ANOVA, one aspect of this modeling is how the factors are coded. Coding of factors refers to the system used for assigning numbers to the levels of the factors. Coding begins with creation of a vector of codes corresponding to each factor in the experiment, to represent that factor's main effect. One approach to coding, called effect coding, is used widely in engineering and many other fields. For reasons that will be explained below, effect coding is greatly preferred for analysis of data from factorial optimization trials. Therefore, everything said about factorial experiments in this book and the companion volume assumes that effect coding is to be used in data analysis.

In effect coding, if there are two levels per factor, each main effect is coded by assigning a value of +1 for one setting of each factor and a value of -1 for the other setting. (Although which setting of each factor is assigned which code is arbitrary, interpretation of the effects is usually easier if the setting hypothesized to correspond to a more favorable mean on the outcome variable is assigned +1.) These vectors appear in the three columns under the Main Effects heading in Table 3.6. Compare these three vectors with the corresponding columns in Table 3.2. It is evident that wherever there is a + in Table 3.2, there is a +1 in Table 3.6, and wherever there is a − in Table 3.2, there is a -1 in Table 3.6.

To code the interactions, the vectors for the corresponding main effects are multiplied together. For example, the vector for the $MI \times PEER$ effect is obtained by multiplying the codes for the *MI* and *PEER* main effects. (This is why interactions are sometimes called multiplicative effects.) The vector for the $MI \times PEER \times TEXT$ effect is obtained by multiplying all three main effect vectors together. The effect codes for the interactions are shown in the four rightmost columns in Table 3.6.

The complete regression model for the hypothetical three-factor experiment can be expressed as

$$\widehat{Y} = b_0 + b_1 X_{MI} + b_2 X_{PEER} + b_3 X_{TEXT} + b_4 X_{MI \times PEER} + b_5 X_{MI \times TEXT} \\ + b_6 X_{PEER \times TEXT} + b_7 X_{MI \times PEER \times TEXT}, \tag{3.3}$$

where the *X*s correspond to the vectors in Table 3.6. That is, X_{MI}, X_{PEER}, and X_{TEXT} correspond to the vectors for the three main effects; $X_{MI \times PEER}$, $X_{MI \times TEXT}$, and

3.11 The General Linear Model (GLM) Approach to Classical Factorial... 95

Table 3.6 Effect coding for 2^3 factorial ANOVA

Experimental condition (cell)	Intercept	Main effects			Interactions			
		MI X_{MI}	PEER X_{PEER}	TEXT X_{TEXT}	MI×PEER $X_{MI \times PEER}$	MI×TEXT $X_{MI \times TEXT}$	PEER×TEXT $X_{PEER \times TEXT}$	MI×PEER×TEXT $X_{MI \times PEER \times TEXT}$
1	1	−1	−1	−1	+1	+1	+1	−1
2	1	−1	−1	+1	+1	−1	−1	+1
3	1	−1	+1	−1	−1	+1	−1	+1
4	1	−1	+1	+1	−1	−1	+1	−1
5	1	+1	−1	−1	−1	−1	+1	+1
6	1	+1	−1	+1	−1	+1	−1	−1
7	1	+1	+1	−1	+1	−1	−1	−1
8	1	+1	+1	+1	+1	+1	+1	+1

$X_{PEER \times TEXT}$ correspond to the vectors for the three two-way interactions; and $X_{MI \times PEER \times TEXT}$ corresponds to the vector for the three-way interaction. The *b*s are unstandardized regression coefficients. b_0 is the intercept, which, when effect coding is used, is the grand mean; b_1, b_2, and b_3 are the regression coefficients corresponding to the three main effects; b_4, b_5, and b_6 are the regression coefficients corresponding to the two-way interactions; and b_7 is the regression coefficient corresponding to the three-way interaction. For example, for a subject in experimental condition 4, in which *MI*=no and *PEER* and *TEXT*=yes, the expression for $\widehat{Y}$ would be

$$\widehat{Y} = b_0 + b_1(-1) + b_2(1) + b_3(1) + b_4(-1) + b_5(-1) + b_6(1) + b_7(-1).$$

Straightforward but perhaps somewhat tedious algebra (some of which is reviewed in the Kugler, Dziak, and Trail (2018) chapter in the companion volume) shows that when effect coding is used, the effects estimated correspond directly to the definitions of main effects and interactions presented above, multiplied by a constant scaling factor of ½. For example:

$$b_1 = b_{MI} = \frac{ME_{MI}}{2} \quad \text{and}$$

$$b_4 = b_{MI \times PEER} = \frac{INT_{MI \times PEER}}{2}.$$

Because this scaling factor affects both the estimates and their standard errors, it has no impact on the hypothesis tests. Thus, the hypothesis tests of each regression weight are identical to the corresponding hypothesis tests of main effects and interactions in a standard ANOVA, and they correspond directly to the hypothesis tests that are produced by most ANOVA routines in statistical software. When powering factorial experiments is discussed below, the emphasis will be on the regression weight, specifically the standardized regression weight, as a convenient expression of effect size.

At this point it may be helpful to note two subtly different uses of the term "independent variable." One use pertains to experimental design, and the other pertains to data analysis from a GLM perspective. The factors in a factorial experiment are independent variables, because the investigator manipulates them. In the hypothetical example, there are three factors: *MI*, *PEER*, and *TEXT*. In a GLM, it is common to refer to each variable on the *X* side as an independent variable. Although there are only three factors in the experiment, the model expressed in Eq. (3.3) has seven independent variables, because each of the seven effects in the ANOVA, that is, the three main effects and four interactions, are represented by one of the *X*s in Eq. (3.3). (The intercept is not usually considered an independent variable.)

To avoid confusion, in this book, the term independent variable will be reserved for (i) the manipulated variable in an experiment in the RCT family or (ii) the variables on the *X* side in a regression equation. From this point on the independent variables in a factorial experiment will be referred to as factors.

3.11.2 Dummy Coding

A different approach to coding effects in factorial ANOVA, namely, dummy coding (or reference cell coding), is sometimes used in the social and behavioral sciences. In dummy coding, the settings of a two-level factor are represented by 1 and 0 instead of by +1 and −1. The 1 is usually assigned to the setting that is expected to correspond to a higher mean on the outcome variable, in a manner directly analogous to assignment of effect codes. The vectors for the higher-order effects are constructed by multiplying the corresponding vectors for the first-order effects, again in a manner directly analogous to the way this is done with effect codes. An extensive treatment of the conceptual and statistical differences between effect coding and dummy coding can be found in the Kugler, Dziak, and Trail (2018) chapter in the companion volume.

The difference between using −1 and +1 as codes vs. using 0 and 1 may seem trivial on the surface, but it has huge implications, particularly in designs involving three or more factors. The omnibus F test is the same whether effect coding or dummy coding is used. However, when dummy coding is used, the effects estimated generally do not correspond to the definitions of main effects and interactions presented above, with the exception of the highest-order effect (e.g., the three-way interaction in an experiment with three factors). As an example, the dummy-coded main effect of *MI* is equivalent to a linear combination of the effect-coded main effect of *MI* and several other dummy-coded effects involving *MI*. From this point forward, in this book, the terms main effect and interaction are used only in reference to effect coding.

Because the individual effect estimates are different depending on whether effect coding or dummy coding is used, the corresponding hypothesis tests are different also. In other words, a particular regression weight may be statistically significant when effect coding is used and not significant when dummy coding is used and vice versa. The Kugler, Dziak, and Trail (2018) chapter in the companion volume includes an analysis of the same data set using effect coding and dummy coding, to illustrate how the results can differ.

3.11.3 Why Effect Coding Is Preferred for Analysis of Data from Factorial Optimization Trials

In this book the focus is primarily on effect coding to perform factorial ANOVA. Neither effect coding nor dummy coding is inherently better, but one may be preferable to the other for the purpose of addressing a particular research question. This book takes the position that to analyze the data from a factorial optimization trial for the purpose of selecting components and component settings for inclusion in an intervention, effect coding is preferable for the following reasons.

First, consider a simple 2^k factorial experiment. Suppose the design is perfectly balanced—that is, there are equal ns across experimental conditions. When effect coding is used to analyze the data from this experiment, the standard error is the same for all regression coefficients, irrespective of whether the coefficient represents a main effect or an interaction. It follows that in a perfectly balanced 2^k experiment, power is identical for any regression coefficient of a given size, whether it corresponds to a main effect or an interaction of any order. Although unequal ns across experimental conditions or covariates included in the analytic model will produce some variability in the standard errors, on average, the standard errors will be the same for main effects and interactions. This makes power analysis relatively straightforward. (However, there is more to say about power for interaction effects. This is discussed in Chap. 4.)

By contrast, when dummy coding is used under the same circumstances, the standard error is usually larger for regression coefficients corresponding to higher-order effects (effects that would correspond to higher-order interactions in effect coding) than for regression coefficients corresponding to lower-order effects (effects that would correspond to main effects and lower-order interactions in effect coding). Consequently, when dummy coding is used to analyze data from a factorial experiment, statistical power tends to be relatively smaller for detection of non-zero regression coefficients corresponding to higher-order effects, as compared to those corresponding to lower-order effects.

Second, when effect coding is used, the estimates of the regression coefficients are uncorrelated in a perfectly balanced design. Even when the ns are unequal, the effect estimates are nearly uncorrelated except in severely unbalanced designs. Uncorrelated effects are preferred because they greatly simplify the interpretation of the regression coefficients, that is, the effects in the ANOVA model. Although main effects must always be evaluated in the light of any important interaction effects, interpretation of main effects and interactions is more straightforward when they are uncorrelated. By contrast, when dummy coding is used, the regression coefficients are always correlated and may be substantially correlated, even in data from perfectly balanced factorial experiments.

A word of caution The approach to coding effects that is used in factorial ANOVA can make a big difference in the results obtained and how they should be interpreted. Unfortunately, statistical software and the accompanying documentation are sometimes not very clear about which approach to coding is being used. This is discussed in more detail in the Kugler, Dziak, and Trail (2018) chapter in the companion volume. Whichever approach to coding is preferred for a given analysis, it is essential to ensure that the power analysis and statistical analyses are based on the intended coding system.

3.12 Powering a Factorial Experiment: Main Effects

3.12.1 Power and Factorial Experiments

This section will emphasize power in 2^k designs, because as mentioned above, the efficiency of these designs makes them highly recommended for optimization trials (such as screening experiments in engineering; Montgomery, 2009).

The logic behind how an experiment is powered differs between the RCT and factorial experiments. In RCTs, power is reflected in the per-condition sample size, which will be represented here by a lowercase n. If an RCT has a small per-condition n, it is unlikely to have sufficient power. By contrast (assuming that effect coding is to be used and the data are to be analyzed using an appropriate factorial ANOVA model), power for detection of the main effect of a factor in a balanced factorial experiment is reflected in the N per level of that factor, not the per-condition n. In fact, as will be demonstrated shortly, it is possible for a factorial experiment to have a small per-condition n and nevertheless deliver a high level of statistical power. To see why this is true, it is necessary to understand the basics of how to power a factorial experiment.

The starting point for a power analysis is specification of the size of the effect to be detected. There are several equivalent ways to express the size of main effects and interaction effects for the purpose of power analysis (see Cohen, 1988), including the unstandardized regression coefficient b, Cohen's d, and Cohen's f. This book will express effect sizes primarily in terms of the standardized regression coefficient, represented by β^* (with the asterisk added to avoid confusion with the symbol for the Type II error). In a balanced factorial ANOVA using effect coding, β^* can be expressed as follows:

$$\beta^* = \frac{b}{s_y}, \tag{3.4}$$

where s_y is the standard deviation of the outcome variable y.

When powering a factorial experiment, the focus is on the smallest β^* that the investigator wishes to be able to detect at a particular desired level of power. If power is maintained for detection of a regression weight corresponding to a particular β^* in a factorial experiment, it is automatically maintained for detection of any other effects of the same size or larger. In other words, if the sample size is sufficient to detect a main or interaction effect of, say, $\beta^* = .2$ at a power of .8, it is sufficient to detect all other main or interaction effects that are $\geq .2$ at a power of at least .8. This will be explained further below.

3.12.2 Identifying Effect Sizes for Power Analysis From the Conclusion-Priority and Decision-Priority Perspectives

Factorial experiments are typically powered primarily to detect main effects (although if there are important interactions, they can and should be included in the power analysis; see Chap. 4). This means that an investigator powering a factorial experiment will have to begin by identifying the size of the main effects to be detected, particularly the smallest main effect to be detected. But, how does the investigator identify the main effect sizes upon which to base the power analysis?

The conclusion-priority and decision-priority perspectives, discussed above, suggest different approaches. The investigator who takes the conclusion-priority perspective may be interested in establishing whether there is a main effect that is at least the size prior literature suggests. This investigator will comb the scientific literature for evidence supporting a particular effect size or range of effect sizes and will power the study using an effect size at the middle or low end of the range.

By contrast, the investigator who is in the optimization phase of MOST, and therefore is taking the decision-priority perspective, may want to pose the following question: *What is the minimum main effect a factor needs to exhibit to indicate that the corresponding component should be selected for the screened-in set?* If the investigator decides that a particular effect size is the cutoff for considering the corresponding component for inclusion in the screened-in set, then that effect size can plausibly be used in the power analysis for the optimization trial. A factor may in reality exert an effect that is smaller than the designated cutoff, but in that case the corresponding component is too weak to be included in the screened-in set. The resource management principle suggests that if a component is too weak to be included in the screened-in set, resources need not be devoted to detecting its effect.

3.12.3 Comparison of Two Putatively Active Levels of a Component

Particularly careful thought should be given to selection of an effect size and the desired Type II error rate when the research question is not whether a particular component is more effective than a control, but instead whether a putatively active level of a component can be considered equivalent to another putatively active level of a component. For example, suppose it has previously been established that motivational interviewing is effective using professional staff. However, due to the expense of using professional staff, Dr. B wishes to determine whether trained volunteers can produce comparable outcomes. Dr. B plans to include an *MI* factor in the optimization trial, with the two levels professional and volunteer. If the volunteer is not associated with a detectably worse outcome than the professional, Dr. B will consider including motivational interviewing delivered by trained volunteers in the optimized intervention. This would essentially be an equivalence/noninferiority study and would be subject to all of the same considerations. Readers are referred to Walker and Nowacki (2011) and Kraemer (2011).

3.12.4 Demonstration of Power Analysis for a Factorial Experiment

Let us demonstrate the efficiency of factorial experiments by comparing power analyses conducted under several scenarios. Power analysis for factorial experiments can be performed using a variety of software. Whichever software is selected, it is wise to ensure that it is clear what system is being used to code the effects, because power for detection of a particular effect is likely to be different depending on whether effect coding, dummy coding, or some other system is used. Power analysis for planning a factorial experiment can be conducted using software such as PROC POWER in SAS. The power analyses presented in this book were done using a macro based on PROC POWER called FactorialPowerPlan (available for free download at http://methodology.psu.edu). This macro has been prepared to assist in power analysis for planning factorial experiments in MOST.

Returning to the hypothetical example, suppose the investigator wishes to maintain power of at least .8 to detect the main effects of each of the three factors. The investigator wishes to be able to detect unstandardized regression coefficients of the following sizes:

main effect of *MI*: $b = 2.0$
main effect of *PEER*: $b = 1.5$
main effect of *TEXT*: $b = 2.5$.

Further suppose $s_y = 10$, so applying Eq. (3.4) to obtain the standardized regression coefficients yields:

main effect of *MI*: $\beta^* = .20$
main effect of *PEER*: $\beta^* = .15$
main effect of *TEXT*: $\beta^* = .25$

The FactorialPowerPlan macro indicates that the required sample size is $N = 351$. The usual practice is to round up so that equal cell sizes can be maintained, which in this case requires $N = 352$. This means that the per-level sample size is 176, and $n = 44$ in each of the eight experimental conditions. (It is possible that if a covariate were included in the model, the required sample size would be smaller; the effect of covariates on power is outside the scope of this book.) Table 3.7 illustrates this experiment.

Table 3.7 Experimental conditions and per-condition n in 2^3 factorial design with overall $N = 352$

Experimental condition	MI	PEER	TEXT	n
1	No[a]	No	No	44
2	No	No	Yes	44
3	No	Yes	No	44
4	No	Yes	Yes	44
5	Yes	No	No	44
6	Yes	No	Yes	44
7	Yes	Yes	No	44
8	Yes	Yes	Yes	44

[a]No means not included in intervention; yes means included in intervention

Now suppose the investigator wishes to detect regression coefficients of .15 for all three main effects. Despite this reduction in the expected size of two of the three effects, the sample size necessary to maintain power of .8 remains $N = 352$. This is because with $N = 352$ the per-level N remains 176 for estimation of each of the main effects. If a per-level $N = 176$ provides power of .8 to detect a standardized regression coefficient of .15, it provides power of .8 to detect all of the regression coefficients of .15 and at least that much power to detect any larger regression coefficients.

Essentially, a factorial experiment is powered to detect the smallest main effect of interest. This can be considered a two-sided coin. One side of the coin is that the same sample size will automatically deliver more power to detect any larger main effects. The other side of this coin is that if several main effects are of interest, one of which is smaller than the others, it will be necessary to obtain a large enough sample to achieve sufficient power for the smaller effect, even though this sample size may exceed what is needed to detect the larger effects. Although this is by no means necessary, the most efficient approach is probably to try to arrange the experiment so that the effects on which the power analysis is to be based are all about the same size.

3.12.5 The Negligible Impact on Power of Adding a Factor to a Factorial Experiment

An investigator who wishes to add an arm to an RCT will find it necessary to increase the sample size, usually substantially, to maintain the same level of power. By contrast, it is often possible to add factors to a factorial experiment without the need to add more than a few, or even any, experimental subjects. This is another aspect of the efficiency of factorial experiments.

Suppose Dr. B would like to investigate the performance of an additional component, say mindfulness meditation, by adding a factor to the experiment.

3.12 Powering a Factorial Experiment: Main Effects

Table 3.8 Experimental conditions and per-condition n in 2^4 factorial design with overall $N = 352$

Experimental condition	MI	PEER	TEXT	MIND	n
1	No[a]	No	No	No	22
2	No	No	No	Yes	22
3	No	No	Yes	No	22
4	No	No	Yes	Yes	22
5	No	Yes	No	No	22
6	No	Yes	No	Yes	22
7	No	Yes	Yes	No	22
8	No	Yes	Yes	Yes	22
9	Yes	No	No	No	22
10	Yes	No	No	Yes	22
11	Yes	No	Yes	No	22
12	Yes	No	Yes	Yes	22
13	Yes	Yes	No	No	22
14	Yes	Yes	No	Yes	22
15	Yes	Yes	Yes	No	22
16	Yes	Yes	Yes	Yes	22

[a] No means not included in intervention; yes means included in intervention

Suppose the new factor, *MIND*, has levels no and yes, and the investigator wishes to be able to detect a main effect for this factor corresponding to a standardized regression coefficient $\geq .15$. Table 3.8 shows the new 2^4 experimental design. Suppose resource limitations make it necessary to maintain $N = 352$, so in the new design, there will be $n = 22$ per cell, as indicated in the table.

Will $N = 352$ be adequate to maintain power $= .8$ now that the design has been expanded to include another factor? Although the per-condition n is smaller in the new 2^4 design, the per-level N is identical to what it was in the 2^3 design, and thus statistical power remains identical. The upper panel of Table 3.9 illustrates how the main effect of *MI* would be computed in the 2^4 design. As the table shows, this computation involves the per-level N of 176 for each of the two levels and thus is based on the total N of 352. The lower panel of Table 3.9 illustrates how the main effect of the new factor, *MIND*, would be computed. Again, the effect is estimated based on the total N of 352.

This illustrates how factorial experiments can be tremendously efficient with respect to experimental subjects. Even if several more factors had been added, adequate power would be maintained without having to increase N, as long as the investigator wished to detect regression weights no smaller than the ones originally used to power the experiment. As Tables 3.8 and 3.9 show, the new 2^4 factorial experiment will have 16 experimental conditions. If, say, three more factors had been added to the 2^3 design instead of one, the new 2^6 experiment would have 64 experimental conditions. Assuming the effect sizes the investigator wishes to detect remain the same or larger, power would be maintained with approximately the same N. Although not strictly necessary, if resources permit, it would be a good idea

Table 3.9 Illustration of computation of main effects in 2^4 factorial design with overall $N = 352$[a]

Main effect of *MI*: mean of experimental conditions 9—16 compared to mean of experimental conditions 1—8

Experimental condition	MI	PEER	TEXT	MIND	n
1	No[b]	No	No	No	22
2	No	No	No	Yes	22
3	No	No	Yes	No	22
4	No	No	Yes	Yes	22
5	No	Yes	No	No	22
6	No	Yes	No	Yes	22
7	No	Yes	Yes	No	22
8	No	Yes	Yes	Yes	22
9	Yes	No	No	No	22
10	Yes	No	No	Yes	22
11	Yes	No	Yes	No	22
12	Yes	No	Yes	Yes	22
13	Yes	Yes	No	No	22
14	Yes	Yes	No	Yes	22
15	Yes	Yes	Yes	No	22
16	Yes	Yes	Yes	Yes	22

Main effect of *MIND*: mean of experimental conditions 2, 4, 6, 8, 10, 12, 14 and 16 compared to mean of experimental conditions 1, 3, 5, 7, 9, 11, 13, and 15

Experimental condition	MI	PEER	TEXT	MIND	n
1	No	No	No	No	22
2	No	No	No	Yes	22
3	No	No	Yes	No	22
4	No	No	Yes	Yes	22
5	No	Yes	No	No	22
6	No	Yes	No	Yes	22
7	No	Yes	Yes	No	22
8	No	Yes	Yes	Yes	22
9	Yes	No	No	No	22
10	Yes	No	No	Yes	22
11	Yes	No	Yes	No	22
12	Yes	No	Yes	Yes	22
13	Yes	Yes	No	No	22
14	Yes	Yes	No	Yes	22
15	Yes	Yes	Yes	No	22
16	Yes	Yes	Yes	Yes	22

[a]Note that the means of each level of both MI and MIND are estimated based on $N = 176$
[b]No means not included in intervention; yes means included in intervention

to round up to $N = 384$ to maintain balance. This would result in a per-level N of 192 and $n = 6$.

In summary, in contrast to power in the RCT, power in a factorial experiment is largely a function of the per-level N (all else being equal), not the per-condition n. Thus, although it may be surprising to those who have been trained primarily in the RCT, a factorial experiment with $n = 6$, or even smaller, can be adequately powered if the per-level N is sufficient. In fact, a reasonably accurate informal estimate of the required sample size for a 2^k experiment can be obtained by identifying the smallest regression coefficient to be detected, specifying its size, and looking up the required sample size as if the test statistic is a t-test and this regression coefficient is the only one to be detected. As discussed above, a sample size that provides sufficient power to detect a regression coefficient of a particular size in a 2^k experiment will provide sufficient power to detect any larger regression coefficient in the same experiment.

From one perspective, factorial experiments involving many factors are more efficient than those involving fewer factors, because they obtain more information from subjects. Particularly in situations in which it is expensive, difficult, or time-consuming to recruit and retain subjects, it may be most efficient to include as many factors as possible to examine multiple intervention components in fewer experiments. In fact, if the budget (in terms of money, time, or other resources) can bear the expense of the additional experimental conditions (remembering that the overall N will not increase by an appreciable amount), in one sense, it may be wasteful not to add factors so as to obtain as much information from subjects as possible. However, from another perspective, factorial experiments involving many factors are less efficient, because they require mounting so many more experimental conditions. In a situation where the overhead associated with mounting each additional condition is high even with the overall N fixed, this may be a serious drawback.

A careful comparison of the expenses associated with different experimental design alternatives is an important aspect of following the resource management principle and can be useful to present when justifying a proposed design to a potential funder. Chap. 6 discusses how to compare the relative resource requirements of different experimental designs to select the most efficient one for a particular situation.

3.12.6 *The Large Impact on Power of Adding a Level to a Factor: Why 2^k Factorial Optimization Trials are Recommended*

An experimental factor can have more than two levels, and a single experiment can involve factors with different numbers of levels. The notation that expresses factorial designs can easily accommodate this. A design that has four factors with three levels each would be expressed as a 3^4 design. A design that has two factors with three levels each and two factors with two levels each would be a $2^2 3^2$ design.

However, as mentioned earlier in this chapter, it is recommended that, in general, factorial optimization trials be limited to two levels per factor. This is because 2^k experiments are very efficient. Above it was explained how adding one or more factors to a factorial experiment may not increase the sample size necessary to maintain a desired level of statistical power. However, the situation is different when adding a level to a factor. Adding a level to even one factor in a factorial experiment can greatly increase the overall N needed to maintain power. This is because, as has been discussed above, the per-level N is an important determinant of statistical power. In a 2^k experiment, the overall N required to maintain power at a desired level is twice the required per-level N. If even one factor has three levels, all else being equal the overall N must be increased to three times the per-level N to maintain the same level of power.

To see why, let us return to the example with three factors: *MI*, *PEER*, and *TEXT*, in which approximately $N = 352$ would be required to achieve the desired level of power. Suppose Dr. B decided to take a different approach to investigating the effect of text messaging. Instead of a two-level factor having levels yes and no, Dr. B creates a factor *TEXT* involving three levels: no, low (4 texts per week), and high (3 texts per day). This is now a $2^2 3$ design with 12 experimental conditions. Suppose the following contrasts involving the *TEXT* factor are to be examined: no vs. low and low vs. high. For each of these contrasts, Dr. B wishes to detect a standardized regression coefficient of .15.

Table 3.10 shows the new experimental design with the original sample size, increased slightly to $N = 360$ to maintain balance, distributed across the 12 experimental conditions resulting in $n = 30$. In this design the per-level N for *MI* and *PEER*

Table 3.10 $2^2 3$ factorial experimental design with overall $N = 360$, with shading indicating the main effect of *TEXT*, no vs. low comparison

Experimental condition (cell)	MI	PEER	TEXT	n
1	No[a]	No	No	30
2	No	No	Low	30
3	No	No	High	30
4	No	Yes	No	30
5	No	Yes	Low	30
6	No	Yes	High	30
7	Yes	No	No	30
8	Yes	No	Low	30
9	Yes	No	High	30
10	Yes	Yes	No	30
11	Yes	Yes	Low	30
12	Yes	Yes	High	30

[a]No means not included in intervention; yes means included in intervention

3.12 Powering a Factorial Experiment: Main Effects

is $N = 180$, and power for detection of the *MI* and *PEER* main effects is the same as it was in the 2^k design. However, the per-level N for *TEXT* is $N = 120$, because the overall $N = 360$ is now spread across three levels. Consider the no vs. low comparison on the *TEXT* factor. The mean of the no level would be computed based on the light shaded cells in Table 3.10, and the mean of the low level would be computed based on the dark shaded cells. The unshaded cells would not be involved in the computation of the regression coefficient, although they would be involved in the estimation of the error term. Thus, as Table 3.10 illustrates, the comparison of no vs. low is based on the 240 subjects in those two levels, not the entire sample size of $N = 360$. Similarly, the low vs. high comparison is also based on a subset of 240 subjects, this time those assigned to either the low or high levels. Consequently, the power for these comparisons is not maintained at .8; instead it is about .64. To maintain power of .8 for these comparisons, it would be necessary to increase the sample size by about 50 percent to about $N = 540$. With this larger sample size, the per-level N for the *MI* and *PEER* factors would increase, and so the statistical power for detection of their main effects would increase beyond .8.

In summary, in any balanced factorial experiment—whether the factors have two levels or more than two levels—the sample size per level of a factor is an important determinant of the power available for detection of that factor's main effect. It is a good idea to remember the following two points. The first point is that in general, adding a level to even one factor in a factorial experiment can require a substantial increase in subjects to maintain the same level of power. The second point pertains to experiments in which the number of levels varies across the factors. Assuming all else is equal, that is, effect sizes and α are the same, there will be less statistical power for the tests of comparisons of levels, as well as certain interactions, involving the factors with more levels. This is because when the overall N is spread across more levels, the per-level sample size is smaller.

3.12.7 Recommendations Based on the Resource Management Principle

In engineering and related fields, factorial experiments are frequently designed with two levels per factor and involve numerous factors, because this is usually most efficient. This book recommends that wherever possible, intervention scientists take this approach. In other words, when designing a factorial optimization trial, it is a good idea to use a 2^k design, and to make k as large as possible, that is, include as many factors as is operationally feasible.

What if it is necessary to choose among several possible levels, values, or doses of a component for inclusion in an optimized intervention? If the levels can be ordered quantitatively, a reasonable approach is to start by including only two levels in the factorial experiment: one corresponding to the lowest level that would be considered for inclusion in the intervention and one corresponding to the level expected to produce the best outcome (i.e., the level corresponding to the largest dose, unless a curvilinear dose-response function is hypothesized). This provides a test of the null hypothesis that the strongest level of the component has no effect on the outcome. If this hypothesis is demonstrated to be false, then it makes sense to conduct a follow-up experiment to look at intermediate doses of the component. If this hypothesis is not falsified, then it would be a waste of resources to look at intermediate doses; if the level expected to produce the best outcome demonstrates no effect, a smaller dose is unlikely to demonstrate an effect. To follow this advice in the hypothetical example above, Dr. B would start with two levels of TEXT. The two levels would probably be intensive and no. If the results indicate that this difference is of sufficient magnitude, then follow-up experimentation would be done to determine whether a moderate level of texts, say 4 texts per week, achieves nearly as good an outcome as 3 texts per day. However, if the results of the original experiment indicate that the difference between intensive and no is small, leading to the conclusion that the text support component is not very effective, then the resource management principle suggests it would not be a good use of resources to investigate the intensive vs. moderate comparison.

Now consider a different scenario involving a different question. Suppose an investigator wishes to examine a motivational interviewing component, which can take on the following three levels: no, motivational interviewing via a computer program, and motivational interviewing conducted by a trained interviewer. Let us assume that motivational interviewing conducted by a trained interviewer is expected to produce a better outcome, but motivational interviewing via computer is worth examination because it is relatively inexpensive. Here again, the recommendation is to try to set up the experiment so that it is made up of factors with two levels. There are two alternative strategies that could be used, based on different ideas about the role of motivational interviewing in the intervention.

One of these alternative strategies is appropriate if the two types of support are considered mutually exclusive alternatives, that is, the same content delivered via two different vehicles, only one of which will ultimately be selected for inclusion. This strategy uses essentially the same logic as above. The experiment would include an *MI* factor with two levels: no and conducted by a trained interviewer. If the difference between the two levels is large enough to be of interest, then a subsequent experiment can be conducted to examine whether motivational interviewing via computer is nearly as good as motivational interviewing by a trained interviewer (although, see the discussion above about comparison of two putatively active levels of a component). However, if motivational interviewing by a trained interviewer is ineffective, it makes little sense to try it via computer, given that this approach is expected to be less effective. Instead, it would probably be better to go back to the conceptual model and reconsider whether motivational interviewing should be a candidate for inclusion in the intervention.

The other strategy involves treating the two approaches to motivational interviewing as two independent components, either or both of which could conceivably be included in the optimized intervention. In contrast to the first strategy, this would probably not involve the same material delivered via two different vehicles, unless the conceptual model suggests that this redundancy may be desirable. Instead, motivational interviewing conducted by a trained interviewer and via computer would play two different roles in the intervention. With this strategy there would be two factors in the experiment: one corresponding to interviewer-conducted motivational interviewing, labeled (say) *IMI*, with levels no and yes, and one corresponding to computer-conducted motivational interviewing, labeled (say) *CMI*, with levels no and yes. For this strategy to be appropriate, it would have to make sense from an intervention design point of view to include either or both types of motivational interviewing, so that any of the following combinations would be feasible: neither component, only motivational interviewing conducted by a trained interviewer, only motivational interviewing conducted via computer, and both types of motivational interviewing. This approach would enable estimation of the interaction between the two types of motivational interviewing. An interaction could suggest, for example, that motivational interviewing conducted via computer is more effective when participants are also receiving motivational interviewing conducted by a trained interviewer.

3.12.8 Small N Situations

In some research areas within intervention science, such as treatment of severe autism, communication disorders, or rare diseases, there may be only a small number of experimental subjects available (either due to the rarity of the condition or difficulty in recruitment). Here the pool of subjects available for experimentation may be too small to enable even a moderate level of statistical power in a factorial experiment. However, if the intervention takes place for a sustained period of time and involves considerable one-on-one attention, it may be possible to conduct experimentation in the optimization phase using an N-of-one framework, in which many different experimental conditions are implemented repeatedly on single subjects. Some variation on the micro-randomized trial (Klasnja et al. 2015) approach may be helpful here, or procedures may be adapted from the design of system identification experiments (e.g., Ogunnaike & Ray 1994). More research is needed on how to apply MOST to optimize interventions in small-N situations.

3.13 The Coefficient Correction: Why Is There a 2 in the Denominator of the Two-Way Interaction?

Above the interaction in the factorial ANOVA, for example, the $MI \times PEER$ interaction, was defined as follows:

$$INT_{MI \times PEER} = \frac{(\bar{\mu}_{++\cdot} - \bar{\mu}_{-+\cdot}) - (\bar{\mu}_{+-\cdot} - \bar{\mu}_{--\cdot})}{2}.$$

It seems intuitive that the numerator of the expression for the interaction is the difference between the effect of *MI* on *adhere* when *PEER* is set to yes and when *PEER* is set to no. But what is the reason for dividing this quantity by 2? In this book this denominator, which varies depending on the order of the interaction, is called the coefficient correction.

The explanation provided here borrows heavily from the Kugler, Dziak, and Trail (2018) chapter in the companion volume. Recall the definition of the main effect of, e.g., *MI*:

$$ME_{MI} = \bar{\mu}_{+..} - \bar{\mu}_{-..}.$$

This can be expanded as (see Table 3.2)

$$ME_{MI} = \frac{\mu_{+++} + \mu_{++-} + \mu_{+-+} + \mu_{+--}}{4} - \frac{\mu_{-++} + \mu_{-+-} + \mu_{--+} + \mu_{---}}{4}$$

and equivalently expressed as

$$\begin{aligned}ME_{MI} = &\frac{1}{4}\mu_{+++} + \frac{1}{4}\mu_{++-} + \frac{1}{4}\mu_{+-+} + \frac{1}{4}\mu_{+--} - \frac{1}{4}\mu_{-++}\\ &-\frac{1}{4}\mu_{-+-} - \frac{1}{4}\mu_{--+} - \frac{1}{4}\mu_{---}.\end{aligned} \quad (3.5)$$

In other words, each μ has a coefficient of ¼.

In a similar manner the definition of the interaction (e.g., $MI \times PEER$) can be expanded:

$$INT_{MI \times PEER} = \frac{\left(\frac{(\bar{\mu}_{+++} + \bar{\mu}_{++-})}{2} - \frac{(\bar{\mu}_{-++} + \bar{\mu}_{-+-})}{2}\right) - \left(\frac{(\bar{\mu}_{+-+} + \bar{\mu}_{+--})}{2} - \frac{(\bar{\mu}_{--+} + \bar{\mu}_{---})}{2}\right)}{2}$$

$$INT_{MI \times PEER} = \left(\frac{(\bar{\mu}_{+++} + \bar{\mu}_{++-})}{4} - \frac{(\bar{\mu}_{-++} + \bar{\mu}_{-+-})}{4}\right)$$
$$- \left(\frac{(\bar{\mu}_{+-+} + \bar{\mu}_{+--})}{4} - \frac{(\bar{\mu}_{--+} + \bar{\mu}_{---})}{4}\right)$$

$$\begin{aligned}INT_{MI \times PEER} = &\frac{1}{4}\bar{\mu}_{+++} + \frac{1}{4}\bar{\mu}_{++-} - \frac{1}{4}\bar{\mu}_{-++} - \frac{1}{4}\bar{\mu}_{-+-}\\ &-\frac{1}{4}\bar{\mu}_{+-+} - \frac{1}{4}\bar{\mu}_{+--} + \frac{1}{4}\bar{\mu}_{--+} + \frac{1}{4}\bar{\mu}_{---}.\end{aligned} \quad (3.6)$$

Comparison of Eqs. (3.5) and (3.6) shows that each is a linear combination of cell means, with each mean weighted with coefficient = ¼. Without the coefficient correction, the cell means in the definition of the interaction would be weighted differently from the cell means in the definition of the main effect. The coefficients in the definition of the two-way interaction would be ½, which would mean the main effects and interaction would be on different scales. Thus the coefficient correction ensures that the cell means in all effects in a factorial ANOVA, from the main effect up to the highest-order interaction, are weighted using the same coefficient.

As the order of the interaction increases, the size of the coefficient correction increases accordingly. For example, for a three-way interaction the coefficient correction is 4:

$$INT_{MI \times PEER \times TEXT}$$
$$= \frac{[(\mu_{+++} - \mu_{-++}) - (\mu_{+-+} - \mu_{--+})] - [(\mu_{++-} - \mu_{-+-}) - (\mu_{+--} - \mu_{---})]}{4}.$$

For this three-way interaction, a coefficient correction of 4 produces a linear combination of cell means with each mean weighted using coefficient $= ¼$. The algebraic demonstration of this is left to the reader.

As discussed in Chap. 4, there is an alternative definition of the interaction that does not include the coefficient correction. As long as consistency is maintained, either definition is appropriate. However, using the definition with the coefficient correction greatly simplifies the correspondence between the definition of the interaction and the regression coefficient and, by extension, the translation of raw effects into effect sizes for power analysis.

3.14 Summary: The Efficiency of the Factorial Experiment and the Efficiency of the RCT

In this chapter the different characteristics of factorial experiments and the RCT have been compared, including the different approaches to powering the two types of studies. Let us now review why earlier in the chapter it was said that treating a factorial experiment like a multi-arm RCT ignores the signature strengths of the factorial experiment.

The purpose of the RCT is to provide direct comparisons between means of individual experimental conditions. Consider a treatment-control comparison in a two-arm RCT. Here, all else being equal, the n in these two conditions drives the statistical power available for this treatment-control comparison. By contrast, the purpose of a factorial experiment is to provide comparisons between the means associated with the levels of each factor. Except for estimation of the highest-order interaction, these means are based on aggregates of experimental conditions, as shown, for example, in Tables 3.3 and 3.9. In other words, each effect estimate is based on all the subjects in each level being compared. For this reason, a factorial experiment can enjoy excellent statistical power with a per-condition n that would be much too small to provide adequate power for an RCT. For example, Table 3.8 illustrates a well-powered factorial experiment with $n = 22$. Depending on the circumstances, a factorial experiment can have good power even with per-condition ns in the single digits. This can be counterintuitive to scientists who have been trained primarily in the RCT.

Any power analysis is done assuming the data are to be analyzed in a particular way. If the data are analyzed in some manner other than the one upon which the power analysis was based, the desired level of power may not be maintained. Put a different way, statistical power is partly determined by how the research questions are framed and the approach to data analysis that the framing of the questions dictates. For example, suppose an investigator has conducted the experiment depicted in Table 3.8 but is not interested in the main effect of each factor and interactions between factors. Instead, the investigator wants to determine which combinations of levels of the factors are significantly better than experimental condition 1, in which all of the factors are set to the no level. Framing the research question in this manner transforms an adequately powered factorial experiment into a badly underpowered multi-arm RCT.

Thus, the RCT and the factorial experiment are both efficient but for different applications. The RCT is most efficient at directly comparing means of experimental conditions. It is not an efficient way of assessing which of the components making up the treatment condition has a detectable effect, and it cannot be used to determine whether any of them interact. Its signature strength is establishing whether one experimental condition, for example, a treatment condition representing a particular combination of intervention components, is detectably different from another, such as a control. By contrast, the factorial experiment is most efficient at estimation of main effects and interactions. It is not an efficient way of directly comparing means of experimental conditions. Its signature strength is investigating which components have detectable effects and which interact with each other.

In MOST, each type of experimental design is reserved for the research question it addresses most efficiently. Factorial experimental designs are used, among others (see Chaps. 1 and 8), in the optimization phase of MOST because they represent an efficient way to weed out ineffective components and, from those that remain, identify the components and component levels that will make up the optimized intervention. The RCT is used in the evaluation phase because it is the most efficient way to compare the optimized intervention to a suitable control or comparator.

3.15 What's Next?

This chapter has provided an introduction to the factorial experiment. The 2^k experiment was emphasized, because generally this is the most efficient approach to examination of the effectiveness of intervention components. To keep the amount of material manageable, this chapter focused more on main effects than on interactions, even though interactions are critically important. Interactions are the primary topic of the next chapter, Chap. 4.

References

Caldwell, L. L., Smith, E. A., Collins, L. M., Graham, J. W., Lai, M., Wegner, L., ... Jacobs, J. (2012). Translational research in South Africa: Evaluating implementation quality using a factorial design. *Child & Youth Care Forum, 41*, 119–136.

Cohen, J. (1988). *Statistical power analysis for the behavioral sciences* (2nd ed.). Hillsdale, NJ: Lawrence Erlbaum Associates.

Collins, L. M., Dziak, J. R., & Li, R. (2009). Design of experiments with multiple independent variables: A resource management perspective on complete and reduced factorial designs. *Psychological Methods, 14*, 202–224.

Collins, L. M., & Kugler, K. C. (Eds.). (2018). *Optimization of multicomponent behavioral, biobehavioral, and biomedical interventions: Advanced topics*. New York, NY: Springer.

Fisher, R. A. (1925). *Statistical methods for research workers*. Edinburgh, Scotland: Oliver and Boyd.

Fisher, R. A. (1971). *The design of experiments*. New York, NY: Hafner Publishing.

Goldstein, H. (2011). *Multilevel statistical models* (4th ed.). London, UK: Wiley.

Kirk, R. E. (2013). *Experimental design: Procedures for the behavioral sciences* (4th ed.). Los Angeles, CA: SAGE.

Klasnja, P., Hekler, E. B., Shiffman, S., Boruvka, A., Almirall, D., Tewari, A., & Murphy, S. A. (2015). Microrandomized trials: An experimental design for developing just-in-time adaptive interventions. *Health Psychology, 34*(S), 1220.

Kraemer, H. C. (2011). Another point of view: superiority, noninferiority, and the role of active comparators. *The Journal of Clinical Psychiatry, 72*, 1350–1352.

Kugler, K. C., Dziak, J. J., & Trail, J. (2018). Coding and interpretation of effects in analysis of data from a factorial experiment. In L. M. Collins & K. C. Kugler (Eds.), *Optimization of multicomponent behavioral, biobehavioral, and biomedical interventions: Advanced topics* (forthcoming). New York, NY: Springer.

McClure, J. B., Derry, H., Riggs, K. R., Westbrook, E. W., John, J. S., Shortreed, S. M., ... An, L. C. (2012). Questions about quitting (Q 2): Design and methods of a Multiphase Optimization Strategy (MOST) randomized screening experiment for an online, motivational smoking cessation intervention. *Contemporary Clinical Trials, 33*, 1094–1102.

Michie, S., Abraham, C., Whittington, C., McAteer, J., & Gupta, S. (2009). Effective techniques in healthy eating and physical activity interventions: A meta-regression. *Health Psychology, 28*, 690–701.

Montgomery, D. C. (2009). *Design and analysis of experiments* (7th ed.). Hoboken, NJ: Wiley.

Murray, D. M. (1998). *Design and analysis of group-randomized trials* (Vol. 29). New York, NY: Oxford University Press.

Nahum-Shani, I., & Dziak, J. J. (2018). Multilevel factorial designs in intervention development. In L. M. Collins & K. C. Kugler (Eds.), *Optimization of multicomponent behavioral, biobehavioral, and biomedical interventions: Advanced topics* (forthcoming). New York, NY: Springer.

Ogunnaike, B. A., & Ray, W. H. (1994). *Process Dynamics, Modeling, and Control*. Oxford, UK: Oxford University Press.

Piper, M. E., Fiore, M. C., Smith, S. S., Fraser, D., Bolt, D. M., Collins, L. M., ... Baker, T. B. (2016). Identifying effective intervention components for smoking cessation: A factorial screening experiment. *Addiction, 111*, 129–141.

Raudenbush, S. W., & Bryk, A. S. (2002). *Hierarchical linear models: Applications and data analysis methods* (2nd ed.). Thousand Oaks, CA: SAGE.

Schlam, T. R., Fiore, M. C., Smith, S. S., Fraser, D., Bolt, D. M., Collins, L. M., ... Loh, W. Y. (2016). Comparative effectiveness of intervention components for producing long-term abstinence from smoking: A factorial screening experiment. *Addiction, 111*(1), 142–155.

Walker, E., & Nowacki, A. S. (2011). Understanding equivalence and noninferiority testing. *Journal of General Internal Medicine, 26*, 192–196.

Wu, C. F. J., & Hamada, M. S. (2009). *Experiments: Planning, analysis, and optimization* (2nd ed.). Hoboken, NJ: Wiley.

Chapter 4
Interactions Between Components and Moderation of Component Effects

Abstract Interactions, introduced in Chap. 3, merit their own chapter for several reasons. First, the interaction is both a subtle and an important concept, one that takes more than a few pages to explore thoroughly. Second, it is helpful to note the difference between synergistic and antagonistic interactions and the implications of this difference for decision-making in the multiphase optimization strategy (MOST). Third, how to power a factorial experiment for detection of interactions between factors can be confusing unless the investigator is aware that there are two different definitions of the interaction in use in the scientific literature. Fourth, investigators may wish to explore whether the effects of experimental factors are moderated by observed variables (e.g., subject gender). Readers should be familiar with the material in Chaps. 1, 2, and 3.

Contents

4.1	Introduction	116
4.2	Interactions and Moderation	117
4.3	Definition of the Interaction Effect in Factorial ANOVA	118
	4.3.1 Review of the Classical Definition of the Interaction	118
4.4	Interpreting Interactions by Plotting	119
	4.4.1 Plots of Means Where There Is No Interaction	119
	4.4.2 Synergistic and Antagonistic Interactions	123
	4.4.3 Synergistic Two-Way Interactions	123
	4.4.4 Antagonistic Two-Way Interactions	125
	4.4.5 Higher-Order Interactions	128
4.5	The Role of Main Effects and Interactions in Decision-Making: Effect Hierarchy, Effect Sparsity, and Effect Heredity	130
	4.5.1 A Brief Reminder	131
4.6	A Decision-Priority Perspective on Interactions	132
	4.6.1 Interactions and Decision-Making	132
	4.6.2 Implications for Powering Factorial Experiments	133
4.7	What Is the Statistical Power for Detection of Interaction Effects as Compared to Main Effects?	134

	4.7.1	Power Is Identical for Main Effects and Interactions With Identical Regression Weights in 2^k Experiments, but Not Necessarily in Other Experiments	134
	4.7.2	Cohen and Fleiss Use Different Definitions of "Interaction"	135
	4.7.3	The Ability to Detect Interactions Depends Partly on the Size of the Interaction, but Who Knows What Effect Size to Expect?	138
4.8	Moderation of the Effects of Factors by Observed Variables (Effect Modification)		138
	4.8.1	Examining Moderation Based on Naturally Varying Observed Moderators	139
	4.8.2	Examining Moderation by Including the Moderator as a Factor in the Experiment	140
	4.8.3	Moderation of Component Effects and the Continual Optimization Principle ..	142
4.9	What's Next?		142
References			143

4.1 Introduction

Chapter 3 reviewed the factorial design and explained how this experimental design can be used in an optimization trial to examine the effects of individual intervention components. In a factorial optimization trial, there is an experimentally manipulated factor corresponding to each candidate component. Based on the resulting data, the main effect of each factor can be estimated, along with interactions between factors. An interaction between two factors is present when the effect of one factor is different depending on the level of the other. In a factorial optimization trial, this indicates that the performance of one component varies depending on the presence or level of another component.

In this chapter statistical and conceptual aspects of interactions are discussed to provide a necessary foundation for the material in the rest of the book. Interactions must be considered carefully when the investigator makes decisions about the choice of design for the optimization trial (Chaps. 5 and 6) and, after the optimization trial is completed, which components and component levels to include in the optimized multicomponent behavioral, biobehavioral, or biomedical intervention (Chap. 7).

The discussion in this chapter is limited to models for normal outcome variables. Interactions can be estimated when other link functions are used, such as the logistic, but the interpretation of the results is more complicated and requires particularly careful consideration (e.g., Karaca-Mandic, Norton, & Dowd, 2012). It is important to note that any main effect or interaction that is significant when one link function is used may or may not be significant when a different link function is used. Linear transformations of variables do not affect the interpretation of effects and associated hypothesis tests in an ANOVA, but nonlinear transformations, such as squaring or log transformations, do affect the interpretation of both main effects and interactions, as well as tests of hypotheses involving these effects.

In this chapter it is assumed that the factors are coded so that the higher level is the more expensive or intense level, and the outcome variable is coded so that a larger value is more desirable. In practice coding this way is not necessary, but it makes interpretation of results more straightforward. The 2^k factorial design is emphasized because, as has been stated repeatedly in this book, it is highly efficient for optimization trials.

4.2 Interactions and Moderation

To assist in the discussion of the interaction effect in factorial ANOVA, let us return to the hypothetical example presented in the previous chapters, in which Dr. B is examining a set of components for possible inclusion in an intervention to reduce viral load in HIV-positive individuals who drink heavily. For simplicity, the discussion in this chapter will be confined to the first three components: motivational interviewing, peer mentoring, and text messaging. Dr. B is to conduct a factorial experiment to examine these components. The factors correspond directly to the components, with the first factor abbreviated *MI*, the second abbreviated *PEER*, and the third abbreviated *TEXT*. Each factor takes on two levels: either no (not included in the intervention) or yes (included). The primary outcome variable is number of days of adherence to ART in the 30-day period following the conclusion of the intervention, and it will be abbreviated *adhere*.

Statistical interactions between two variables are often conceptualized as moderation of one variable's effect by the other variable, particularly in the behavioral sciences. There is a conceptual distinction between interactions involving only experimental factors and those that involve one or more observed variables (see Chap. 2). Although it would not be incorrect to say that one experimental factor moderates the effect of another, in this book, the term moderation is reserved for the special case that occurs when an observed (as opposed to manipulated) variable—such as depression—interacts with an observed or experimental factor. This is sometimes called effect modification, particularly in the epidemiology literature. When two experimental factors interact, either factor can be considered the moderator. It is equivalent to say that the effect of *skills* on *adhere* is different depending on the level of *mind* and to say that the effect of *mind* is different depending on the level of *skills*. By contrast, when an observed variable and an experimental factor interact, the observed variable is usually considered the moderator of the effect of the experimental factor. From this point forward, in this book, the term moderation will be reserved for interactions involving both observed variables and experimental factors, that is, for effect modification, to highlight this conceptual distinction.

4.3 Definition of the Interaction Effect in Factorial ANOVA

4.3.1 Review of the Classical Definition of the Interaction

In this section the definition of the interaction used in this book, presented in Chap. 3, is reviewed for convenience. Let us start with a two-way interaction, say between *MI* and *PEER*. As was discussed in Chap. 3, according to a widely used definition of the interaction (e.g., Montgomery, 2008), two factors interact if the effect of one is different depending on the level of the other. In a 2^k experiment, a two-way interaction is half of the difference in the effect of a particular factor across the levels of a second factor, averaging over all other factors (Montgomery, 2008). In the example, the $MI \times PEER$ interaction is defined as

$$INT_{MI \times PEER} = \frac{(\bar{\mu}_{++\cdot} - \bar{\mu}_{-+\cdot}) - (\bar{\mu}_{+-\cdot} - \bar{\mu}_{--\cdot})}{2}, \qquad (4.1)$$

where $\bar{\mu}_{++\cdot}$ is the average of the cell means corresponding to experimental conditions where both of the first two factors, *MI* and *PEER*, are set to the higher (i.e., +) level; the dot in the third position in the subscript indicates that the average is taken over the levels of *TEXT*; and the 2 in the denominator is the coefficient correction (explained in Chap. 3). The numerator of the $MI \times PEER$ interaction is the difference between the effect of *MI* on *adhere* when *PEER* is set to yes and when *PEER* is set to no; that is, the difference between the effect of *MI* on *adhere* when the intervention includes peer mentoring and when the intervention does not include peer mentoring, averaging over the levels of *TEXT*. If the effect of *MI* on *adhere* is different when peer mentoring is provided—for example, if the effect of *MI* on *adhere* is larger on average when peer mentoring is also provided, which would make the first parenthetical expression in the numerator of Eq. (4.1) larger than the second—then *MI* and *PEER* interact. If the effect of *MI* on *adhere* is the same irrespective of whether or not peer mentoring is provided, the two parenthetical expressions in the numerator of Eq. (4.1) are identical and the difference between them is zero, in other words, *MI* and *PEER* do not interact.

A three-way interaction occurs when a two-way interaction is different depending on the level of a third factor. In the example, the $MI \times PEER \times TEXT$ interaction is defined as

$$INT_{MI \times PEER \times TEXT}$$
$$= \frac{[(\mu_{+++} - \mu_{-++}) - (\mu_{+-+} - \mu_{--+})] - [(\mu_{++-} - \mu_{-+-}) - (\mu_{+--} - \mu_{---})]}{4}.$$
$$(4.2)$$

There are no bars over the μs in the above expression because the example involves only three factors, so the means are not averaged over settings of any additional factors. If there were additional factors in the experiment, there would be bars over the μs, indicating that the means were averaged over those factors. Each of the expressions in brackets in the numerator represents the difference between the effect of *MI* on *adhere* when *PEER* is set to yes and when *PEER* is set to no. The left-hand expression in brackets represents this quantity when *TEXT* is set to yes, and the right-hand expression in brackets represents this quantity when *TEXT* is set to no. The 4 in the denominator is the coefficient correction. If the difference between the effect of *MI* on *adhere* when *PEER* is set to yes and when *PEER* is set to no varies depending on the level of *TEXT*, then there is an *MI* × *PEER* × *TEXT* interaction. If the difference between the effect of *MI* on *adhere* when *PEER* is set to yes and when *PEER* is set to no is the same at each level of *TEXT*, then there is no *MI* × *PEER* × *TEXT* interaction.

Four-way interactions, five-way interactions, and so on are defined using straightforward extensions of the definitions above.

4.4 Interpreting Interactions by Plotting

Plotting marginal means is a good way to interpret the main effects and, particularly, interactions that are estimated in a factorial ANOVA. It is a good idea to plot model predicted marginal means, that is, the means computed based on the regression model that has been fit to the data, rather than the observed marginal means based on the raw data. In particular, when a pretest or other covariates have been included in the model, the predicted means are adjusted for the covariate, making interpretation more straightforward. (However, comparing results for model predicted and observed means can be a useful diagnostic.) This section will present several plots of hypothetical results (i.e., based on artificial data generated for pedagogical purposes) to demonstrate how to interpret such plots. The role of interaction plots in decision-making in the optimization phase of MOST will be discussed in detail in Chap. 7.

4.4.1 Plots of Means Where There Is No Interaction

To interpret plots of interactions properly, it is helpful to be familiar with plots in which no interaction is evident. The data displayed in Fig. 4.1 were generated with the hypothetical example in mind, using this model:

$$adhere = 14 + 8X_{MI}.$$

In this model there is a main effect of *MI* and no other effects. Figure 4.1 shows plots of the marginal means for the *MI* and *PEER* factors, collapsing across *TEXT*. In Fig. 4.1a *PEER* has been placed on the x-axis. The dashed line shows the means for the no and yes levels of *PEER* when *MI* is set to no, and the solid line shows the corresponding means for the two levels of *PEER* when *MI* is set to yes.

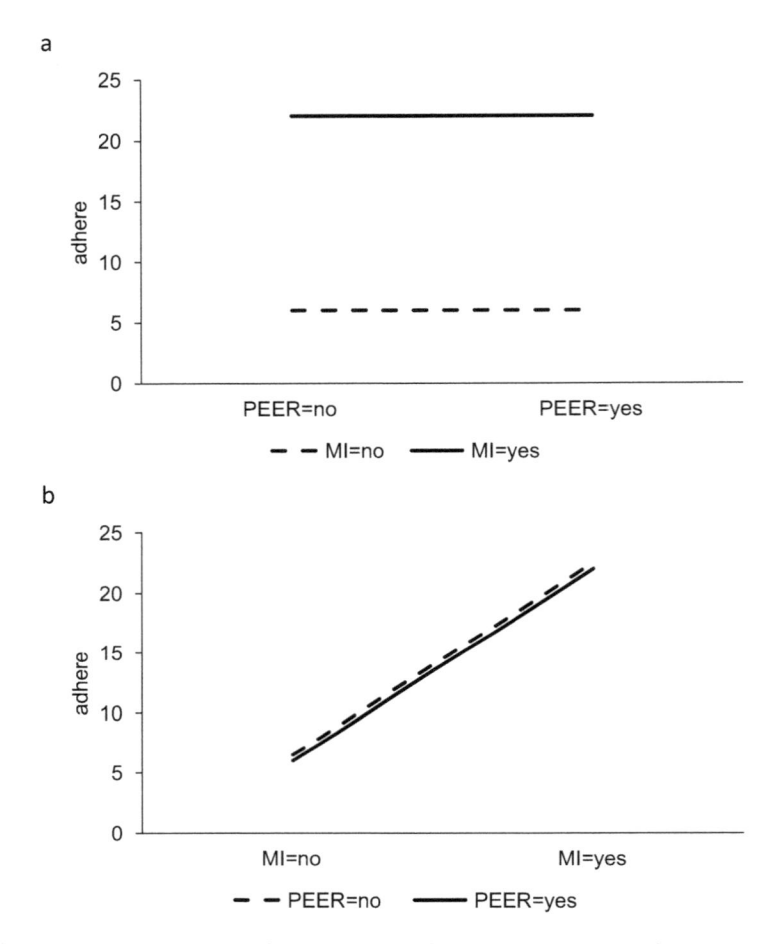

Fig. 4.1 Plots of marginal means for two two-level factors. In these data *MI* demonstrates a main effect; *PEER* does not demonstrate a main effect; there is no *MI* × *PEER* interaction. (**a**) Plot with *PEER* on the *x*-axis. (**b**) Plot of same data with *MI* on the *x*-axis

The main effect of *MI* is evident in Fig. 4.1a; the gap between the solid and dashed lines indicates that the means are higher for *MI* = yes as compared to *MI* = no. It is also evident that there is no main effect of *PEER*, because both the solid and dashed lines are horizontal. This reflects the lack of a difference in the means corresponding to the two levels of *PEER*.

It is arbitrary which factor is placed on the *x*-axis, although sometimes one way of arranging the graph is more helpful in interpreting the data or making decisions than another. Figure 4.1b shows the same data plotted with *MI* on the *x*-axis instead of *PEER*. Here the dashed line corresponds to *PEER* = no, and the solid line corresponds to *PEER* = yes. Because there is no main effect of *PEER*, the solid and dashed lines are identical whether *PEER* = no or *PEER* = yes. In this figure, the slope of the lines represents the main effect of *MI*.

4.4 Interpreting Interactions by Plotting

Table 4.1 Hypothetical cell means of *adhere* (averaged over the levels of *TEXT*)

	PEER	
MI	No	Yes
No	2	10
Yes	18	26

Although it may be counterintuitive, one cell or marginal mean that appears very large or very small is not necessarily evidence for the presence of an interaction. Such a pattern of results can easily be due to main effects alone. Consider a different situation in which the example experiment was conducted, and produced the cell means in Table 4.1 corresponding to the various combinations of the levels of *MI* and *PEER*, averaged over the levels of *TEXT*. The table shows clearly that the experimental condition in which subjects are provided with both motivational interviewing and text messaging has the best outcome. This outcome is much better than the outcome for any of the other conditions shown in Table 4.1. Given that the condition in which both factors are at the yes level represents a substantial increase in *adhere*, does this mean there is an interaction between *MI* and *PEER*?

In fact, the data appearing in Table 4.1 were generated using the following regression model:

$$Y = 14 + 8X_{MI} + 4X_{PEER}.$$

In other words, the only effects in this model are the intercept ($b_0 = 14$), a coefficient representing the main effect of *MI* ($b_{MI} = 8$), and a coefficient representing the main effect of *PEER* ($b_{PEER} = 4$). There is no *MI* × *PEER* interaction. This can be seen by examining the table carefully. If there were an interaction, the effect of *MI* would be different depending on the level of *PEER*. But the effect of *MI* when *PEER* = no is 16 days of adherence (18–2); the effect of *MI* when *PEER* = yes is also 16 days of adherence (26–10). (Recall from Chap. 3 that the main effect of *MI* is $2b_{MI}$.)

Figure 4.2a contains a plot of these data with *PEER* on the *x*-axis. The main effect of *MI* is demonstrated by the gap between the solid and dashed lines. The main effect of *PEER* is demonstrated by the slope of the lines. The fact that the two lines are parallel is important. According to the definition in Eq. (4.1), there is an interaction if the difference between the no and yes levels of *PEER* varies depending on whether *MI* is at the no or yes level. The difference between the no and yes levels of *PEER* at each level of *MI* is represented by the relative position of the lines in Fig. 4.2a. If these lines are parallel, it means the difference between the no and yes levels of *PEER* is the same at each level of *MI*. Thus parallel lines on a graph of marginal means of two factors indicate that the factors do not interact.

Figure 4.2b shows the same marginal means plotted with *MI* on the *x*-axis. The plot looks very similar. There is still a sizeable gap between the solid and dashed lines, but it is smaller than the gap in Fig. 4.2a. This is because in this figure the gap represents the main effect of *PEER*, which is smaller than the main effect of *MI*. Similarly, the slope of the lines is steeper than the slope in Fig. 4.2a, because the main effect this slope represents in Fig. 4.2b, that of *MI*, is larger.

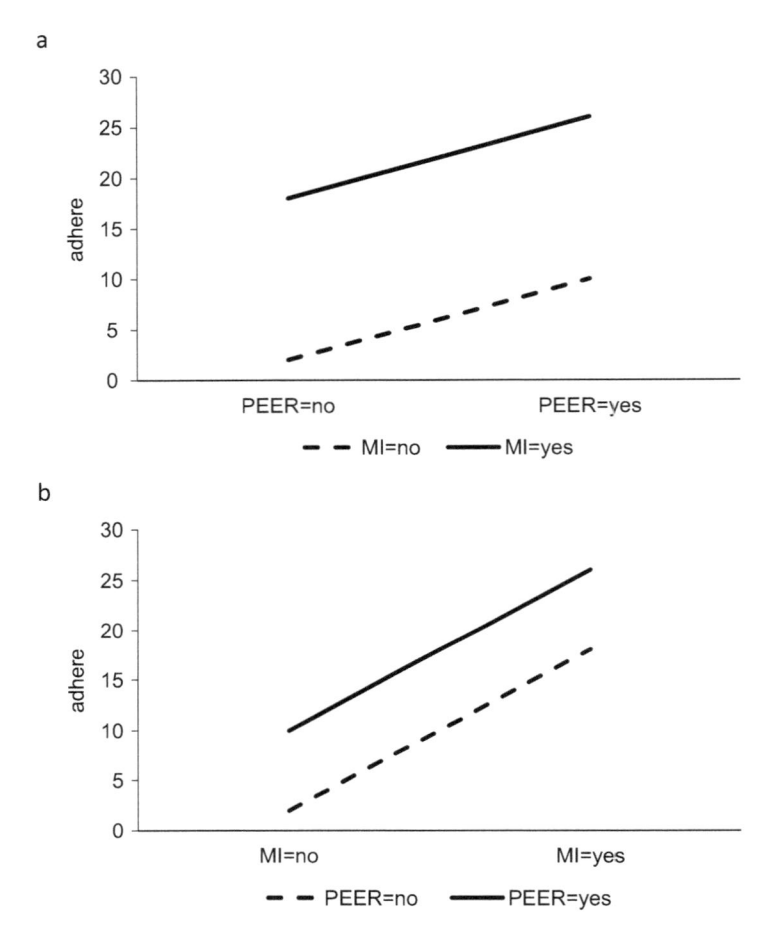

Fig. 4.2 Plots of marginal means for two two-level factors. In these data both *MI* and *PEER* demonstrate main effects; there is no *MI* × *PEER* interaction. (**a**) Plot with *PEER* on the *x*-axis. (**b**) Plot of same data with *MI* on the *x*-axis

Sometimes for planning purposes, it is desirable to determine whether an interaction is expected. It may be helpful to try conducting the following exercise: Carefully review the conceptual model. Then make a table similar to Table 4.1, initially leaving the cell means blank. Do not think about whether there may be main effects or interactions, simply fill in the cells with means that are plausible according to the hypotheses made based on the conceptual model. (This is one example of many where a well-specified conceptual model is essential in MOST; see Chap. 2.) Finally, examine or plot the means to see whether they reflect an interaction.

4.4.2 Synergistic and Antagonistic Interactions

The main effect estimate of a factor provides a sense of the expected average performance of the corresponding candidate component. The estimate of an interaction involving two or more factors provides a sense of whether the combination of the corresponding components is expected to produce a net result that is better or worse than would have been expected based solely on the individual performances of the components. As will be seen in Chap. 7, for this reason, interactions play an important role in the decision-making that is necessary to arrive at the optimized intervention.

In this book a distinction will be drawn between synergistic and antagonistic interactions. This distinction is made in fields such as chemistry and pharmacy, which, like the behavioral, biobehavioral, and biomedical sciences, often involve combining multiple components to achieve a desired effect. A synergistic interaction is defined as follows:

> *In a synergistic interaction, the combined effect of two or more factors is more favorable than would be expected based solely on the main effects.*

Under the circumstances assumed at the outset of this section, namely, that the factors are coded so that the higher level is represented with +1 and a larger value on the outcome variable is more desirable, the coefficient corresponding to a synergistic interaction will have a positive sign.

An antagonistic interaction is defined as follows:

> *In an antagonistic interaction, the combined effect of two or more factors is less favorable than would be expected based solely on the main effects.*

Under the circumstances assumed in this section (i.e., everything coded in a positive direction), the regression coefficient corresponding to an antagonistic interaction will have a negative sign. Note that depending on how the factors and outcome variable are coded, it is possible for an interaction considered synergistic to have a negative sign and for an interaction considered antagonistic to have a positive sign. It is helpful to be clear on the direction of effects and to try to keep it simple if possible.

Synergistic and antagonistic interactions will be explained further and illustrated below.

4.4.3 Synergistic Two-Way Interactions

In a synergistic two-way interaction, the net effect of the factors when they are both set to the higher level is greater than the sum of the two main effects. Consider a situation in which two factors each exert a main effect, and there is a synergistic interaction involving the two factors, as in the following model:

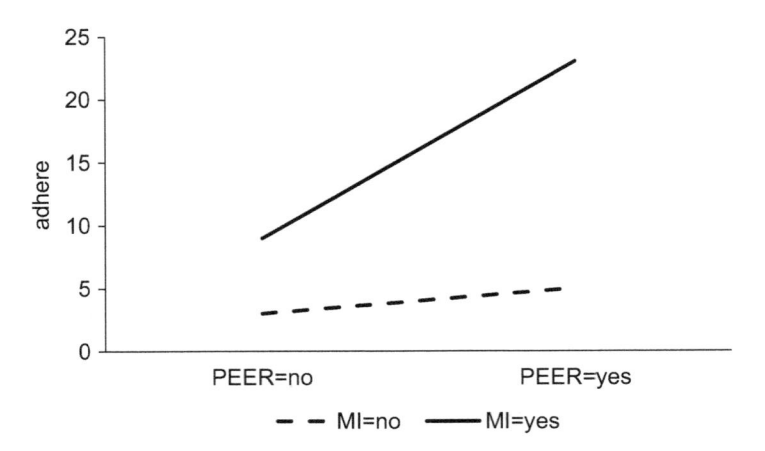

Fig. 4.3 Plot of marginal means for two two-level factors. In these data there is a main effect of *MI*, a main effect of *PEER*, and a synergistic *MI* × *PEER* interaction

$$adhere = 10 + 6X_{MI} + 4X_{PEER} + 3X_{MI \times PEER}.$$

Marginal means generated by this model are plotted in Fig. 4.3. This figure is similar to Fig. 4.2, in that the main effects of both *MI* and *PEER* are evident. An important difference is that the lines are not parallel in Fig. 4.3, indicating the presence of an interaction. Figure 4.3 shows that at both levels of *MI* the difference between the no and yes levels of *PEER* is positive. This difference is relatively small when *MI* = no and considerably larger when *MI* = yes. The outcome for the condition in which both factors are set to yes is better than would be expected based solely on the main effects of *MI* and *PEER*, reflecting the synergistic interaction between the two factors.

Now consider this regression model, which contains one main effect and a synergistic interaction:

$$adhere = 14 + 8X_{MI} + 3X_{MI \times PEER}$$

(note it is implicit that $b_{PEER} = 0$). Figure 4.4 is a plot of the corresponding marginal means. The main effect of *MI* is evident in the gap between the dashed line representing the no level of this factor and the solid line representing the yes level. The lack of a main effect of *PEER* may be harder to see in the figure, because to see this requires mentally averaging the solid line and the dashed line. The midpoint of the dashed and solid lines when *PEER* = no and when *PEER* = yes are identical; this means that on average, there is no difference between the two levels of *PEER*, which is consistent with $b_{PEER} = 0$. The lines are not parallel, indicating the presence of an interaction. In this example, although there is no main effect of *PEER*, there is a difference between the no and yes levels of *PEER* at each individual level of *MI*. When *MI* = no, providing peer mentoring is associated with a worse outcome than not providing peer mentoring. This means that on average,

4.4 Interpreting Interactions by Plotting

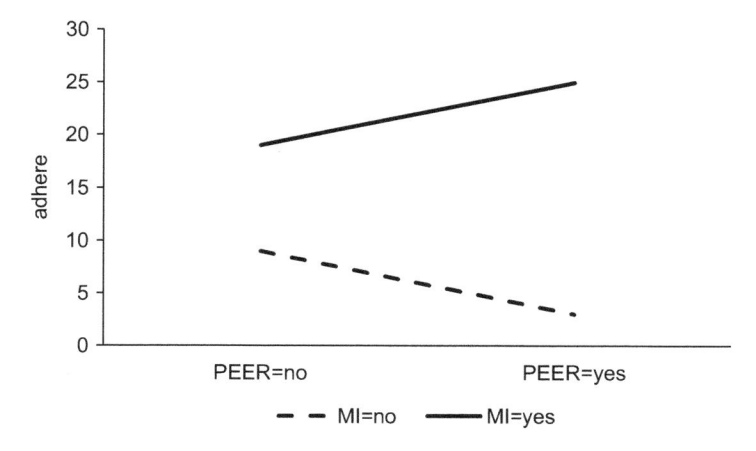

Fig. 4.4 Plot of marginal means for two two-level factors. In these data there is a main effect of *MI* and a synergistic *MI* × *PEER* interaction

providing peer mentoring alone without motivational interviewing leads to a worse outcome than providing neither component (in this hypothetical example). When *MI* = yes, providing peer mentoring in addition to motivational interviewing produces a better outcome than would be expected based on the main effects alone. (The data could have been plotted with *MI* on the *x*-axis; the interpretation would have been identical.)

4.4.4 Antagonistic Two-Way Interactions

In an antagonistic interaction involving two factors, the net effect of the factors when they are both set to the higher level is *less* than the sum of their main effects. The following regression model contains one main effect and an antagonistic interaction:

$$adhere = 14 + 10X_{MI} - 3X_{MI \times PEER}.$$

Figure 4.5 is a plot of the corresponding marginal means. As in Fig. 4.4, the main effect of *MI* is evident in the gap between the dashed and solid lines, and the lack of a main effect of *PEER* can be seen by mentally combining the solid line and the dashed line and noting that on average *adhere* is about the same when *PEER* = no and *PEER* = yes. One way to think about the interaction is that the effect of *MI* is different depending on the level of *PEER*. (Equivalently, it could be stated that the effect of *PEER* is different depending on the level of *MI*, even though there is no overall main effect of *PEER*.) The interaction involving *MI* and *PEER* is antagonistic because when these factors are both set to the higher level, *adhere* is smaller than would be expected if there were no interaction.

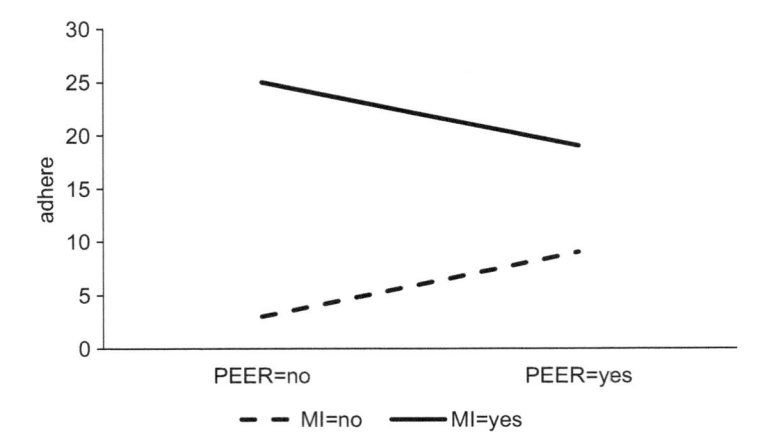

Fig. 4.5 Plot of marginal means for two two-level factors. In these data there is a main effect of *MI* and an antagonistic *MI* × *PEER* interaction

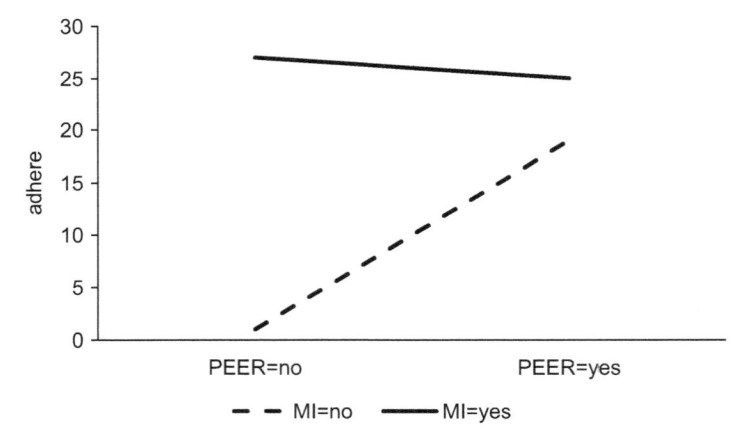

Fig. 4.6 Plot of marginal means for two two-level factors. In these data there is a main effect of *MI*, a main effect of *PEER*, and an antagonistic *MI* × *PEER* interaction

Now consider a situation in which there are two main effects and an antagonistic interaction, such as the following:

$$adhere = 18 + 8X_{MI} + 4X_{PEER} - 5_{MI \times PEER}.$$

The marginal means corresponding to this model are depicted in Fig. 4.6. The figure shows that when *MI* = no, setting *PEER* to yes is associated with a large increase in *adhere*, whereas when *MI* = yes, setting *PEER* to yes is associated with a slight decrease in *adhere*. In this case, the experimental condition in which both factors are set to yes is not associated with the best outcome on *adhere*. Instead, the experimental condition in which *MI* is set to yes and *PEER* is set to no is associated with the most favorable outcome.

4.4 Interpreting Interactions by Plotting

However, even in the presence of an antagonistic interaction between two factors, it is possible for the experimental condition in which both factors are set to the higher level to be associated with the best outcome. In other words, an antagonistic interaction between two factors does not necessarily mean it would be a mistake to include both of the corresponding factors in the intervention (or, depending on what the levels represent, to include them both set to the higher level). Consider this regression model:

$$adhere = 14 + 8X_{MI} + 4X_{PEER} - 2X_{MI \times PEER}$$

with marginal means graphed in Fig. 4.7. The $MI \times PEER$ interaction is evident because the lines are not parallel; the effect of *PEER* varies depending on the level of *MI*. In this case, the difference in *adhere* between the higher and lower levels of *PEER* is more pronounced when $MI =$ no, but the difference is still in the desired direction when $MI =$ yes. Moreover, the experimental condition in which both factors are set to yes produces the largest value of *adhere*.

As Fig. 4.7 illustrates, sometimes an antagonistic interaction simply indicates that when two or more factors are combined, there are diminishing returns, even though both of the factors have an effect in the desired direction and the experimental condition in which both are set to the higher level is associated with the most favorable outcome. In the example, perhaps when motivational interviewing and peer mentoring are combined, there is a modest and unavoidable amount of overlap in the message being delivered by the two components; after all, they will be focused on the same subjects of reducing alcohol use and increasing adherence to the ART treatment regimen. Both are contributing to increasing *adhere*, but when the two are combined, the whole is slightly less than the sum of the parts. The antagonistic interaction reflects this. The implications of antagonistic interactions for making

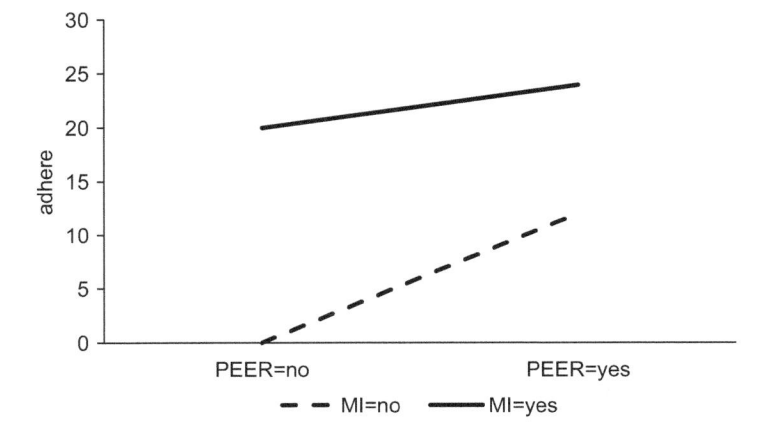

Fig. 4.7 Plot of marginal means for two two-level factors. In these data there is a main effect of *MI*, a main effect of *PEER*, and an antagonistic $MI \times PEER$ interaction. Although the interaction is antagonistic, the outcome is most favorable in the experimental condition in which both factors are set to yes

decisions about which components and component levels to include in an optimized intervention are discussed in Chap. 7.

4.4.5 Higher-Order Interactions

Higher-order interactions can also be designated synergistic or antagonistic. Suppose there is a main effect of *MI*, a main effect of *PEER*, and a three-way interaction involving *MI*, *PEER*, and *TEXT*. An example of such a model is

$$adhere = 14 + 8X_{MI} + 4X_{PEER} + 2X_{MI \times PEER \times TEXT}.$$

The sign of the coefficient corresponding to the interaction is positive, so the interaction is synergistic. Means based on this model are graphed in Fig. 4.8. Recall

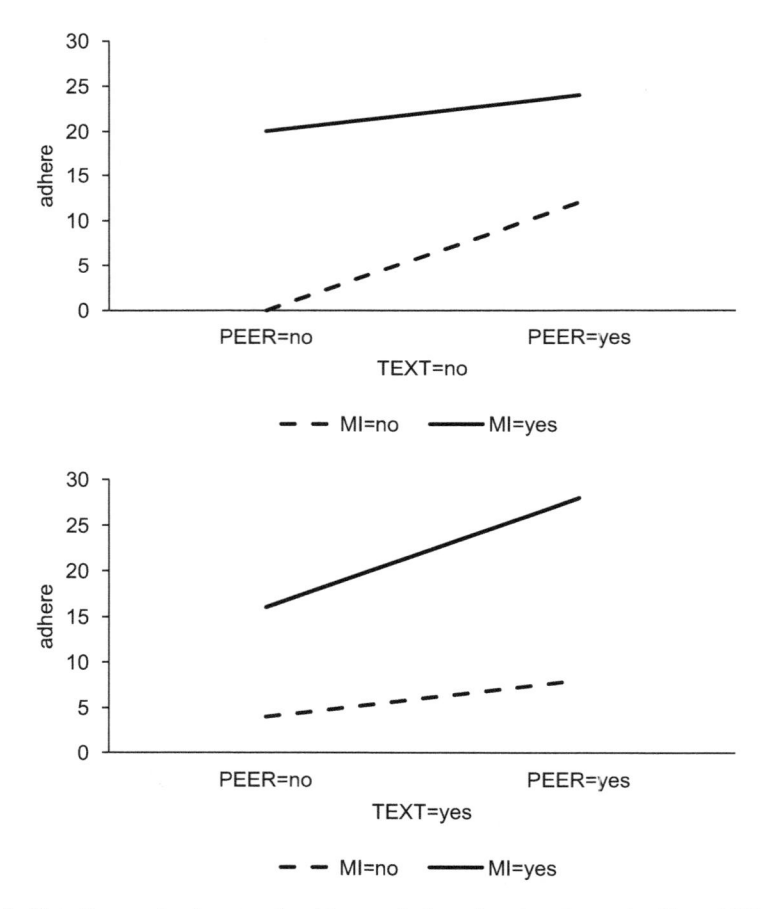

Fig. 4.8 Plot of means for three two-level factors. In these data there is a main effect of *MI*, a main effect of *PEER*, and a synergistic *MI* × *PEER* × *TEXT* interaction

4.4 Interpreting Interactions by Plotting

from Chap. 3 that a three-way interaction occurs when a two-way interaction is different depending on the level of a third factor. Three-way interactions are usually graphed by using a separate panel for the two-way interaction at each level of the third factor. It is arbitrary which factor is assigned to which axis or selected to be the third factor.

When both *MI* and *PEER* are set to yes, the outcome when *TEXT* is set to yes is better than when it is set to no. Thus, although *TEXT* has no main effect, in combination with the higher levels of the other two factors, it is associated with more days of adherence to the ART treatment regimen; in other words, the most favorable outcome occurs when all three factors are set to yes.

Now consider a similar scenario, except the three-way interaction is antagonistic:

$$adhere = 14 + 8X_{MI} + 4X_{PEER} - 2X_{MI \times PEER \times TEXT}.$$

Means for this model are graphed in Fig. 4.9. In this case the outcome for the condition in which *MI*, *PEER*, and *TEXT* are all set to yes is less favorable than the

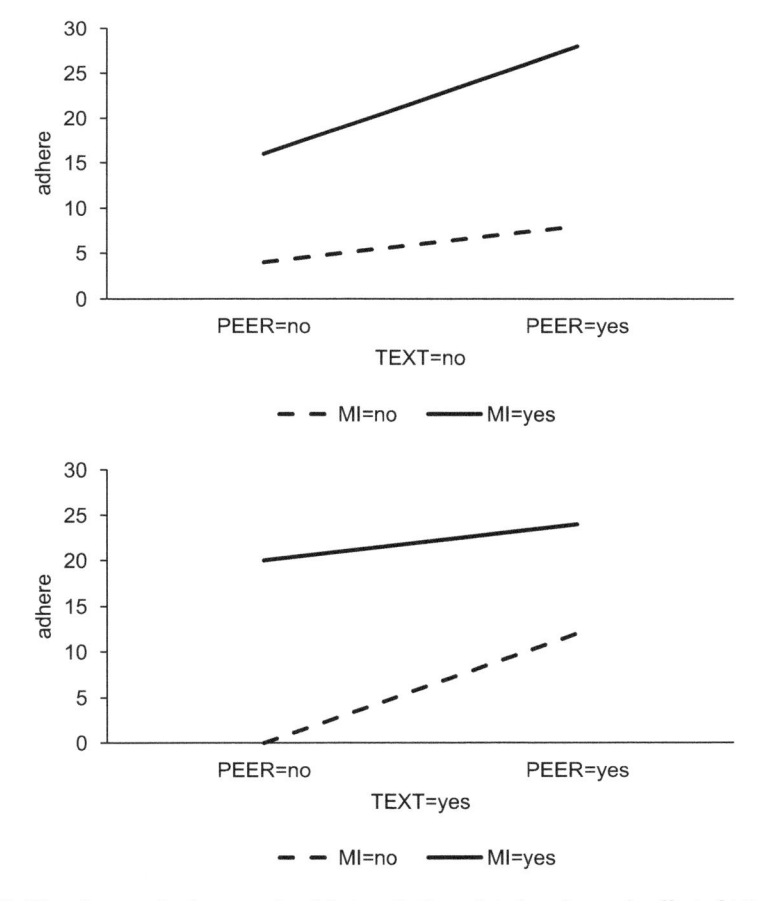

Fig. 4.9 Plot of means for three two-level factors. In these data there is a main effect of *MI*, a main effect of *PEER*, and an antagonistic *MI* × *PEER* × *TEXT* interaction

outcome for the condition in which *MI* and *PEER* are set to yes, but *TEXT* is set to no, reflecting the antagonistic three-way interaction.

4.5 The Role of Main Effects and Interactions in Decision-Making: Effect Hierarchy, Effect Sparsity, and Effect Heredity

The purpose of an optimization trial in MOST is to provide the information needed to make decisions about which components and component levels to select to comprise the optimized intervention. Experiments are conducted for essentially the same purpose in other fields such as engineering, although of course instead of optimizing interventions, scientists in these fields are optimizing manufacturing processes and the like.

This book recommends an approach for interpreting and making use of the results of factorial experiments that has been developed in these fields. This approach is based on the principles of effect hierarchy, effect sparsity, and effect heredity. In this section these concepts are approached from an intervention science perspective. The brief review presented here borrows heavily from the excellent discussion in Wu and Hamada (2009).

The effect hierarchy principle states that main effects are the most important effects scientifically. It follows, then, that the optimized intervention should be made up primarily of components and component levels that have demonstrated they have an effect on average. The effect hierarchy principle also states that as the order of an effect increases, its likely scientific importance decreases. The effect sparsity principle (also called the Pareto principle) says that of the many effects estimated in a factorial ANOVA, only a few will be scientifically important. Effect heredity states that an interaction is considered important only if at least one of the "parent" factors involved in the interaction has an important main effect. (How the term "important" is defined in decision-making in MOST is discussed in Chap. 7.)

Taken together, the three principles suggest that decision-making starts with main effects. In other words, if a factor demonstrates an important main effect, the corresponding component (or higher component level) is considered for inclusion in the optimized intervention. In the example, suppose only *MI* demonstrated an important main effect. Then motivational interviewing would tentatively be considered for inclusion in the optimized intervention, and *PEER* and *TEXT* would tentatively not be considered. Then, because main effects may not tell the whole story if there are important interactions, the initial decisions would be reconsidered in the light of any important interactions. Suppose *PEER* did not demonstrate an important main effect, but it had an important synergistic interaction with *MI*. Based on this interaction, consideration would be given to including peer mentoring in the optimized intervention even though *PEER* did not have an important main effect. The three principles also suggest that relatively few of the many interactions are likely to be important, with lower-order effects more likely to be important than higher-order effects. The use of these principles in making decisions about which components and component levels to select for the optimized intervention is discussed in Chap. 7.

It should be emphasized that these principles are general guidelines that are always overruled by the conceptual model (see Chap. 2). They are never invoked with respect to a particular effect if the conceptual model specifies that the effect is scientifically important. For example, an important interaction between two factors predicted by the conceptual model would be examined carefully whether or not the factors demonstrate main effects.

The idea of starting decision-making with main effects may seem strange, particularly in the light of advice given by some authors who suggest that it may be inadvisable to interpret main effects when interactions are present (e.g., Myers & Montgomery, 1995) or who advocate starting with the highest-order interactions and working down. Here a few points are raised to defend the perspective taken in this book.

First, the argument made in this section pertains only to experiments that are to be used to select components and component levels for inclusion in an optimized intervention. The perspective taken here is strictly decision-priority; in a conclusion-priority situation, it may be advisable to take a different approach.

Second, this book does not advocate ignoring important interactions; on the contrary, the decision-making approach outlined in Chap. 7 stresses that main effects must be interpreted carefully whenever substantial interactions are present. The approach of starting by selecting components/component levels that correspond to factors that have demonstrated an important main effect is consistent with engineering-based approaches to decision-making based on the results of a factorial experiment. This approach is suitable for MOST because the purpose of the factorial optimization trial is first and foremost to identify strong components that show an effect on average across a variety of combinations of levels of other components.

Third, this approach makes sense only when the factorial ANOVA is based on effect coding (see Chap. 3). When effect coding is used, the effect estimates are uncorrelated in a balanced experiment and nearly uncorrelated even if there is some unbalance. This helps provide more conceptual separation between main effects and interactions, although this does not mean any important effect can ever be ignored. When dummy coding is used, there may be substantial correlations between main effects and interactions even in a balanced experiment, making it difficult to interpret individual effects.

4.5.1 A Brief Reminder

It is a good idea to be clear about what definitions of effects are being used and in particular not to confuse effect-coded estimates that are consistent with the classical definition with those produced by a dummy-coded ANOVA. In fact, it is unfortunate that the terms "main effect" and "interaction" are commonly used to refer to regression coefficients from both effect-coded and dummy-coded analyses in factorial ANOVA, when in most cases these quantities are not the same. The distinction between effect-coded and dummy-coded effects is not always made clear in the output of statistical packages (see the Kugler, Dziak, and Trail (2018) chapter in the companion volume; Collins & Kugler, 2018) and in published literature, particularly in the social and behavioral sciences, even though this distinction is critical to the interpretation of any results. Therefore, when an "interaction" is reported, sometimes it is not clear whether or not the effect is an interaction according to the classical definition. This remark is not intended to be a critique of the use of dummy coding to conduct factorial ANOVA; either an effect-coded interaction or a dummy-coded higher-order effect may be of scientific interest. Rather, this is intended to caution readers that software documentation and scientific reporting practices sometimes make it difficult to determine which type of effect is being considered.

4.6 A Decision-Priority Perspective on Interactions

As is emphasized throughout this book, the investigator working within the MOST framework is ultimately interested in using the results of an optimization trial to make decisions about which components and component levels are to be selected for the optimized intervention. Therefore, it makes sense to view interaction effects from a decision-priority perspective. The decision-priority and conclusion-priority perspectives are contrasted in Chap. 3. Briefly, a decision-priority perspective on interpretation of experimental results is taken when the primary objective is to inform decision-making about the composition of the optimized intervention. By contrast, a conclusion-priority perspective is taken when the primary objective is to draw scientific conclusions about the putative effects of the factors in the experiment.

Recall that, as explained in Chap. 1, the results of an optimization trial are always used to select a level of a component. When the levels to choose from are yes and no, selecting the yes level means the component is included in the optimized intervention, and selecting the no level means it is not included. When the levels to choose from are, e.g., high and low, the component will be included in the intervention whether the high or low level is selected. This is why this book refers to selection of components and component levels for an intervention, rather than just selection of components.

4.6.1 Interactions and Decision-Making

As was touched on in Chap. 2 and will be discussed in detail in Chap. 7, decisions about selection of components/component levels for inclusion in the optimized intervention rely heavily on modeling the outcome, Y, using a regression model

4.6 A Decision-Priority Perspective on Interactions

derived from experimental results. When a general linear model approach is taken to factorial ANOVA, and effect coding is used, interaction terms can be viewed as accounting for variance in Y that remains after the main effects are removed (e.g., Cohen, 1988; Jaccard, 1998). In other words, from this perspective, interactions tell "the rest of the story" about the outcome Y that is not told by the main effects. Consider a regression model representing data from an experiment with two factors, Factor A and Factor B. If there is no interaction between the two factors, this model is a good representation of the data:

$$\widehat{Y} = b_0 + b_1 X_A + b_2 X_B,$$

where b_1 and b_2 represent the main effects of Factor A and Factor B, respectively. If there is an important interaction between Factor A and Factor B (what is meant by "important" is discussed in Chap. 7), the main effects do not tell the whole story about $\widehat{Y}$, and therefore the above expression would provide a poor basis for decision-making. Instead, an additional term is needed:

$$\widehat{Y} = b_0 + b_1 X_A + b_2 X_B + b_3 X_{A \times B}.$$

This additional term expresses the idea that the combined effect of Factor A and Factor B is different from what would be expected based solely on the main effects, and therefore, to "tell the story" of the outcome and make decisions on that basis, the interaction term is necessary. Similarly, a three-way interaction can be viewed as accounting for variance in Y that remains after the main effects and two-way interactions are removed and so on.

4.6.2 Implications for Powering Factorial Experiments

Taking a decision-priority perspective has some implications for how to power factorial experiments for detection of interactions. From the decision-priority perspective, a factorial experiment should be powered to detect any regression weight, whether it corresponds to a main effect or an interaction, that is likely to make a difference in decision-making. Thus if an interaction will not make a difference in decision-making, the resource management principle suggests that resources need not be devoted to detecting it. On the other hand, if an interaction is critical for decision-making, resources must be devoted to assure there is enough power to have a reasonable chance of detecting it. Thus to power a factorial experiment, the investigator must decide what is the smallest size interaction that is large enough to make a difference in decision-making and ensure that the sample size is large enough to provide sufficient power to detect an interaction of at least that size, given the selected Type I error rate. (The role of interactions in decision-making is discussed briefly below and in more detail in Chap. 7.)

4.7 What Is the Statistical Power for Detection of Interaction Effects as Compared to Main Effects?

This section may be skipped without loss of continuity. Those who read this section should first read Sect. 3.14 in Chap. 3, which explains the coefficient correction, that is, why there is a 2 in the denominator of the definition of a two-way interaction, a 4 in the denominator of the definition of the three-way interaction, and so on.

The purpose of this section is to help clarify the issue of power for detection of interactions between factors in factorial experiments. Power for detection of interactions between experimental factors and observed (i.e., nonexperimental) variables, in other words, for detection of moderation (effect modification), will be discussed later in this chapter.

This section is included because there have been two contradictory statements in the literature on statistical power. Some literature (e.g., Cohen, 1988) has stated that in any particular 2^k factorial experiment analyzed using effect coding, there is identical power to detect any regression coefficient of size b, irrespective of whether it corresponds to a main effect or an interaction. Other literature (e.g., Fleiss, 1986) has stated that all else being equal, there is less power available for detection of interactions than for main effects. Clearly both Cohen (1988) and Fleiss (1986) are unimpeachable sources of information. How can both be correct?

4.7.1 Power Is Identical for Main Effects and Interactions With Identical Regression Weights in 2^k Experiments, but Not Necessarily in Other Experiments

First, let us clarify that Cohen's (1988) statement that power is identical for detection of main effects and interactions pertains only to 2^k experiments, that is, factorial experiments in which each of the k factors has two (and only two) levels. One example of the efficiency of 2^k experiments is how they preserve power for estimation of interactions. This property is not shared by other factorial experiments (e.g., 3^k experiments). In a detailed explanation of power for interactions in factorial experiments, Cohen wrote:

> ...*for any given size of effect... and significance criterion... the power of the interaction tests in a factorial design will, on the average, be smaller than that of the main effects (excepting 2^k designs, where they will be the same)*. (p. 374; emphasis added)

Cohen demonstrates that, all else being equal, power to detect an interaction tends to diminish as the number of levels in the factors involved in the interaction increases. This is consistent with what was said in Chap. 3 in the present book, where it was explained that given a fixed N, power for detection of any main effects of factors with more than two levels will be smaller compared to power for detection of main effects of factors with only two levels. In general, all else being equal, a

4.7 What Is the Statistical Power for Detection of Interaction Effects as... 135

factorial experiment that includes even one factor with more than two levels is more resource intensive than a 2^k experiment, because it requires a larger N to maintain a comparable level of statistical power for many of the effects estimated. This is one reason why the 2^k design is recommended for optimization trials.

4.7.2 Cohen and Fleiss Use Different Definitions of "Interaction"

However, even if the conversation is confined to 2^k experiments, Cohen (1988) and Fleiss (1986) still represent different perspectives on power to detect interaction effects. The different perspectives are due to the use of different definitions of the interaction and, by extension, the interaction effect size.

In the italicized statement above, Cohen expressed effect size in terms of the regression coefficient, as does the present book. In Chap. 3 it was explained that the regression coefficient is always exactly half of the corresponding main effect or interaction. This is true provided that the definitions of main effects and interactions used in this book are used. In other words, according to the definitions of interactions used in this book (e.g., Eqs. (4.1) and (4.2) above), if there is sufficient power to detect a particular regression coefficient, there is by definition sufficient power to detect the corresponding main effect or interaction.

However, Fleiss (1986) and others (see Jaccard, 1998) define the interaction without the use of the coefficient correction, that is, as only the numerator of the expressions in Eqs. (4.1) and (4.2). In other words, the coefficient correction of 2 in the denominator of the two-way interaction, 4 in the denominator of the three-way interaction, and so on is omitted. Let us refer to interactions defined this way as INT^*, so the $MI \times PEER$ interaction is

$$INT^*_{MI \times PEER} = (\bar{\mu}_{++.} - \bar{\mu}_{-+.}) - (\bar{\mu}_{+-.} - \bar{\mu}_{--.}), \qquad (4.3)$$

and the $MI \times PEER \times TEXT$ interaction is

$$INT^*_{MI \times PEER \times TEXT} = [(\mu_{+++} - \mu_{-++}) - (\mu_{+-+} - \mu_{--+})] \\ - [(\mu_{++-} - \mu_{-+-}) - (\mu_{+--} - \mu_{---})]. \qquad (4.4)$$

Some readers may find the INT^* definition more intuitive, but it has a less straightforward relation to the regression coefficient. Consider the relations among $INT_{MI \times PEER}$, $INT^*_{MI \times PEER}$ and $b_{MI \times PEER}$:

$$b_{MI \times PEER} = \frac{INT_{MI \times PEER}}{2} = \frac{[(\mu_{++.} - \mu_{-+.}) - (\mu_{+-.} - \mu_{--.})]/2}{2} = \frac{INT^*_{MI \times PEER}/2}{2}$$
$$= \frac{INT^*_{MI \times PEER}}{4}$$

Thus an effect size for a two-way interaction by the *INT** definition translates to a regression coefficient that is half as large as what will be obtained starting with effect size of identical magnitude by the *INT* definition. This discrepancy grows as the order of the interaction increases. It is straightforward to show that

$$b_{MI \times PEER \times TEXT} = \frac{INT_{MI \times PEER \times TEXT}}{2} = \frac{INT^*_{MI \times PEER \times TEXT}}{8}.$$

Because, as discussed in Chap. 3, the coefficient correction becomes larger as the order of the interaction increases, therefore the divisor applied to *INT** becomes larger.

Of course, no matter what definition is used for the interaction, the regression coefficient corresponding to a main effect, say ME_{PEER}, is half of that main effect:

$$b_{PEER} = \frac{ME_{PEER}}{2}.$$

Thus when an interaction effect is expressed in terms of the *INT* definition (e.g., Eq. (4.1) or Eq. (4.2)), if a main effect and an interaction are said to have the same expected size, the two effects correspond to the same regression weight. Therefore, all else being equal, they will be associated with the same level of statistical power. By contrast, when an interaction is expressed in terms of the *INT** definition (e.g., Eq. (4.3) or Eq. (4.4)), a main effect and interaction said to have the same expected size do not correspond to the same regression weight. The regression coefficient corresponding to the interaction will always be smaller in absolute value and therefore will be associated with less statistical power. This discrepancy becomes wider as the order of the interaction increases. Neither definition is right or wrong, but for the sake of clarity in conducting power analysis for detection of interaction effects, writing scientific reports concerning interactions, and even conversing about interactions, it is critical to know which definition is being referred to.

Example Suppose Dr. B is conducting a power analysis to plan the example 2^3 factorial experiment, which includes the factors *MI*, *PEER*, and *TEXT*. For each factor, Dr. B wants to be able to detect a main effect corresponding to a raw difference between means of 3 or an unstandardized regression coefficient of $b = 1.5$. A power analysis has shown that to maintain power $= 0.8$ for detection of all main effects with this effect size, $N = 351$ is required.

Now suppose Dr. B wants to be able to detect an $MI \times PEER$ interaction that is the same size as the main effects, that is, an interaction corresponding to a raw size of 3. What sample size is required to maintain the same level of power to detect this interaction? Will additional experimental subjects be required? To answer this question, it is necessary to translate the interaction to a regression coefficient

4.7 What Is the Statistical Power for Detection of Interaction Effects as...

(or some other expression of effect size). To translate the interaction to a regression coefficient, it is necessary to know whether Dr. B is expressing the $MI \times PEER$ interaction the way Cohen would, in terms of INT, or the way Fleiss would, in terms of INT^*.

Suppose Dr. B has the INT definition in mind. Then the $MI \times PEER$ interaction translates directly to the unstandardized regression coefficient:

$$b_{MI \times PEER} = \frac{INT_{MI \times PEER}}{2} = \frac{3}{2} = 1.5.$$

Because within a balanced effect-coded 2^k factorial ANOVA power is identical for detection of any regression weight of a particular size, no additional subjects are required to detect the interactions. The experiment has been planned to provide power $= 0.8$ for detection of any $b \geq 1.5$. $N = 351$ is sufficient to provide power $= 0.8$ for detection of main effects and interactions that are expected to have $b \geq 1.5$.

By contrast, suppose Dr. B has the INT^* definition in mind. When the INT^* definition of the interaction is used, the $MI \times PEER$ interaction translates to a smaller regression coefficient, because

$$b_{MI \times PEER} = \frac{INT^*_{MI \times PEER}/2}{2} = \frac{3/2}{2} = .75.$$

In general, all else being equal, b (or any other effect size metric used in power analysis) is smaller when the INT^* definition is used than when the INT definition is used. For this reason, when ME and INT^* are equal, generally additional subjects will be required to maintain power for detecting the interaction as compared to detecting the main effect. In contrast to the example above in which the INT definition was used, here the sample size must be approximately quadrupled to maintain power $= 0.8$ for detection of the $MI \times PEER$ interaction (Fleiss, 1986). This increase to $N = 1404$ is required because the investigator wishes to maintain power for detection of a smaller effect. In a 2^k experiment, there is nothing special about interactions in this respect; exactly the same increase in sample size would have been required if a main effect of $b = 0.075$ were to be detected.

In summary, power analysis for factorial experiments can be approached using the Cohen definition or the Fleiss definition of the interaction. Of course, which definition is used is arbitrary, as long as it is clear which definition is being used, and the definition is applied consistently (Dziak, Nahum-Shani, & Collins, 2012). This book uses the definitions of interactions in Eqs. (4.1) and (4.2) above, in other words, INT rather than INT^* is used as the definition of the interaction. This is consistent with the decision-priority perspective, because this perspective emphasizes building a regression model to predict the outcome, and the INT definition corresponds more closely to the regression weight.

4.7.3 The Ability to Detect Interactions Depends Partly on the Size of the Interaction, but Who Knows What Effect Size to Expect?

So far it has been shown that the answer to the question *Are interaction effects harder to detect than main effects?* may be *yes* or *no*, depending on whether one views interactions using the Fleiss definition or the Cohen definition, respectively. But there is a third answer: *It is impossible to say*. This is because at this writing relatively little is known, theorized, or even conjectured about the expected size of interactions in intervention science by either the *INT* or *INT** definitions. Intervention science has had relatively few empirical findings that pertain to interactions between the components of a multicomponent intervention (or to main effects of individual components either), because the field has seldom conducted experiments that enable investigation of these effects. Similarly, theory has for the most part been silent on whether particular intervention components might be expected to interact. So in this sense, it is impossible to say whether interactions are likely to be harder to detect than main effects—at this time at least—because the anticipated size of interactions is unknown. If more experiments are conducted that permit examination of interactions, a body of knowledge will be accumulated, and the field eventually will gain a better sense of the expected size of interaction effects.

4.8 Moderation of the Effects of Factors by Observed Variables (Effect Modification)

It may be of interest to examine whether a particular observed (unmanipulated) variable moderates the effect of an experimental factor, in other words, whether effect modification has occurred. For example, as Fig. 2.3 shows, Dr. B hypothesizes that the effect of the motivational interviewing component is moderated by depression. One hypothesis is that those with a clinical diagnosis of depression will be less receptive to motivational interviewing (recall that for simplicity not all of the components in Fig. 2.3 are included in the experiment discussed in this chapter).

Suppose Dr. B plans to conduct a 2^3 experiment involving the factors *MI*, *PEER*, and *TEXT* and also wishes to examine whether a clinical diagnosis of depression at pretest is a moderator. Statistically, moderation of the effect of an experimental factor by an observed variable is modeled in the regression equation by including an interaction term involving the factor and the observed variable. Suppose depression is represented by *DEP*, a categorical variable with two levels, depressed and not depressed (decided via clinical diagnosis). Then investigation of Dr. B's hypothesis would involve including a term for the main effect of *DEP* and the interaction term *DEP* × *MI*.

The concept of moderation is the essentially same in a factorial experiment as in a randomized controlled trial (RCT). However, in an RCT, there is only one manipulated variable, the treatment/control assignment. By contrast, moderation in a factorial experiment opens up many possibilities. The investigator has the opportunity to examine moderation of the main effect of each factor. Although the original hypothesis involved only the *MI* factor, Dr. B may also wish to examine whether *DEP* moderates the effects of the other two factors. Dr. B could also examine moderation of interactions involving two or more factors. For example, moderation of a two-way interaction would be coded as a three-way interaction involving the original two variables and the moderator.

Below two general approaches to incorporating moderation of the effects of the factors into a factorial experiment are discussed. One involves relying on natural variation in the moderator; the other involves controlling the variation in the moderator to treat it like a factor in the experiment.

4.8.1 Examining Moderation Based on Naturally Varying Observed Moderators

This is the more commonly used approach to examining moderation. In this approach, the pretest data on whether or not each individual has a clinical diagnosis of depression data, that is, *DEP*, would be collected. After the experiment was conducted, analyses would be done to examine whether *DEP* moderates, in other words, interacts with, the experimental factors.

The starting point for conducting this moderation analysis based on natural variation in the moderators is computation of the appropriate product terms. First the main effects of any categorical moderator variables are coded, and numeric moderator variables are centered. A thorough discussion of this, along with approaches for conducting regression analyses involving interactions, can be found in the classic book by Aiken and West (1991). Categorical moderators can be either dummy-coded or effect-coded; of the two approaches, effect coding is more consistent with the approach advocated in this book for analysis of data from a factorial ANOVA. It may make interpretation more straightforward to standardize numeric moderator variables to z-scores (which, of course, also centers them). Then product terms to represent the interactions are created as needed.

In a balanced factorial experiment, the distribution of each level of each factor is identical across the levels of each other factor. However, the distributions of naturally varying moderators are unlikely to be identical across the levels of the factors in the experiment. Although on average properly conducted random assignment will produce cells with approximately identical means and proportions on uncontrolled variables in the long run, in any particular experiment, there is likely to be some deviation from this ideal in practice. Stratified random assignment can help in a limited way. As will be discussed in Chap. 6, the purpose of stratified random assignment is to ensure that each experimental condition has the same distribution of a particular variable. For example, if 20% of the sample is depressed, stratified random assignment will ensure that these depressed individuals constitute about 20% of the subjects in each experimental condition. The use of stratification must be carefully thought through, because it is not practical to stratify on more than one or (at most) two of the many possible moderators when conducting randomization.

As was discussed in Chap. 3, when effect coding is used to analyze data from a balanced factorial experiment, the parameter estimates in the ANOVA are uncorrelated. Whenever an experiment is unbalanced, some degree of correlation among the effect estimates is introduced. Thus effect estimates involving any naturally occurring variables, whether main effects or interactions, are likely to be correlated with other effects in the model and may be substantially correlated. Such correlations are likely to be smaller if categorical moderators are effect-coded rather than dummy-coded and quantitative moderators are centered (Aiken & West, 1991), but these measures will not eliminate the correlations.

This imbalance has two implications. First, when there is imbalance involving a moderator variable, all else being equal the standard error associated with the regression weights for any effect estimates involving the moderator will be larger, and statistical power will be smaller. Thus, power for detection of a naturally varying moderator can be small. Second, when effect estimates are correlated, adding or removing an effect from the regression equation may change the remaining effect estimates. The more correlated the effect estimates are, the more they will change when an effect is added or removed. Thus care must be taken in building a regression model that involves moderators.

4.8.2 Examining Moderation by Including the Moderator as a Factor in the Experiment

Depending on how important the hypothesis is that a variable will have a moderating effect, and particularly if the presence or absence of moderation will have an impact on the decisions about which components and component levels to include in the optimized intervention, then the investigator may wish to build the moderator into the factorial experiment as one of the factors rather than relying on naturally occurring variation in the moderator as described above. Moderator variables typically cannot be directly manipulated, but their variation can be controlled via subject recruitment and random assignment, enabling the inclusion of the moderator as a factor in the experiment.

Suppose Dr. B views the hypothesis that *DEP* moderates the effect of *MI* as very important. If the effect of *MI* appears to be less in those who are clinically depressed, Dr. B plans to conduct additional research in the future to identify one or more components that can be added to the intervention to increase its effectiveness in clinically depressed participants. To build the variable *DEP* into the factorial experiment, Dr. B would recruit subjects so that 50% are depressed and 50% are not. Then

4.8 Moderation of the Effects of Factors by Observed Variables (Effect Modification) 141

Table 4.2 Experimental conditions in 2^4 experiment including *DEP* as a factor

Experimental condition	MI	PEER	TEXT	DEP
1	No[a]	No	No	Not depressed
2	No	No	No	Depressed
3	No	No	Yes	Not depressed
4	No	No	Yes	Depressed
5	No	Yes	No	Not depressed
6	No	Yes	No	Depressed
7	No	Yes	Yes	Not depressed
8	No	Yes	Yes	Depressed
9	Yes	No	No	Not depressed
10	Yes	No	No	Depressed
11	Yes	No	Yes	Not depressed
12	Yes	No	Yes	Depressed
13	Yes	Yes	No	Not depressed
14	Yes	Yes	No	Depressed
15	Yes	Yes	Yes	Not depressed
16	Yes	Yes	Yes	Depressed

[a]No means not included in intervention; yes means included in intervention

random assignment would be conducted separately for each subgroup (see Chap. 6 for more about random assignment), essentially replicating the experiment within each subgroup. In a sense, the experiment is now a 2^4 factorial experiment, with DEP added as a factor. This is illustrated in Table 4.2. In data analysis, *DEP* may be treated like any other factor. Hypothesis tests can be performed on the main effect of *DEP* or on interactions of any order involving *DEP* and any of the other factors.

The advantage of this approach as compared to relying on naturally occurring variation in the moderator is that it enables the moderator to be examined in a highly efficient factorial experiment. All else being equal, this approach will typically provide more power for detection of the moderation effect than relying on naturally occurring variation in the moderator. The approach is typically practical for only a small number of moderators—one or two at the most. Careful thought must be given to powering the experiment to detect the anticipated interactions between the moderator and the experimental factors. Before planning an experiment that will include a moderator as a factor, it is critical to be clear about whether the definition of interaction under consideration is *INT* or *INT**, because this will have a huge impact on the power analysis and planned sample size.

Including a moderator as a factor in an experiment is the most straightforward if the moderator has a limited number of categories, preferably two, or can reasonably be categorized (although see Aiken & West, 1991, for a discussion of the loss in power that can accompany categorization). Response surface experiments may be considered if there are one or more quantitative factors and it is deemed inadvisable to divide them into categories. Such experiments are outside the scope of this book; the reader is referred to Wu and Hamada (2009).

4.8.3 Moderation of Component Effects and the Continual Optimization Principle

Examination of moderation of component effects is consistent with the continual optimization principle of MOST (see Chap. 1), which states that optimization of an intervention is an ongoing process of incremental improvement. As Fig. 1.1 illustrates, once the evaluation phase of MOST has been completed, the investigator returns to the preparation phase and begins the process of developing a better intervention. A finding that a component's effects are moderated suggests that the component works better for certain individuals, or in certain settings, or under certain circumstances, than others. A worthwhile goal for improving an intervention is to develop an intervention strategy that works well across the board.

For example, suppose in Dr. B's experiment, secondary analyses conducted after the optimization phase indicated that motivational interviewing worked well for non-depressed individuals but did not work well for depressed individuals. This finding could form the basis for a new cycle of MOST. As always, the new cycle would start with the preparation phase (see Fig. 1.1). This would include refinement of the conceptual model to include a hypothesized explanation of why motivational interviewing did not work well for depressed individuals yet worked well for the other participants. The refined conceptual model would help point to a strategy for developing an intervention that is as effective for depressed participants as it is for non-depressed participants. Dr. B can seek either a different approach to motivational interviewing that will be effective with depressed participants or an entirely different strategy other than motivational interviewing for establishing appropriate health beliefs in depressed participants. Either could form the basis for an adaptive intervention (see Chap. 8 in this volume and Almirall, Nahum-Shani, Wang, and Kasari (2018) and Rivera, Hekler, Savage and Downs (2018) in the companion volume), in which depressed and non-depressed participants would be assigned different versions of the motivational interviewing component or entirely different components. Alternatively, Dr. B could try to develop a new, more robust component that would work equally well in depressed and non-depressed participants. Dr. B would probably want to test additional new components in the optimization phase of this round of MOST, to capitalize on the efficiency of the factorial experiment. Dr. B would build a new optimized intervention, and then proceed to the evaluation phase, in which the effectiveness of the newly optimized intervention would be examined in an RCT.

4.9 What's Next?

This chapter and the previous chapter have reviewed the factorial experiment, particularly the 2^k experiment, which is highly efficient for use in the optimization phase of MOST. Chapter 3 emphasized the general concept of the factorial experiment and main effects, whereas this chapter has discussed the concept of the interaction and moderation of factor effects. In both chapters it was demonstrated that although factorial experiments make very efficient use of experimental subjects, they can require implementation of a large number of experimental conditions. In the next chapter, Chap. 5, reduced factorial designs that require fewer experimental conditions are discussed. Under some circumstances, such designs, particularly fractional factorial designs, can be very economical. Chap. 6 reviews how to compare the economy of various experimental designs under consideration.

References

Aiken, L. S., & West, S. G. (1991). *Multiple regression: Testing and interpreting interactions*. Thousand Oaks, CA: Sage.

Cohen, J. (1988). *Statistical power analysis for the behavioral sciences*. Hillside, NJ: Lawrence Earlbaum Associates.

Collins, L. M., & Kugler, K. C. (Eds.). (2018). *Optimization of multicomponent behavioral, biobehavioral, and biomedical interventions: Advanced topics*. New York, NY: Springer.

Dziak, J. J., Nahum-Shani, I., & Collins, L. M. (2012). Multilevel factorial experiments for developing behavioral interventions: Power, sample size, and resource considerations. *Psychological Methods, 17*, 153–175.

Fleiss, J. L. (1986). *The design and analysis of clinical experiments*. New York, NY: Wiley.

Jaccard, J. (1998). *Interaction effects in factorial analysis of variance* (Vol. 118). Thousand Oaks, CA: Sage.

Karaca-Mandic, P., Norton, E. C., & Dowd, B. (2012). Interaction terms in nonlinear models. *Health Services Research, 47*, 255–274.

Kugler, K. C., Dziak, J. J., & Trail, J. (2018). Coding and interpretation of effects in analysis of data from a factorial experiment. In L. M. Collins & K. C. Kugler (Eds.), *Optimization of multicomponent behavioral, biobehavioral, and biomedical interventions: Advanced topics* (forthcoming). New York, NY: Springer.

Montgomery, D. C. (2008). *Design and analysis of experiments* (7th ed.). Hoboken, NJ: Wiley.

Myers, R. H., & Montgomery, D. C. (1995). *Response surface methodology: Process and product optimization using designed experiments*. New York, NY: Wiley.

Wu, C. F. J., & Hamada, M. S. (2009). *Experiments: Planning, analysis, and optimization* (2nd ed.). Hoboken, NJ: Wiley.

Chapter 5
Balanced and Unbalanced Reduced Factorial Designs

Abstract In this chapter, balanced and unbalanced reduced factorial designs for use in optimization of multicomponent behavioral, biobehavioral, and biomedical interventions are discussed. These designs, which include a carefully selected subset of the experimental conditions in a corresponding complete factorial, can be more efficient and economical than complete factorials when implementation of experimental conditions is expensive or logistically challenging. However, there are important trade-offs that the investigator must make in exchange for this efficiency and economy. Reduced factorial designs are not for every situation, but when used appropriately and strategically, they can make excellent use of limited research resources. Readers should be familiar with the material in all previous chapters, particularly Chaps. 3 and 4.

Contents

5.1	Introduction to Balanced Reduced Designs (i.e., Fractional Factorial Designs)	146
5.2	Hypothetical Example of a Fractional Factorial Design	147
	5.2.1 Some Interesting Aspects of Fractional Factorial Designs	149
	5.2.2 Notation for Fractional Factorial Designs	151
5.3	What's the Catch? Aliasing	151
	5.3.1 A Small Example of Combining of Effects	152
	5.3.2 Confounding and Aliasing	155
5.4	Overview of Rationale and Strategy	155
	5.4.1 Rationale: Targeting Resources to Scientifically Important Effects	155
	5.4.2 Strategy of Selecting a Design That Deliberately Aliases Effects	157
5.5	More About the Strategy of Deliberately Aliasing Effects	158
5.6	Clear and Strongly Clear Effects	159
5.7	Two Key Characteristics of Fractional Factorial Designs	160
5.8	Trade-Offs to Consider When Selecting a Fractional Factorial Design	161
	5.8.1 Resolution and Resource Requirements	162
	5.8.2 Bundling of Effects and Resource Requirements	164
5.9	Aliasing: What to Look For	165

© Springer International Publishing AG 2018
L. M. Collins, *Optimization of Behavioral, Biobehavioral, and Biomedical Interventions*, Statistics for Social and Behavioral Sciences,
https://doi.org/10.1007/978-3-319-72206-1_5

5.10 Selecting a Fractional Factorial Design Using Software 166
   5.10.1 Overview: Different Approaches to Using Software to Select
           a Fractional Factorial Design 166
   5.10.2 Selecting a Design by Specifying the Desired Number
           of Experimental Conditions 166
   5.10.3 Selecting a Design by Specifying a Minimum Desired Resolution 170
   5.10.4 Selecting a Design by Specifying Both the Desired Number
           of Experimental Conditions and a Desired Resolution 170
   5.10.5 Selecting a Design by Specifying Which Effects Are in Which Categories .. 171
5.11 The Term "Experimental Design" as Used by Statisticians and as Used
      by Intervention Scientists 172
   5.11.1 Switching Factor Level Labels 173
   5.11.2 Permuting the Order in Which Factors Are Listed 174
   5.11.3 When the Investigator Desires to Omit a Particular
           Experimental Condition 175
5.12 Fractional Factorial Designs With Clustered Data 177
5.13 Following the Resource Management Principle When Considering a Fractional
      Factorial Design 178
   5.13.1 General Recommendations 179
   5.13.2 Some Specific Recommendations 181
   5.13.3 Resources for Further Study 181
5.14 Unbalanced Reduced Factorial Designs, Interactions, and Aliasing 181
   5.14.1 Familiar Experimental Designs as Unbalanced Reduced Factorial Designs .. 182
   5.14.2 Aliasing in Individual Experiments and MACEs 184
   5.14.3 What Is the Effect of a Factor? 187
5.15 A Few Points to Remember About Fractional Factorial Designs 189
5.16 What's Next 190
References 190

5.1 Introduction to Balanced Reduced Designs (i.e., Fractional Factorial Designs)

Chapter 3 discussed how factorial designs make very efficient use of experimental subjects. Under many circumstances, it is even possible to add one or more factors to a factorial design without needing to add subjects to maintain power. Thus, factorial experiments can provide a great deal of scientific information based on relatively few subjects. However, although factorial experiments tend to be economical in terms of subject requirements, they can require implementation of many experimental conditions. The number of conditions in a factorial experiment increases rapidly as factors are added. A 2^4 factorial experiment has 16 conditions; adding one more factor to make it a 2^5 experiment doubles the number of conditions to 32; adding another factor to make it a 2^6 experiment doubles the number of conditions again, to 64; and so on. Thus an investigator who wishes to conduct, say, a 2^6 experiment may have sufficient subjects to power the experiment, but insufficient resources to manage 64 experimental conditions.

Reduced factorial designs (Collins, Dziak, & Li, 2009) involve a subset of the experimental conditions in a standard, or complete, factorial design. Reduced factorial designs may be balanced or unbalanced. Fractional factorial designs (Collins

5.2 Hypothetical Example of a Fractional Factorial Design

et al., 2009; Wu & Hamada, 2009) are a type of balanced reduced factorial design. They provide an alternative for investigators who wish to take advantage of the modest sample size requirements of factorial experiments but have the resources to implement only a limited number of experimental conditions. In a fractional factorial design, only a fraction of the experimental conditions in the corresponding complete factorial design are run—hence the name fractional factorial.

This chapter will rely on the hypothetical example described in Chap. 1, in which Dr. B wishes to develop an intervention to reduce viral load in HIV-positive individuals who drink heavily. Based on prior scientific literature, clinical experience, and a well-specified conceptual model (reviewed in detail in Chap. 2), Dr. B has identified five intervention components that are hypothesized to be critical in helping HIV-positive individuals reduce their drinking and improve their adherence to anti-retroviral therapy (ART). They are (a) motivational interviewing, to engage the participant in the process of examining his or her alcohol use and ART adherence; (b) peer mentoring, to provide contact with a positive role model; (c) text messaging, to provide access to a support system; (d) mindfulness meditation, to improve mental health; and (e) behavioral skills training, to improve behavioral skills for managing alcohol use and ART. Suppose there is currently no standard of care for the first four components, but physicians and HIV clinics routinely give HIV-positive individuals who drink heavily and have poor ART adherence a workbook to complete to help improve their behavioral skills. Dr. B is considering two levels for each of these components. Each of the first four components can be set either to no (not included in the intervention) or yes (included). The behavioral skills component can be at either a low level, consisting of the workbook only (i.e., standard of care), or a high level, consisting of the workbook plus training delivered by a behavioral skills counselor. The primary outcome variable is the number of days of adherence to ART in the 30-day period following the conclusion of the intervention and will be abbreviated *adhere*.

5.2 Hypothetical Example of a Fractional Factorial Design

Suppose Dr. B is now in the optimization phase of the multiphase optimization strategy (MOST) and would like to conduct a factorial experiment to examine the performance of the five components described above. The components, the corresponding experimental factors, and the levels of each factor are shown in Table 5.1. Table 5.2 lists all of the 32 experimental conditions that would be included in a 2^5 factorial design. A power analysis (see Chap. 3 for a discussion of how to power factorial experiments) indicates that power $\geq .8$ to detect an effect of $\beta^* \geq .15$ would be obtained with $N = 351$, which is rounded up to $N = 352$ to achieve equal cell sizes of $n = 11$.

However, Dr. B has some concerns about implementation of 32 experimental conditions. This experiment will require careful management to ensure that each subject receives the exact treatment he or she has been randomly assigned. This means making sure that no subject mistakenly receives a component or component

Table 5.1 Factors and levels in hypothetical 2^5 factorial experiment

Component	Factor name	Lower level	Higher level
Motivational interviewing	*MI*	No[a]	Yes
Peer mentoring	*PEER*	No	Yes
Text messaging	*TEXT*	No	Yes
Mindfulness meditation	*MIND*	No	Yes
Behavioral skills training	*SKILLS*	Low (workbook only)	High (workbook plus training)

[a]No means not included in intervention; yes means included in intervention

Table 5.2 Experimental conditions in hypothetical 2^5 factorial design

Experimental condition	*MI*	*PEER*	*TEXT*	*MIND*	*SKILLS*	Per-condition n[a]
1	No[b]	No	No	No	Low	11
2	No	No	No	No	High	11
3	No	No	No	Yes	Low	11
4	No	No	No	Yes	High	11
5	No	No	Yes	No	Low	11
6	No	No	Yes	No	High	11
7	No	No	Yes	Yes	Low	11
8	No	No	Yes	Yes	High	11
9	No	Yes	No	No	Low	11
10	No	Yes	No	No	High	11
11	No	Yes	No	Yes	Low	11
12	No	Yes	No	Yes	High	11
13	No	Yes	Yes	No	Low	11
14	No	Yes	Yes	No	High	11
15	No	Yes	Yes	Yes	Low	11
16	No	Yes	Yes	Yes	High	11
17	Yes	No	No	No	Low	11
18	Yes	No	No	No	High	11
19	Yes	No	No	Yes	Low	11
20	Yes	No	No	Yes	High	11
21	Yes	No	Yes	No	Low	11
22	Yes	No	Yes	No	High	11
23	Yes	No	Yes	Yes	Low	11
24	Yes	No	Yes	Yes	High	11
25	Yes	Yes	No	No	Low	11
26	Yes	Yes	No	No	High	11
27	Yes	Yes	No	Yes	Low	11
28	Yes	Yes	No	Yes	High	11
29	Yes	Yes	Yes	No	Low	11
30	Yes	Yes	Yes	No	High	11
31	Yes	Yes	Yes	Yes	Low	11
32	Yes	Yes	Yes	Yes	High	11

[a]Overall $N = 352$ provides power $\geq .8$ to detect effect size of $\beta^* \geq .15$
[b]No means not included in intervention; yes means included in intervention

5.2 Hypothetical Example of a Fractional Factorial Design

level that is not a part of the assigned treatment and that no subject fails to receive a component or component level that is a part of the assigned treatment. Dr. B's assessment is that given the number of staff members that can be hired with the available resources, it is feasible to conduct an experiment with 352 subjects, but 32 different conditions is too many to handle. After careful consideration, Dr. B concludes that 16 conditions would be manageable.

Assuming Dr. B wishes to stay with a factorial experiment, there are two options. One is to drop a factor, leaving four factors, and conduct a 2^4 factorial experiment, which would have 16 experimental conditions. The other is to conduct a fractional factorial experiment. Fractional factorial designs are a special case of factorial designs. A fractional factorial design will enable Dr. B to retain all five factors yet reduce the number of experimental conditions. Later in this chapter the advantages and disadvantages of complete and fractional factorial designs will be directly compared. First, let us explore the idea of conducting a fractional factorial experiment.

5.2.1 Some Interesting Aspects of Fractional Factorial Designs

There are a variety of fractional factorial designs that could be selected for Dr. B's experiment. The process of selecting a fractional factorial design, including how the exact set of experimental conditions in a particular fractional factorial design is arrived at, will be discussed later in this chapter. For now, suppose Dr. B has selected the fractional factorial design in Table 5.3. Examination of Table 5.3 reveals some interesting aspects of fractional factorial designs.

First, every experimental condition in the fractional factorial design in Table 5.3 is also found in the complete factorial design shown in Table 5.2. In other words, the conditions in Table 5.3 are a subset of those in Table 5.2. Every fractional factorial design is made up of a subset of conditions from a complete factorial design that involves the same factors.

Second, the design in Table 5.3 has 16 experimental conditions, which is exactly half of the experimental conditions of the design in Table 5.2. Every fractional factorial design consists of a specific fraction of the conditions from the corresponding complete factorial. In this case the fraction is ½, but, as will be demonstrated below, in other cases the fraction may be smaller. A fractional factorial design is often referred to as, for example, a half-fraction or a quarter-fraction design, to indicate that it involves one-half or one-quarter, respectively, of the experimental conditions in the corresponding complete factorial.

Third, like the complete factorial design in Table 5.2, the fractional factorial design in Table 5.3 is balanced; each level of each factor appears the same number of times. (The reader is encouraged to examine Table 5.3 to verify this. For more about why balance is important in factorial designs, see Chap. 3.) Balanced fractional factorial designs are as efficient in terms of use of subjects as complete factorial

Table 5.3 2^{5-1} Fractional factorial design

Experimental condition from Table 5.2	MI	PEER	TEXT	MIND	SKILLS	Per-condition n^a
2	No[b]	No	No	No	High	22
3	No	No	No	Yes	Low	22
5	No	No	Yes	No	Low	22
8	No	No	Yes	Yes	High	22
9	No	Yes	No	No	Low	22
12	No	Yes	No	Yes	High	22
14	No	Yes	Yes	No	High	22
15	No	Yes	Yes	Yes	Low	22
17	Yes	No	No	No	Low	22
20	Yes	No	No	Yes	High	22
22	Yes	No	Yes	No	High	22
23	Yes	No	Yes	Yes	Low	22
26	Yes	Yes	No	No	High	22
27	Yes	Yes	No	Yes	Low	22
29	Yes	Yes	Yes	No	Low	22
32	Yes	Yes	Yes	Yes	High	22

[a]Overall $N = 352$ provides power $\geq .8$ to detect effect size of $\beta^* \geq .15$
[b]No means not included in intervention; yes means included in intervention

designs. In addition, in a balanced complete or fractional factorial design, when the data are properly analyzed and effect coding is used, individual effect estimates are uncorrelated. However, as will be explained below, the interpretation of the individual effect estimates is a bit different in fractional factorial designs than in complete factorial designs.

Fourth, as compared to the design in Table 5.2, the one in Table 5.3 has a larger per-condition n, yet the two designs have exactly the same statistical power. Someone comparing the designs in Tables 5.2 and 5.3 and viewing the comparison from an RCT perspective might mistakenly conclude that the fractional factorial design has more power or, put another way, assume that a fractional factorial design requires a smaller sample size to maintain the same level of power, because it has fewer experimental conditions. This mistaken conclusion might be made because adding conditions to or removing them from an RCT usually changes the sample size requirements, often dramatically. However, this is generally not so in balanced factorial experiments. As discussed in Chap. 3, in a factorial experiment power is determined by the per-level N, not the per-condition n. The designs in Tables 5.2 and 5.3 have exactly the same statistical power because, although the per-condition n is different, the per-level N in the two designs is identical. In fact, when using software to power a balanced factorial experiment, it is usually unnecessary to specify whether the design is complete or fractional. Thus, the motivation for considering a fractional factorial experiment is not to reduce the required sample size. Rather, it is to reduce the burden and expense associated with implementing experimental conditions.

5.3 What's the Catch? Aliasing

Fifth, unlike the design in Table 5.2, the design in Table 5.3 does not include a so-called "control" condition in which all of the factors are set to the lowest level. In other words, the condition in which *PEER*, *MI*, *TEXT*, and *MIND* are set to no and *SKILLS* is set to low is not represented in this design. This may be counterintuitive to someone accustomed to an RCT perspective. Recall the discussion in Chap. 3 about how in a factorial experiment individual conditions are never directly compared to each other. Although a factorial experiment may include a condition in which every factor is set to the lowest level, this condition does not serve as the control for the entire experiment. Instead, each factor has one level that serves as the control for that factor. For example, the main effect of *MI* is estimated by comparing the mean of the first half of the conditions, that is, those in which *MI* is set to no, to the mean of the second half of the conditions, in which *MI* is set to yes. In this comparison, the conditions in which *MI* is set to no, taken together, serve as the control for the *MI* factor. The control for each of the other factors is made up of a different subset of conditions. Thus, a single condition in which all of the factors are set to the lowest level is not necessary to maintain experimental control in a fractional factorial design. This is why it is possible to conduct a fractional factorial experiment without a condition corresponding to a "control" and still draw strong causal inferences about each factor.

5.2.2 Notation for Fractional Factorial Designs

Chapter 3 introduced the standard exponential notation for factorial designs. For example, a 2^3 factorial design involves three factors, each with two levels, and there is a total of $2^3 =$ eight experimental conditions.

This notation is extended for fractional factorial designs and contains a lot of information in a compact form. The fractional factorial design in Table 5.3 is referred to as a 2^{5-1} factorial. This indicates that the design involves five factors, each with two levels; there is a total of $2^{5-1} = 2^4 = 16$ experimental conditions; and the fractional factorial design is a $2^{-1} = \frac{1}{2} =$ one-half fraction of the corresponding complete factorial design.

5.3 What's the Catch? Aliasing

The design in Table 5.3 is extremely efficient. Like any balanced factorial design, its sample size requirements are modest in relation to the amount of scientific information it provides—but it requires implementation of only half of the experimental conditions that would be required by a complete factorial experiment. The reader may be wondering, *what's the catch?*

Fractional factorial designs are efficient, but there are important trade-offs that must be weighed when considering their use. When conditions are removed from a factorial experiment, it becomes impossible to estimate as many effects as can be estimated using data from a complete factorial experiment. As a result, the main effects of the experimental factors become combined with some interactions between experimental factors, and some interactions between experimental factors become combined with other interactions between experimental factors. This combining of effects in fractional factorial designs is called *aliasing*. When effects are aliased in a fractional factorial experiment, they cannot be disentangled without implementing additional experimental conditions.

As will be explained below, it is known exactly which effects will be aliased in every fractional factorial design. In the design in Table 5.3, each main effect is combined with one four-way interaction, and each two-way interaction is combined with one three-way interaction. As a specific example, the main effect of *MI* is combined with the *PEER* × *TEXT* × *MIND* × *SKILLS* interaction. Thus when this aliased effect is estimated, whether it is attributable primarily to the main effect of *MI*, primarily to the *PEER* × *TEXT* × *MIND* × *SKILLS* interaction, or to a combination of two cannot be determined empirically without conducting additional experimentation. It is even possible that one of the effects making up the aliased effect has a positive sign and the other has a negative sign, producing a net effect at or near zero.

5.3.1 A Small Example of Combining of Effects

The reason why aliasing of effects occurs in a fractional factorial design can be seen by examining the effect codes for the design. It is difficult to look at all 32 vectors of effect codes for a 2^5 experiment, so combination of effects in a fractional factorial design will be demonstrated here by using the effect codes for a much smaller experiment involving only *MI*, *PEER*, and *TEXT*. The conditions in this 2^3 experimental design are shown in Table 5.4.

A fractional factorial design, in this case a half-fraction design, can be created by including only experimental Conditions 1, 4, 6, and 7. The conditions in the fractional factorial design are shaded in Table 5.4. (This fractional factorial design is used for illustrative purposes only, because the set of effect codes is small and therefore easy to examine. In contrast to the fractional factorial design in Table 5.3, the fractional factorial design in Table 5.4 has some undesirable properties and would be unlikely to be useful in the behavioral, biobehavioral, or biomedical sciences. This will be explained later in the chapter.)

The vectors of effect codes for the full 2^3 design are shown in block (a) of Table 5.5. The effect codes corresponding to the four conditions in the fractional factorial design appear in block (b) in Table 5.5 and then are repeated in blocks (c)–(f) to illustrate some points to be made below.

Let us examine the effect codes corresponding to the experimental conditions in the fractional factorial design. Suppose Dr. B had run these four experimental conditions and was analyzing the data. Imagine Dr. B mistakenly thought it was

5.3 What's the Catch? Aliasing

Table 5.4 2^3 Factorial experimental design with conditions in a 2^{3-1} factorial design shaded

Experimental condition (cell)	MI	PEER	TEXT
1	No[a]	No	No
2	No	No	Yes
3	No	Yes	No
4	No	Yes	Yes
5	Yes	No	No
6	Yes	No	Yes
7	Yes	Yes	No
8	Yes	Yes	Yes

[a]No means not included in intervention; yes means included in intervention

possible to estimate all eight effects that could have been estimated based on a full 2^3 factorial experiment and therefore used all eight vectors of effect codes and attempted to fit this regression equation:

$$\hat{Y} = b_0 + b_1 X_{MI} + b_2 X_{PEER} + b_3 X_{TEXT} + b_4 X_{MI \times PEER} + b_5 X_{MI \times TEXT} + b_6 X_{PEER \times TEXT} + b_7 X_{MI \times PEER \times TEXT}$$

where b_0 is the intercept; b_1, b_2, and b_3 correspond to the main effects; and the remaining coefficients correspond to the interactions in the model. Dr. B's attempt to estimate this model containing the intercept and all seven regression coefficients would fail, because several of the vectors of effect codes are perfectly collinear.

To understand why, look at block (c) in Table 5.5. Two columns in this block are shaded: the column corresponding to the intercept and the column corresponding to the three-way interaction. Because Conditions 2, 3, 5, and 8 have been removed from the design, based on the remaining experimental conditions, these two columns become identical, except for sign. Thus in this fractional factorial design, the vector corresponding to the intercept and the vector corresponding to the three-way interaction are perfectly correlated (although they are negatively correlated, they are perfectly collinear) and therefore indistinguishable from a statistical point of view. The same is true for the columns corresponding to the *MI* main effect and the *PEER* × *TEXT* interaction (block (d) in Table 5.5); those corresponding to the *PEER* main effect and the *MI* × *TEXT* interaction (block (e)); and those corresponding to the *TEXT* main effect and the *MI* × *PEER* interaction (block (f)).

The appropriate way to conduct the analysis of data from this fractional factorial design is to remove the redundant vectors and fit a new regression equation with only the remaining vectors. What happens to the effects that are removed, and how should the effects in the new regression equation be interpreted? The effects that are removed from the previous equation do not simply vanish. Instead, the effects that were in the previous equation become combined to create new effects. The new regression equation can be expressed as follows:

Table 5.5 Effect codes for 2^3 factorial ANOVA

Experimental condition	Intercept	Main effects			Interactions			
		MI	PEER	TEXT	MI × PEER	MI × TEXT	PEER × TEXT	MI × PEER × TEXT
		X_{MI}	X_{PEER}	X_{TEXT}	$X_{MI \times PEER}$	$X_{MI \times TEXT}$	$X_{PEER \times TEXT}$	$X_{MI \times PEER \times TEXT}$

(a) Effect codes for all conditions in 2^3 factorial ANOVA

1	+1	−1	−1	−1	+1	+1	+1	−1
2	+1	−1	−1	+1	+1	−1	−1	+1
3	+1	−1	+1	−1	−1	+1	−1	+1
4	+1	−1	+1	+1	−1	−1	+1	−1
5	+1	+1	−1	−1	−1	−1	+1	+1
6	+1	+1	−1	+1	−1	+1	−1	−1
7	+1	+1	+1	−1	+1	−1	−1	−1
8	+1	+1	+1	+1	+1	+1	+1	+1

(b) Effect codes for conditions in the 2^{3-1} fractional factorial design

1	+1	−1	−1	−1	+1	+1	+1	−1
4	+1	−1	+1	+1	−1	−1	+1	−1
6	+1	+1	−1	+1	−1	+1	−1	−1
7	+1	+1	+1	−1	+1	−1	−1	−1

(c) Shading shows that the intercept is aliased[a] with the MI × PEER × TEXT interaction

1	+1	−1	−1	−1	+1	+1	+1	−1
4	+1	−1	+1	+1	−1	−1	+1	−1
6	+1	+1	−1	+1	−1	+1	−1	−1
7	+1	+1	+1	−1	+1	−1	−1	−1

(d) Shading shows that the MI main effect is aliased with the PEER × TEXT interaction

1	+1	−1	−1	−1	+1	+1	+1	−1
4	+1	−1	+1	+1	−1	−1	+1	−1
6	+1	+1	−1	+1	−1	+1	−1	−1
7	+1	+1	+1	−1	+1	−1	−1	−1

(e) Shading shows that the PEER main effect is aliased with the MI × TEXT interaction

1	+1	−1	−1	−1	+1	+1	+1	−1
4	+1	−1	+1	+1	−1	−1	+1	−1
6	+1	+1	−1	+1	−1	+1	−1	−1
7	+1	+1	+1	−1	+1	−1	−1	−1

(f) Shading shows that the TEXT main effect is aliased with the PEER × MI interaction

1	+1	−1	−1	−1	+1	+1	+1	−1
4	+1	−1	+1	+1	−1	−1	+1	−1
6	+1	+1	−1	+1	−1	+1	−1	−1
7	+1	+1	+1	−1	+1	−1	−1	−1

[a]Aliased means code vectors are identical or identical if one vector is multiplied by −1. Aliasing is indicated by shading

$$\hat{Y} = b_{0, MI \times PEER \times TEXT} + b_1 X_{MI, PEER \times TEXT} + b_2 X_{PEER, MI \times TEXT} \\ + b_3 X_{TEXT, PEER \times MI}. \tag{5.1}$$

In this equation b_0 corresponds to the combination of the intercept and the three-way interaction (to see why, refer to block (c) in Table 5.5); b_1 corresponds to the combination of the *MI* main effect and the *PEER* × *TEXT* interaction (block (d)); b_2 corresponds to the combination of the *PEER* main effect and the *MI* × *TEXT* interaction (block (e)); and b_3 corresponds to the combination of the *TEXT* main effect and the *MI* × *PEER* interaction (block (f)).

5.3.2 Confounding and Aliasing

This combining of effects is referred to in the literature by two different terms: confounding and aliasing. Although in most of the scientific literature these terms are used interchangeably, in this book a distinction is drawn between confounding and aliasing. The term confounding is reserved for inadvertent combining of effects, such as what occurs in observational studies where there is no experimental control and in experiments in which control has been breached by poor implementation or some kind of disruption. For example, if assignment to experimental conditions was carelessly done and instead of random assignment men primarily were given motivational interviewing and women primarily were not given motivational interviewing, sex and *MI* would be confounded. According to this definition, confounding is usually detected after the fact (if it is detected at all).

The term aliasing is reserved for combining of effects that occurs as a result of deliberate experimental design decisions made by the investigator. For example, an investigator who has selected the experimental design in Table 5.3 has deliberately chosen to alias *MI* with the *PEER* × *TEXT* × *MIND* × *SKILLS* interaction. Because the investigator can determine exactly which of the effects of any manipulated variables will be aliased in any experimental design, information about aliasing can and should be used strategically to select an appropriate experimental design. If the selection is done thoughtfully, fractional factorial designs can be an effective way to follow the resource management principle.

It will be shown later in this chapter that fractional factorial designs are not the only experimental designs used in intervention science that involve aliasing.

5.4 Overview of Rationale and Strategy

5.4.1 Rationale: Targeting Resources to Scientifically Important Effects

The rationale for the use of fractional factorial designs is that when resources are limited, whatever resources are available should be devoted primarily to estimating scientifically important effects, rather than to estimating unimportant effects. But which effects are scientifically important?

If the work of the preparation phase of MOST has been done thoughtfully, by the time the optimization phase is reached, it is straightforward to determine whether an effect is scientifically important in a given research endeavor. If an effect has been specified in the conceptual model, it is scientifically important; if it has not been specified in the conceptual model, it is not scientifically important. This is one example of the critical role the conceptual model plays in MOST (see Chap. 2); it can form the basis for determination of the experimental design used in the optimization phase. For example, the conceptual model described in Chap. 2 includes an interaction between the mindfulness meditation and behavioral skills training components. Thus any fractional factorial design selected to examine the five intervention components in this conceptual model must be sure to alias this two-way interaction with one or more other effects that are unimportant scientifically and are expected to be negligible in size. This model does not specify any higher-order interactions, so it may be reasonable to assume that the three-way and four-way interactions and the five-way interaction are expected to be negligible in size. However, it is possible for effects to be unimportant scientifically yet expected to be sizeable.

In some areas, theory may not be advanced enough to provide sufficient detail in the conceptual model about which effects are scientifically important and which are likely to be large even if not scientifically important. In this case, the scientist can rely on the effect hierarchy, effect sparsity, and effect heredity principles (Wu & Hamada, 2009), which were discussed in Chap. 4. These three principles have been used by statisticians to justify the development of fractional factorial design theory and are also applied by engineers and other scientists to guide the selection of fractional factorial designs. The principles are invoked only when the conceptual model does not provide sufficient guidance; anything specified in the conceptual model overrules them.

The effect hierarchy principle states that the main effects are most likely to be scientifically important, followed by two-way interactions, followed by three-way interactions, etc. The higher the order of the effect, the less likely it is to be important. The effect sparsity principle, also known as the Pareto principle (Box & Meyer, 1986), specifies that out of the large number of effects estimated in a factorial experiment, the number of important ones is small. According to the effect heredity principle, in order for an interaction to be considered scientifically important, at least one of its parent main effects must be significant. In this book a modified version of effect heredity will be used. This modified version states that an interaction is most likely to be scientifically important if at least one of its parent main effects is significant; otherwise, it should be viewed with extreme caution. These principles will be discussed again in Chap. 7 in relation to making decisions about which components and component levels comprise the optimized intervention.

Taken together, these three principles suggest that in the absence of an a priori rationale for taking a different perspective, it is reasonable to devote resources and attention to the limited subset of effects in a factorial experiment that are likely to be the most important scientifically. A basic principle underlying the use of fractional factorial designs is that lower-order effects, that is, main effects and two-way interactions, are in general more scientifically interesting and important than higher-order effects, such as 4-way and 5-way interactions. Thus in every fractional factorial design, the subset of scientifically important effects includes the main effects. Often two-way interactions are also designated important, and in some cases the subset may include a few carefully selected three-way interactions. In general, if in addition to main effects and two-way interactions, more than a few three-way interactions or any four-way or interactions or higher are considered scientifically important, a complete factorial is probably a better choice than a fractional factorial. Choosing between a complete and fractional factorial will be discussed further later in this chapter and in Chap. 6.

5.4.2 Strategy of Selecting a Design That Deliberately Aliases Effects

Once the investigator considering using a fractional factorial design has determined which effects are scientifically important, the next step is to seek a design that deliberately aliases each of these important scientific effects only with effects that (a) are not scientifically important *and* (b) can be assumed to be negligible in size. The effect estimates based on such a design will be a combination of one scientifically important effect along with one or more (depending on the design selected) other effects. The logic is that because these other effects are assumed to be negligible in size, the combined estimates can be attributed primarily to the scientifically important effects.

For example, suppose the 2^{3-1} fractional factorial experiment in the example above is run and Eq. (5.1) is fit to the resulting data. As described above, b_1 will then correspond to the combination of the *MI* main effect and the *PEER* × *TEXT* interaction. b_1 can be interpreted as the main effect of *MI* if it can be assumed that the *PEER* × *TEXT* interaction is negligible in size and scientifically unimportant. In general, one would not make this assumption about two-way interactions; this is intended as an illustrative example.

In some fractional factorial designs, there may be a limited number of effects that are scientifically unimportant but cannot be assumed to be negligible in size. When selecting a fractional factorial design, these should not be aliased with scientifically important effects, but they can be aliased with other scientifically unimportant effects.

Recall that Chap. 3 introduced the difference between two perspectives on scientific research: conclusion-priority and decision-priority. In the conclusion-priority perspective, drawing scientific conclusions is the primary objective, whereas the primary objective in the decision-priority perspective is making decisions about which components to include in an intervention. When the conclusion-priority perspective is taken, it may be important to have very precise estimates of effects. In this case, a complete factorial design, in which effects are not aliased, is probably a better choice than a fractional factorial. However, when the decision-priority perspective is taken, achieving a high level of precision may not be as important as conserving resources. All that is needed is sufficient precision to enable correct decision-making. Here a fractional factorial design may merit serious consideration.

5.5 More About the Strategy of Deliberately Aliasing Effects

When selecting a fractional factorial design, the primary objective is to conserve resources while maintaining reasonably accurate estimation of scientifically important effects. As discussed above, all fractional factorial designs alias effects, that is, combine two or more effects that cannot be disentangled without additional experimentation. The overall strategy of selecting a fractional factorial design involves aliasing effects of scientific importance only with effects that are assumed negligible in size, so that the resulting effect estimate can be attributed primarily to the scientifically important effect.

A complete factorial design is always the starting point for selection of a fractional factorial design. It is helpful to divide the effects in the complete factorial design into three categories, which are summarized in Table 5.6: (i) scientifically important, (ii) scientifically unimportant and expected to be negligible in size, and (iii) scientifically unimportant but potentially may be sizeable. Category (i) always contains the main effects, may contain some or all of the two-way interactions, and occasionally may include a few three-way interactions. Category (ii), which is usually the largest, typically contains all interactions involving four or more factors, most or all of the three-way interactions, and possibly some two-way interactions. Category (iii) contains effects that are not scientifically important, yet may not be negligible in size. Category (iii) should be used sparingly and with discretion; it is usually the smallest of the three categories and may even be empty.

Then a design is sought that aliases the effects that are of scientific interest (i.e., category (i)) *only* with effects that are assumed to be negligible in size (i.e., category (ii)) and *never* with other effects that are of scientific interest or any effects that may

Table 5.6 Summary of categorization of effects for selection of a fractional factorial design with k factors

Category	Label	Usually contains	Can be aliased with
i	Scientifically important	Main effects, some two-way interactions, very few three-way interactions	Category ii only
ii	Scientifically unimportant, can be assumed negligible	Two-way interactions up to the k-way interaction	Category i, ii, iii
iii	Scientifically unimportant, may be sizeable	Some two-way interactions, possibly very few three-way interactions	Category ii or iii

be sizeable (i.e., category (iii)). This is why it is so important to divide the effects that are not of scientific interest into those that are expected to be negligible and those that may not be negligible; those that may not be negligible definitely must not be aliased with scientifically important effects! Effects in categories (ii) and (iii) may be aliased with each other.

In general, all else being equal, when more effects can be assumed to be scientifically uninteresting and also negligible in size, designs that are less expensive to conduct because they require fewer experimental conditions are available. When more effects are placed in categories (i) and (iii), that is, declared to be either of scientific importance or potentially sizeable, more conditions are required, and therefore the available designs will be more expensive to conduct. Another way to state this is that if the investigator views many effects as scientifically uninteresting and is willing to assume that all but a few of these effects are negligible in size, opportunities to use highly efficient designs open up. However, such strong assumptions are not always justified. If the investigator views many effects as scientifically important or potentially sizeable even if they are not scientifically important, it will be necessary to use a design involving more experimental conditions. To take this to its logical extreme, if many or all higher-order interactions are deemed scientifically important or potentially sizeable, it will be necessary to conduct a complete factorial experiment.

5.6 Clear and Strongly Clear Effects

The term alias (or aliasing) structure refers to how the aliasing is set up in a particular experimental design. Within the alias structure of a particular design, some effects may be referred to as "clear" or "strongly clear." An effect is referred to as clear if it is aliased only with interactions involving three or more factors and strongly clear if it is aliased only with interactions involving four or more factors (Wu & Chen, 1992). This book takes the position that for applications in the behavioral, biobehavioral, and biomedical sciences, any fractional factorial design selected should have an alias structure that keeps the main effects clear at a minimum and ideally keeps them strongly clear.

When the 2^{3-1} design depicted in the shaded rows of Table 5.4 was first presented in this chapter, it was pointed out that this design has some undesirable properties; it was included solely as a convenient way to illustrate the concept of aliasing. Now that some more information about fractional factorial designs has been reviewed, it is time to discuss these undesirable properties.

The 2^{3-1} design in Table 5.4 does not keep the main effects clear. Instead, it aliases each main effect with one of the two-way interactions. To be able to draw conclusions about main effects based on this design, it would be necessary to assume that all of the two-way interactions are negligible and place them in category (ii). Even when the conceptual model does not specify any two-way interactions, it is usually not a good idea to select a design that aliases main effects and two-way interactions.

Fortunately, there are many fractional factorial designs that require assumptions that are much more plausible (although none with only three factors). In the 2^{5-1} design in Table 5.3, each main effect is aliased with one four-way interaction. Thus, in this design, interpretation of the main effects requires the assumption that the four-way interactions are negligible. This is a weaker and more plausible assumption than the assumption that the two-way interactions are negligible.

5.7 Two Key Characteristics of Fractional Factorial Designs

Fractional factorial designs are often referred to by two key characteristics: the design's *fraction* and its *resolution*. These characteristics are known for every fractional factorial design, and an investigator may wish to consider only designs with a particular fraction or resolution.

The fraction of a design is a straightforward concept. As mentioned previously, each fractional factorial design involves a fraction of the experimental conditions that appear in the corresponding complete factorial. Thus, a fractional factorial design can be a half-fraction, quarter-fraction, etc. of the corresponding complete factorial. It is important to note the fraction, because its denominator determines how many effects will be combined with each effect of scientific interest. It may be helpful to think of the effects in a fractional factorial design as being bundled (Chakraborty, Collins, Strecher, & Murphy, 2009). In a half-fraction design, each bundle includes two effects; in a quarter-fraction design, each bundle has four effects; and so on. It is easy to tell what a particular design's fraction is; it is necessary only to know the number of experimental conditions in the design and in the corresponding full factorial.

The term resolution is a bit more abstract than the term fraction. Resolution is a characteristic of the alias structure of a fractional factorial design. Above, effects were referred to as clear if they are aliased with interactions involving three or more factors and strongly clear if they are aliased with interactions involving four or more factors. The resolution of a design is a shorthand method of stating how clear the effects are in its alias structure.

Resolution is designated with an integer, usually a Roman numeral. This Roman numeral is one larger than the lowest-order interaction with which any main effect in the design is aliased. For example, in a Resolution V fractional factorial design, the main effects are aliased with four-way interactions or higher, that is, they are strongly clear. In other words, no main effect is aliased with a two-way or three-way interaction. For a design to be designated Resolution V, it is also necessary that the two-way interactions be aliased with three-way interactions or higher; no two-way interaction is aliased with another two-way interaction or, as has already been stated, with a main effect.

In a Resolution IV fractional factorial design, the main effects are aliased with three-way interactions or higher. This means the main effects are at least clear; some may be strongly clear. No main effect is aliased with a two-way interaction. Two-way interactions are aliased with other two-way interactions or higher-order interactions.

Unlike the fraction of a design, a design's resolution usually cannot be determined by a glance at the experimental conditions alone, except perhaps by design experts. A design's resolution can be determined mathematically or via examination of the alias structure. However, usually it is not necessary for the investigator to do this. As will be discussed below, the easiest way to select a fractional factorial design is by using software, which will return the resolution of any design or even enable the investigator to select a design of a particular minimum resolution.

The Roman numeral indicating the resolution of a fractional factorial design is usually included in the design notation as a subscript, for example, 2_V^{5-1} (the design depicted in Table 5.3) or 2_{III}^{3-1} (the shaded part of Table 5.4).

5.8 Trade-Offs to Consider When Selecting a Fractional Factorial Design

When selecting a fractional factorial design for use in the optimization phase of MOST, the investigator hopes to identify a design that, although it is less expensive than the corresponding complete factorial design because it has a reduced number of experimental conditions, is likely to lead to the same decisions about which components and component levels make up the optimized intervention. The economy afforded by a fractional factorial design enables the investigator to examine the same number of intervention components while expending fewer resources or, alternatively, to use the same level of resources to examine more components.

In exchange for the reduction in cost associated with a fractional factorial experiment, the investigator assumes some risk that the results may not lead to the same decisions as the results of a complete factorial experiment. This can happen if an effect that is assumed to be negligible—and therefore has been aliased with an effect of scientific interest—in reality is large enough to disrupt decision-making. For example, suppose an investigator decided to use the 2^{3-1} fractional factorial design depicted in the shaded rows in Table 5.4. To estimate the main effect of *MI*, this requires the assumption that the *PEER* × *TEXT* interaction is negligible in size. Now suppose that in reality, the main effect of *MI* is negligible, and the *PEER* × *TEXT* interaction is large. The estimate of the combined effect would be large, and the investigator would mistakenly attribute this to the main effect of *MI*. This would probably result in the selection of *MI* into the screened-in set, which would be an error because *MI* does not have a strong main effect.

In selecting a fractional factorial design, it is important to keep this risk at or below justifiable levels. In general, as one reduces the resources that are required by a fractional factorial design, one assumes greater risk of making some incorrect decisions; conversely, reduced risk usually must be purchased by expending resources. The investigator must find a design that offers a reduction in cost sufficient to offset an increased level of risk. Only the investigator can determine what degree of cost savings is sufficient and what level of risk is justifiable.

In this section two trade-offs that must be made by investigators considering fractional factorial designs are discussed. One trade-off is between the number of experimental conditions in a design and the design's resolution. The other trade-off is between the number of experimental conditions and the number of effects per bundle. These trade-offs must be considered carefully when selecting a design according to the resource management principle, because they bear directly on the risk of making an incorrect decision about selection of components into the screened-in set.

Often choosing whether to use a complete or fractional factorial design and what type of fractional factorial design to select begins with consideration of the number of experimental conditions in the complete factorial and the potential reduction in number of experimental conditions that could be afforded by a fractional factorial design. Table 5.7 contains a summary of some of the design options that are available for experiments with four to seven factors. Only designs that keep the main effects clear or strongly clear, that is, are Resolution IV or higher, are shown in this table. (For a comprehensive treatment of the wide array of fractional factorial designs, see Wu and Hamada, 2009.)

5.8.1 Resolution and Resource Requirements

Table 5.7 shows two different fractional factorial designs with six factors. One of these designs is a Resolution VI, and the other is a Resolution IV. The Resolution VI design aliases main effects with five-way interactions and higher, whereas the Resolution IV design aliases main effects with three-way interactions and higher. According to the effect hierarchy principle, higher-order interactions are less likely to be large than lower-order interactions, and therefore it is a safer bet to assume that five-way interactions will be negligible than to assume that three-way interactions will be negligible. Thus, the Resolution VI design requires less stringent assumptions than the Resolution IV design; in other words, the investigator is less likely to draw the wrong conclusion about the significance of an effect because the effect was aliased with one mistakenly assumed to be negligible. For this reason, the Resolution VI design might be preferred. However, the preferred aliasing in the Resolution VI design comes with a cost. The Resolution VI design is more resource-intensive, because it requires 32 experimental conditions, and the Resolution IV design requires 16. There is often a direct trade-off between resolution and cost, because given the same number of factors, a higher-resolution design that makes less stringent assumptions typically requires more experimental conditions.

5.8 Trade-Offs to Consider When Selecting a Fractional Factorial Design

Table 5.7 Some fractional factorial designs that may be useful in optimization of multicomponent behavioral, biobehavioral, and biomedical interventions

Complete factorial	Number of experimental conditions in complete factorial	Fractional factorial	Number of experimental conditions in fractional factorial	Number of effects in each bundle	Resolution	Lowest order interaction that is aliased with Main effects	Lowest order interaction that is aliased with 2-way interactions
2^4	16	2^{4-1}	8	2	IV	3-way	2-way
2^5	32	2^{5-1}	16	2	V	4-way	3-way
2^6	64	2^{6-1}	32	2	VI	5-way	4-way
2^6	64	2^{6-2}	16	4	IV	3-way	2-way
2^7	128	2^{7-1}	64	2	VII	6-way	5-way
2^7	128	2^{7-2}	32	4	IV	3-way	2-way
2^7	128	2^{7-3}	16	8	IV	3-way	2-way

5.8.2 Bundling of Effects and Resource Requirements

All else being equal, designs with fewer experimental conditions cost less to implement than designs with more experimental conditions. Thus strictly from a cost perspective, a quarter-fraction design would be preferred to a half-fraction design, because it has fewer experimental conditions. Here again there is a trade-off. As discussed above, although smaller-fraction designs are less expensive, they bundle more effects together. For example, in a quarter-fraction design four effects are bundled together, whereas in a half-fraction design only two effects are bundled together. This means that in a quarter-fraction design an effect of primary scientific interest, say a main effect, is bundled with three other effects, whereas in a half-fraction design the same effect is bundled with only one other effect.

Consider the last two rows of Table 5.7. Two fractional factorial designs for seven factors are listed there. Both are Resolution IV designs, in which main effects are aliased with three-way interactions and higher, but one is a quarter-fraction requiring 32 experimental conditions, whereas the other is an eighth-fraction requiring 16 experimental conditions. Even though both designs are the same resolution and the eighth-fraction design is clearly less expensive, it may not be the better choice, because the eighth-fraction design requires stronger assumptions.

Suppose the investigator is comfortable assuming that the three-way interactions and higher are negligible in size but acknowledges the possibility that this assumption may be wrong in one or more cases. Further suppose this investigator assigns a 1% probability that any single effect assumed negligible is actually sizeable and that these probabilities are independent. Now let us compare the quarter-fraction and the eighth-fraction designs. Consider a particular main effect. According to the binomial theorem, in a quarter-fraction design, there would be a 2.97% probability that at least one of the three effects bundled with this main effect is sizeable. In an eighth-fraction design, this probability increases to 6.79%, because now seven effects are bundled with the main effect, any one of which has a 1% probability of being sizeable. In a half-fraction, the probability would be 1%, because only one effect is bundled with each effect of scientific interest.

In summary, designs involving smaller fractions of the corresponding complete factorial are less expensive. However, they involve larger bundles of aliased effects, and therefore, as compared to larger-fraction designs, are associated with an increased probability that within a given bundle at least one effect assumed negligible will in fact be sizeable. In this sense, the risk of making an incorrect decision about the significance of an effect of scientific interest may be greater in designs involving smaller fractions than in designs that have the same resolution but larger fractions.

5.9 Aliasing: What to Look For

Recall in Sect. 5.4 it was mentioned that when considering a fractional factorial design, it is helpful to categorize the effects in the complete factorial design as (see Table 5.6) (i) scientifically important, (ii) scientifically unimportant and expected to be negligible in size, and (iii) scientifically unimportant but potentially sizeable.

When selecting a fractional factorial design, it is important to identify a design that aliases effects into desirable bundles only. There are two types of desirable bundles of effects. The first type includes at most one effect of scientific interest, that is, at most one effect drawn from category (i), with any remaining effects in the bundle scientifically unimportant and negligible in size, that is, drawn from category (ii). For example, in a half-fraction design, if a bundle includes an effect from category (i), the other effect in the bundle should be from category (ii). In a quarter-fraction, a bundle might include one effect from category (i) accompanied by three effects from category (ii). This type of bundle is desirable because when a single effect of scientific importance is bundled only with effects that are scientifically unimportant and negligible in size, it can be concluded that the observed bundled effect is attributable primarily to the important effect.

The second type of desirable bundle may be made up of any combination of effects drawn from categories (ii) and (iii). Bundles made up entirely of category (ii) effects, entirely of category (iii) effects, or a combination of category (ii) and category (iii) effects are acceptable. Even though this type of bundle contains no effects of scientific importance, it is desirable because it avoids aliasing a scientifically important effect with another effect that may be large.

As mentioned above, the overall aliasing in a fractional factorial design is referred to as the alias (or aliasing) structure. Fractional factorial designs with undesirable alias structures, that is, alias structures that include undesirable effect bundles, should be avoided. Undesirable bundles are those that include an effect from category (i) along with one or more effects from either category (i) or category (iii). In other words, undesirable bundles include more than one scientifically important effect, or include a scientifically important effect along with a scientifically unimportant but potentially sizeable effect. An undesirable effect bundle "buries" one or more scientifically important effects with other sizeable effects. Thus, when an undesirable effect bundle is estimated, it is impossible to attribute the observed effect to a single scientifically important effect.

Keeping track of the different types of effects may seem daunting, but as will be demonstrated below, the use of software makes the job much easier. Although fractional factorial design software can often be used without the tedium of specifying which effects have been placed in which category, it is always necessary to review the alias structure of any design under consideration to confirm that it is appropriate before making a final decision. Thus having a clear idea about which effects are assigned to which categories is essential for selection of a fractional factorial design, irrespective of which method is used to identify candidate designs for consideration.

5.10 Selecting a Fractional Factorial Design Using Software

This section will discuss in specific terms how an investigator can select a fractional factorial design. Designs can be selected using books, such as Wu and Hamada (2009), but the easiest way is to use software. Design software all follows the same basic logic: The user specifies certain desired characteristics of a fractional factorial design, and the software returns a design with the desired characteristics, provided that one exists. The user should then review the aliasing structure to confirm that it is acceptable.

Suppose Dr. B would like to conduct a factorial experiment to examine the performance of the five components listed in Table 5.1. The corresponding complete factorial design appears in Table 5.2. Below the use of PROC FACTEX in SAS to select a fractional factorial design for Dr. B under a variety of scenarios is demonstrated. Design routines can also be found in R and Minitab.

5.10.1 Overview: Different Approaches to Using Software to Select a Fractional Factorial Design

Two general approaches to using software to select a fractional factorial design are discussed in this section. In the first approach, the user specifies the desired number of experimental conditions in the fractional factorial design and/or the desired resolution of the design. The software returns an experimental design that may alias any effects as long as the specified criteria are met. In the second approach, the user specifies which effects are in categories (i), (ii), and (iii). The software returns a design that aliases each effect in category (i) only with effects in category (ii). The first approach requires relatively little user input, but there is no guarantee that the design includes only bundles that the investigator will consider desirable; the alias structure must be reviewed carefully to determine whether or not it is acceptable. The second approach requires more user input, but if the categorization is accurately specified the software returns a design that is guaranteed to include only desirable bundles, if such a design exists.

Suppose Dr. B has identified the main effects and two-way interactions as the effects of primary scientific importance and is comfortable assuming that all of the three-way interactions and up are negligible.

5.10.2 Selecting a Design by Specifying the Desired Number of Experimental Conditions

Suppose Dr. B has determined that implementing 16 experimental conditions is feasible. Dr. B can identify a design with 16 experimental conditions by providing the following statements in SAS.

5.10 Selecting a Fractional Factorial Design Using Software

```
PROC FACTEX;
FACTORS MI PEER TEXT MIND SKILLS;
SIZE DESIGN=16;
MODEL RESOLUTION=MAXIMUM;
EXAMINE ALIASING(5) DESIGN;
RUN;
```

The second line lists the names of the factors; this indicates how many factors are to be included in the design and provides labels for the output. (It is not necessary to specify the number of levels here because the factors in Dr. B's experiment all have two levels, which is the SAS default. PROC FACTEX can handle complicated designs with varying numbers of levels per factors, but these are outside the scope of this book.) The third line directs PROC FACTEX to provide a design with 16 experimental conditions. In the fourth line, RESOLUTION = MAXIMUM specifies that the design provided should be the maximum resolution obtainable with 16 conditions. The fifth line requests two different items to be output. ALIASING (5) requests the program to list the alias structure up to the five-way interaction. DESIGN requests the effect code vectors for the main effects. The effect code vectors indicate which levels of each factor should be included in each experimental condition.

The above SAS statements will return the effect code vectors for the main effects that are shown in Table 5.8. If the no/low level of each factor is substituted for the -1 s and the yes/high level of each factor is substituted for the 1 s, the result is the 2_V^{5-1} fractional factorial design in Table 5.3. (Although it is not shown in the above code, in some software, including SAS, the user can include labels for the levels of the factors and output a readable table that shows the labels for the levels in each experimental condition, similar to Table 5.3).

Table 5.8 Main effect code vectors for 2^{5-1} fractional factorial design in Table 5.3

MI	PEER	TEXT	MIND	SKILLS
−1	−1	−1	−1	1
−1	−1	−1	1	−1
−1	−1	1	−1	−1
−1	−1	1	1	1
−1	1	−1	−1	−1
−1	1	−1	1	1
−1	1	1	−1	1
−1	1	1	1	−1
1	−1	−1	−1	−1
1	−1	−1	1	1
1	−1	1	−1	1
1	−1	1	1	−1
1	1	−1	−1	1
1	1	−1	1	−1
1	1	1	−1	−1
1	1	1	1	1

Table 5.9 Alias structure of 2_V^{5-1} design in Table 5.3

Effect	Aliased with
Intercept	MI × PEER × TEXT × MIND × SKILLS
Main effect MI	PEER × TEXT × MIND × SKILLS
Main effect PEER	MI × TEXT × MIND × SKILLS
Main effect TEXT	MI × PEER × MIND × SKILLS
Main effect MIND	MI × PEER × TEXT × SKILLS
Main effect SKILLS	MI × PEER × TEXT × MIND
MI × PEER	TEXT × MIND × SKILLS
MI × TEXT	PEER × MIND × SKILLS
MI × MIND	PEER × TEXT × SKILLS
MI × SKILLS	PEER × TEXT × MIND
PEER × TEXT	MI × MIND × SKILLS
PEER × MIND	MI × TEXT × SKILLS
PEER × SKILLS	MI × TEXT × MIND
TEXT × MIND	MI × PEER × SKILLS
TEXT × SKILLS	MI × PEER × MIND
MIND × SKILLS	MI × PEER × TEXT

The alias structure for this design is provided in Table 5.9. As the table shows, each bundle in this half-fraction design includes two effects. This is a Resolution V design, so no main effects are aliased with any effects of lower order than a four-way interaction. In this design the five-way interaction is aliased with the intercept; each main effect is aliased with a different four-way interaction; and each two-way interaction is aliased with a different three-way interaction.

Is this design made up only of desirable bundles? This depends on which effects the investigator has identified as being of primary scientific interest and what assumptions he or she is willing to make. It was stated above that Dr. B sees the main effects and two-way interactions as being of primary scientific interest and is comfortable with the assumption that the three-way interactions and up are negligible in size. Each of the bundles listed in Table 5.9 aliases an effect of scientific interest with one of the effects assumed to be negligible; no bundle aliases any main effect or two-way interaction with an effect that is thought to be sizeable. Therefore, the aliasing in this design would most likely be acceptable to Dr. B.

Dr. B can investigate the possibility of an even less expensive fractional factorial design by requesting a design with eight experimental conditions, simply by using the same code as above but specifying

```
SIZE DESIGN=8;
```

The result is the 2_{III}^{5-2} design shown in Table 5.10a. This design is different from the one in Table 5.3 in two important ways. First, whereas the design in Table 5.3 is a half-fraction, this one is a quarter-fraction. Second, the design in Table 5.3 is a Resolution V, whereas this one is a Resolution III. (Note that although the design in Table 5.10a is a subset of the complete factorial, it is not a subset of the design in Table 5.3).

5.10 Selecting a Fractional Factorial Design Using Software

Table 5.10 Experimental conditions and alias structure of 2_{III}^{5-2} fractional factorial design

(a) Experimental conditions

Experimental condition from Table 5.2	MI	PEER	TEXT	MIND	SKILLS
2	No[a]	No	No	No	High
7	No	No	Yes	Yes	Low
11	No	Yes	No	Yes	Low
14	No	Yes	Yes	No	High
20	Yes	No	No	Yes	High
21	Yes	No	Yes	No	Low
25	Yes	Yes	No	No	Low
32	Yes	Yes	Yes	Yes	High

(b) Alias structure

Effect	Aliased with		
Intercept	$MI \times MIND \times SKILLS$	$PEER \times TEXT \times SKILLS$	$MI \times PEER \times TEXT \times MIND$
Main effect MI	$MIND \times SKILLS$	$PEER \times TEXT \times MIND$	$MI \times PEER \times TEXT \times SKILLS$
Main effect PEER	$TEXT \times SKILLS$	$MI \times TEXT \times MIND$	$MI \times PEER \times MIND \times SKILLS$
Main effect TEXT	$PEER \times SKILLS$	$MI \times PEER \times MIND$	$MI \times TEXT \times MIND \times SKILLS$
Main effect MIND	$MI \times SKILLS$	$MI \times PEER \times TEXT$	$PEER \times TEXT \times MIND \times SKILLS$
Main effect SKILLS	$MI \times MIND$	$PEER \times TEXT$	$MI \times PEER \times TEXT \times MIND \times SKILLS$
$MI \times PEER$	$TEXT \times MIND$	$MI \times TEXT \times SKILLS$	$PEER \times MIND \times SKILLS$
$MI \times TEXT$	$PEER \times MIND$	$MI \times PEER \times SKILLS$	$TEXT \times MIND \times SKILLS$

[a] No means not included in intervention; yes means included in intervention

The alias structure for this design is shown in Table 5.10b. As the table shows, each bundle in this quarter-fraction design includes four effects. Because this is a Resolution III design, main effects and two-way interactions are aliased. For example, the main effect of *MI* is aliased with the *MIND* × *SKILLS* interaction, the *PEER* × *TEXT* × *MIND* interaction, and the *MI* × *PEER* × *TEXT* × *SKILLS* interaction. Although an experiment with this design would be relatively inexpensive to implement, Dr. B would probably find its alias structure unacceptable because it aliases main effects and two-way interactions. (Quarter-fraction designs can be Resolution IV or higher, but this is possible only for designs that involve more than five factors. Several such designs are listed in Table 5.7.)

5.10.3 Selecting a Design by Specifying a Minimum Desired Resolution

Suppose Dr. B wants to find a design that is at least Resolution IV. To find this design Dr. B can submit the following code:

```
PROC FACTEX;
FACTORS MI PEER TEXT MIND SKILLS;
MODEL RESOLUTION=4;
SIZE DESIGN=MINIMUM;
RUN;
```

The third line specifies that a design of Resolution IV or higher is to be returned. The fourth line specifies that the Resolution IV (or higher) design with the smallest number of experimental conditions is to be returned.

In this case, the design that meets the specified criteria is the half-fraction Resolution V design shown in Table 5.3. (There is no 2_{IV}^{5-1} design.)

5.10.4 Selecting a Design by Specifying Both the Desired Number of Experimental Conditions and a Desired Resolution

Now suppose Dr. B likes the idea of a fractional factorial design with only eight experimental conditions, but the Resolution III design in Table 5.10 is unacceptable because it aliases main effects and two-way interactions. Dr. B would like to identify a design that has only eight conditions but is at least Resolution IV. The code to look for this kind of design would be

5.10 Selecting a Fractional Factorial Design Using Software

```
PROC FACTEX;
FACTORS MI PEER TEXT MIND SKILLS;
SIZE DESIGN=8;
MODEL RESOLUTION=4;
RUN;
```

In this case, SAS would return a message saying "No such design found." As was discussed in the preceding section, a Resolution IV design does not exist for 2^5 experiments. The next best option would be the 16-condition Resolution V design discussed previously, but that design is eliminated because it does not meet the criterion of eight or fewer experimental conditions. The result is that no design meeting both of the specified criteria is found.

5.10.5 Selecting a Design by Specifying Which Effects Are in Which Categories

It was stated above that Dr. B has identified main effects and two-way interactions as being of primary scientific interest and is comfortable assuming that the rest of the effects are negligible in size. Instead of requesting a design with a particular resolution or number of experimental conditions, it is possible to specify which effects are in categories (i), (ii), and (iii) and let the software return a recommended design on that basis.

```
PROC FACTEX;
FACTORS MI PEER TEXT MIND SKILLS;
MODEL ESTIMATE = (MI PEER TEXT MIND SKILLS MI | PEER | TEXT | MIND | SKILLS @2);
SIZE DESIGN=MINIMUM;
RUN;
```

ESTIMATE = ( ) in the third line lists all the effects that are of scientific importance, in this case the main effects and two-way interactions. In other words, the third line lists all of the effects in category (i). (MI | PEER | TEXT | MIND | SKILLS @2 is an efficient way to specify all two-way interactions in SAS.) Unless instructed otherwise (illustrated below), SAS assumes all effects not specified in category (i) are assigned to category (ii).

SAS will search for an experimental design that does not alias any of the effects in category (i) with each other. In the fourth line, DESIGN = MINIMUM serves the same purpose as above, that is, it instructs SAS to return the fractional factorial design with the smallest number of experimental conditions, out of the set of designs that meets the stated criteria. In this case, SAS returns the 2_V^{5-1} design depicted in Table 5.3.

Now suppose Dr. B is concerned that there may be two nonnegligible three-way interactions that are not scientifically important. They are the *TEXT* × *MIND* × *SKILLS* interaction and the *MI* × *PEER* × *TEXT* interaction. Then the SAS code would be as follows.

```
PROC FACTEX;
FACTORS MI PEER TEXT MIND SKILLS;
MODEL ESTIMATE = (MI PEER TEXT MIND SKILLS MI|PEER|TEXT|MIND|SKILLS @2)
NONNEGLIGIBLE = (TEXT*MIND*SKILLS MI*PEER*TEXT);
SIZE DESIGN=MINIMUM;
RUN;
```

By declaring that the *TEXT* × *MIND* × *SKILLS* and *MI* × *PEER* × *TEXT* interactions are nonnegligible, this code places them in category (iii) rather than category (ii). This means that these interactions cannot be aliased with any effects of scientific importance. Recall the statement made above that in general, when more effects are placed in category (iii), that is, flagged as potentially sizeable, the available designs tend to involve more experimental conditions. Let us see whether moving these two effects from category (ii) to category (iii) makes a difference in the design returned by SAS.

When the above code is run, SAS returns the complete 2^5 factorial experiment, because this is the only design that will not alias the *TEXT* × *MIND* × *SKILLS* and *MI* × *PEER* × *TEXT* interactions with either a main effect or a two-way interaction. Thus moving these two effects from category (ii) to category (iii) has a huge impact on the number of experimental conditions in the design that is returned by the software. In other situations, particularly where there are more than five factors, declaring one or two three-way interactions as members of category (iii) may not affect the availability of designs with fewer experimental conditions. However, to maximize economy it is a good idea to keep category (iii) as small as is reasonably possible. Of course, any effects specified by the conceptual model to be scientifically important or otherwise expected to be sizeable must always be categorized appropriately.

5.11 The Term "Experimental Design" as Used by Statisticians and as Used by Intervention Scientists

Statisticians on the one hand and behavioral, biobehavioral, and biomedical scientists on the other tend to use the term "experimental design" slightly differently. It is a good idea to be aware of this when using experimental design software like PROC FACTEX in SAS. It may also be helpful for statisticians and behavioral, biobehavioral, and biomedical scientists to keep this in mind when they communicate about experimental design.

5.11 The Term "Experimental Design" as Used by Statisticians...

To a statistician, an experimental design is defined primarily by its statistical characteristics. By contrast, to most scientists who are not statisticians, an experimental design is defined by the set of experimental conditions it comprises. With fractional factorial (and other) designs, two (or more) different sets of experimental conditions will represent different designs from a scientist's point of view, but they may have identical statistical characteristics and therefore constitute one experimental design from a statistician's point of view.

This is worth noting for practical reasons. Suppose PROC FACTEX has been used to identify a fractional factorial design. The design has the desired resolution and desired number of experimental conditions, but there is some small detail it would be helpful to fine-tune. Sometimes this can be accomplished by making minor changes in the program's input that do not change the statistical characteristics of the design that is returned, but result in a different set of experimental conditions that may be slightly more practical or more acceptable. Two examples of this are reviewed next.

5.11.1 Switching Factor Level Labels

From a statistical point of view, which level of a factor is labeled no or yes, or low or high, is arbitrary. Let us return to the 2^{5-1} design in Table 5.3. It was mentioned earlier that this design does not have (or need) a traditional control condition, that is, an experimental condition in which all of the factors are set to the lower level. Now return to the vectors of effect codes in Table 5.8. Previously the experimental conditions were determined by assigning each level of each factor to either the -1 or the 1 level of the effect code vectors. How this is done is arbitrary, provided it is done consistently. There is no reason why the higher level of a factor could not be assigned to -1 and the lower level assigned to 1. Therefore, it is possible to reverse how this assignment was done for the *SKILLS* factor, so that the level previously labeled low is now labeled high and vice versa.

The resulting experimental design is shown in Table 5.11. This design contains the condition in which all the factors are set to the lower level, and it does not include a condition in which all of the factors are set to the higher level. In fact, the design in Table 5.11 is the complement of the one shown in Table 5.3; it is made up of the half of the experimental conditions from the complete factorial that are omitted to obtain the design in Table 5.3. From a statistical perspective, this is exactly the same design as the one in Table 5.3. It even has exactly the same alias structure. However, most scientists would consider this a different design, because it is made up of a different set of experimental conditions.

Table 5.11 2^{5-1} fractional factorial design with identical statistical characteristics as the design in Table 5.3, but containing no condition in which all five factors are set to the higher level

Experimental condition from Table 5.2	MI	PEER	TEXT	MIND	SKILLS
1	No[a]	No	No	No	Low
4	No	No	No	Yes	High
6	No	No	Yes	No	High
7	No	No	Yes	Yes	Low
10	No	Yes	No	No	High
11	No	Yes	No	Yes	Low
13	No	Yes	Yes	No	Low
16	No	Yes	Yes	Yes	High
18	Yes	No	No	No	High
19	Yes	No	No	Yes	Low
21	Yes	No	Yes	No	Low
24	Yes	No	Yes	Yes	High
25	Yes	Yes	No	No	Low
28	Yes	Yes	No	Yes	High
30	Yes	Yes	Yes	No	High
31	Yes	Yes	Yes	Yes	Low

[a]No means not included in intervention; yes means included in intervention

5.11.2 Permuting the Order in Which Factors Are Listed

Another way to obtain a different set of experimental conditions is simply by permuting the order in which the factors are listed in the code. Recall the lines of code that Dr. B used to obtain a fractional factorial design with eight experimental conditions:

```
PROC FACTEX;
FACTORS MI PEER TEXT MIND SKILLS;
SIZE DESIGN=8;
MODEL RESOLUTION=MAXIMUM;
RUN;
```

The design returned by this code is shown in Table 5.10. Suppose Dr. B permuted the order of the factors in the code as follows:

```
PROC FACTEX;
FACTORS MI MIND SKILLS PEER TEXT;
SIZE DESIGN=8;
MODEL RESOLUTION=MAXIMUM;
RUN;
```

The only difference is in the second line, in which the names of the factors are reordered. Of course the order of the names of the factors is completely arbitrary and does not have any impact on the statistical properties of the design that is returned. The experimental conditions in this new 2_{III}^{5-2} design are listed in Table 5.12a. As the table shows, this design is made up of a different subset of conditions from the complete factorial (Table 5.1). The alias structure is shown in Table 5.12b. Perusal of this table reveals that main effects are still aliased with two-way interactions, but each main effect is aliased with different interactions as compared to Table 5.10.

From the point of view of a scientist, the designs in Tables 5.10 and 5.12 are different, because they contain different experimental conditions, and the aliasing is somewhat different. A scientist might prefer one of these designs to the other. From the point of view of a statistician, however, the designs in Tables 5.10 and 5.12 have identical properties. How can two designs that alias different effects be identical from a statistical point of view?

When statisticians work out the properties of experimental designs, they typically identify the factors with letters. A five-factor design would have factors A, B, C, D, and E. Suppose in a particular design the main effect of A is aliased with the $C \times D$ interaction. This means that the main effect of the first factor is aliased with the interaction of the third and fourth factors. This will be true irrespective of which factor is designated first, third, and fourth. When the user of PROC FACTEX inputs the names of the factors in a particular order, no empirical data are being read. This input merely specifies that the first factor in the list is designated as A, the second factor in the list is designated as B, and so on. Changing which factors are first, third, and fourth does not alter the fact that the main effect of the first factor is aliased with the interaction of the third and fourth factors and thus does not alter the essential statistical characteristics of the design that is returned.

5.11.3 When the Investigator Desires to Omit a Particular Experimental Condition

Above, two ways of identifying slight variations on fractional factorial designs were discussed. These slight variations are identical from a statistical point of view, but may not be identical from a scientific point of view. Sometimes the motivation for examining different variations on a fractional factorial design is that the investigator seeks a variation that omits a particular condition. For example, it may be useful to select a design that eliminates the "no-treatment" condition. Such a design may be useful in situations where prospective subjects may be reluctant to agree to participate in an experiment if there is a chance they will be randomly assigned to a "control" condition in which subjects receive only the lower level of each factor and thereby "get nothing." By assuring prospective subjects that every experimental condition includes some putatively active treatment (while, of course, making it clear that it has not yet been established whether or not any of the components have a detectable effect), the experimenter may improve the participation rate.

Table 5.12 2_{III}^{5-2} fractional factorial design and alias structure obtained by permuting factor labels

(a) Experimental conditions

Experimental condition from Table 5.2	MI	PEER	TEXT	MIND	SKILLS
5	No[a]	No	Yes	No	Low
8	No	No	Yes	Yes	High
10	No	Yes	No	No	High
11	No	Yes	No	Yes	Low
18	Yes	No	No	No	High
19	Yes	No	No	Yes	Low
29	Yes	Yes	Yes	No	Low
32	Yes	Yes	Yes	Yes	High

(b) Alias structure

Effect	Aliased with
Intercept	$MI \times PEER \times TEXT$	$MIND \times SKILLS \times TEXT$	$MI \times MIND \times SKILLS \times PEER$
Main effect MI	$PEER \times TEXT$	$MIND \times SKILLS \times PEER$	$MI \times MIND \times SKILLS \times TEXT$
Main effect PEER	$MI \times TEXT$	$MI \times MIND \times SKILLS$	$MIND \times SKILLS \times PEER \times TEXT$
Main effect TEXT	$MI \times PEER$	$MIND \times SKILLS$	$MI \times MIND \times SKILLS \times PEER \times TEXT$
Main effect MIND	$SKILLS \times TEXT$	$MI \times SKILLS \times PEER$	$MI \times MIND \times PEER \times TEXT$
Main effect SKILLS	$MIND \times TEXT$	$MI \times MIND \times PEER$	$MI \times SKILLS \times PEER \times TEXT$
$MI \times MIND$	$SKILLS \times PEER$	$MI \times SKILLS \times TEXT$	$MIND \times PEER \times TEXT$
$MI \times SKILLS$	$MIND \times PEER$	$MI \times MIND \times TEXT$	$SKILLS \times PEER \times TEXT$

[a] No means not included in intervention; yes means included in intervention

The ideal starting point for selection of a fractional factorial design is a complete factorial design in which every experimental condition is sensible and feasible. It is usually not a good idea to use a fractional factorial design to eliminate a particular combination of levels of factors that is not sensible and feasible because it is, for example, logistically difficult to implement, expected to be toxic, or overly burdensome for participants. For example, suppose a particular combination of the levels of four factors is expected to be overly burdensome for subjects. In a complete factorial this would probably be reflected in a sizeable four-way interaction involving those factors. In most cases where a sizeable four-way interaction is expected, a fractional factorial design is not a good choice.

5.12 Fractional Factorial Designs With Clustered Data

In Chap. 3 it was mentioned that factorial designs can be used when a cluster structure is present in the data, even when between-cluster randomization is necessary (e.g., randomization of entire schools to experimental conditions). In other words, fractional factorial designs can be used in multilevel situations. Like complete factorial designs, fractional factorial designs may be used with either within-cluster or between-cluster randomization. In fact, fractional factorial designs can be used to address a problem that may arise when cluster randomization to the conditions of a factorial experiment is being considered: the problem of having a sufficiently large sample size to power a factorial experiment but nevertheless lacking enough units to populate the conditions of a factorial experiment.

For example, suppose an investigator developing a school-based drug abuse prevention program wishes to examine five intervention components in a 2^5 factorial experiment. The investigator has access to approximately $N = 3200$ students. Suppose this will provide adequate power in this case, but random assignment at the school level is necessary. The 3200 students are clustered into 25 schools. A 2^5 factorial experiment has 32 experimental conditions, so although there are enough subjects to maintain the desired level of power, there are not enough schools to assign at least one to every experimental condition. In this case, the investigators can consider a fractional factorial experiment, such as a 2^{5-1} design. This design has only 16 conditions, and thus the random assignment will place one or two schools in each condition. (It is much better to use all the schools and have some imbalance than to throw away schools to achieve balance.) This design is workable with as few as one school per condition (Dziak, Nahum-Shani, & Collins, 2012), but with only one school per condition it becomes particularly important to ensure that no schools drop out of the study.

As always, this approach makes sense only if, in the judgment of the investigator, the required assumptions about some effects being negligible in size are reasonable. For more on the topic of factorial and fractional factorial designs when there is a cluster structure, see Dziak et al. (2012); Nahum-Shani, Dziak, and Collins (2017); and the Nahum-Shani and Dziak (2018) chapter in the companion volume (Collins & Kugler, 2018).

5.13 Following the Resource Management Principle When Considering a Fractional Factorial Design

With unlimited resources, most investigators would prefer to conduct complete factorial experiments rather than fractional factorial experiments, so that there would be no need to consider aliasing. However, as is stated repeatedly in this book, in reality resources are nearly always limited. This makes it necessary for investigators to give careful consideration to the resource management principle, that is, how to obtain the greatest scientific yield given whatever resources are available. To follow the resource management principle, it is helpful to have a clear sense of the trade-offs associated with different options.

Above it was mentioned that Dr. B wants to investigate five intervention components but has the resources to run at most only 16 experimental conditions. Suppose Dr. B wants to conduct a factorial experiment and has narrowed the options down to Option 1, a complete 2^4 factorial experiment, and Option 2, a 2_V^{5-1} fractional factorial experiment. Both of these experiments require the same sample size and involve 16 experimental conditions. The risks and potential advantages associated with each option are summarized in Table 5.13.

Option 1, the complete factorial design, requires no assumptions about the anticipated size of higher-order interactions. From one perspective the complete factorial design could be considered safer, because estimates of higher-order interactions will be kept separated from estimates of main effects and lower-order interactions. However, with this design only four components can be examined. If in reality no higher-order effects are sizeable, there is an opportunity cost; use of the complete factorial misses the opportunity to examine a fifth component. On the other hand, if in reality there are sizeable higher-order effects, then the complete factorial has the potential to move science forward faster than the fractional factorial. Option 2, the fractional factorial, requires the assumption that the higher-order effects are negligible in size. If this assumption is correct, Option 2 offers the potential benefit of leveraging resources to move science forward faster than Option 1. However, in

Table 5.13 Risks and potential benefits associated with choosing between 2^4 and 2^{5-1} experimental design

	Possible reality about sizeable higher-order effects	
Design selected	No higher-order effects present	Some higher-order effects present
Option 1 Complete factorial design: 2^4 Enables examination of 4 components	*Risk of missing an opportunity* to examine 5th factor and thereby move science forward	*Potential benefit of moving science forward faster* by not aliasing scientifically important effects with other large effects
Option 2 Fractional factorial design: 2^{5-1} Enables examination of 5 components	*Potential benefit of moving science forward faster* by taking advantage of the opportunity to examine a 5th factor	*Risk of some incorrect decisions* about which components to include in intervention, based on misleading results

addition to the standard risk of a Type I or Type II error, the selection of Option 2 incurs the risk of incorrect decisions made because a putative main effect or two-way interaction (or lack thereof) is in reality attributable to a higher-order interaction.

5.13.1 General Recommendations

Very often investigators know the maximum number of experimental conditions that available resources can support to enable implementation to be done properly, so that each condition is delivered with fidelity and protocol deviations are kept to an absolute minimum. As discussed above, the choice then comes down to two different approaches that involve the same number of experimental conditions: (a) a complete factorial design that enables examination of fewer components with no aliasing of effects or (b) a fractional factorial design that enables examination of more components with aliasing of effects.

This is a difficult decision that can be made easier only by information about the likely size of higher-order interactions. Fractional factorial designs came out of engineering statistics. In engineering and other fields in which there is a long tradition of conducting factorial experiments, there is much theory and empirical information on interactions. This information is helpful to experimenters when they select a design. At this writing such information is, unfortunately, not readily available in most areas of the behavioral, biobehavioral, and biomedical sciences. Today little is known empirically, hypothesized based on theory, or even conjectured about interactions involving components of interventions.

Despite this lack of information, fractional factorial designs merit serious consideration by behavioral, biobehavioral, and biomedical scientists. The following general recommendations may be helpful for investigators considering whether or not to use a fractional factorial design.

A first general recommendation is to lean heavily on the conceptual model (Chap. 2). The importance of a clearly articulated conceptual model cannot be overstated. In the optimization phase, the conceptual model plays a critical role in determining which effects are scientifically important. A good conceptual model is detailed and comprehensive; it sums up and integrates all available information that pertains to the intervention to be optimized. This includes any hypothesized interactions involving any number of intervention components. If an interaction is specified in the conceptual model underlying an intervention under development, this means it is scientifically important and large enough to have an impact on which components will be selected for the optimized intervention. Thus any interaction specified in the conceptual model, such as the interaction between mindfulness meditation and behavioral skills training in the conceptual model underlying the hypothetical example, belongs in category (i). Conversely, if a scientist developing an intervention has not included a particular interaction in the conceptual model, this is the same as specifying that the interaction does not belong in category (i).

However, given how little is currently known about interactions in the behavioral, biobehavioral, and biomedical sciences, it may be prudent to make an exception to this rule for two-way interactions and select designs that keep all two-way interactions clear even if they are not specified in the conceptual model.

The conceptual model is tremendously useful in specifying which effects are of scientific interest and, by default, which are not of scientific interest. However, even a highly detailed conceptual model does not usually specify whether any of the interactions that are not of scientific interest may be large enough to merit placement in category (iii). A second recommendation is to start with the presumption that all interactions that are not considered scientifically important are negligible in size and then consider which specific interactions might be sizeable and why. If there are no good reasons for believing that more than a small number of these interactions may be sizeable, then a fractional factorial design may be a possibility; if there are good reasons to believe that several may be sizeable, a complete factorial may be a better choice. This approach is more thoughtful and ultimately more productive than a vague blanket statement that "any of the higher-order interactions could be large." It may be helpful to remember that when selecting an experimental design, "negligible" does not mean exactly zero. Rather, it means small enough not to change or disrupt decision-making about which components and component levels will be selected for the optimized intervention. In thinking this through, it is critical to remember that all of the effects referred to in this book are defined using effect coding, not dummy coding (see Chaps. 3 and 4 in this volume, and the Kugler, Dziak, and Trail (2018) chapter in the companion volume).

A third recommendation is to weigh resource considerations carefully to determine whether or not a fractional factorial design will be appreciably more economical in a given situation. There are some situations in which a fractional factorial design can offer tremendous savings; in other situations, the savings may be so minor that there is little difference in cost between a complete and fractional factorial design, so it probably would be better to go with a complete factorial. For example, if the intervention is delivered completely electronically and the experiment can be largely automated, then there may be little expense associated with implementing a large number of experimental conditions. Here the costs of a complete and fractional factorial may be about the same, and therefore it may not make sense to consider a fractional factorial. Chapter 6 discusses this in more detail, including how to weigh resource considerations in selecting an experimental design.

A fourth recommendation is that it may be prudent to take a conservative approach to selecting a fractional factorial design until theory has paid more attention to interactions between intervention components and more information is gathered about interactions in empirical research. From this perspective, half-fraction designs, which alias each effect of scientific interest with only one other effect, are a good choice for optimization trials. A conservative approach would also suggest that consideration should be given primarily to designs that are Resolution V or higher; Resolution IV designs should be approached with caution; and Resolution III designs, which alias main effects and two-way interactions, should be avoided.

5.13.2 Some Specific Recommendations

Here two fractional factorial designs are highlighted for consideration by readers of this book. Both of these are half-fraction designs that are Resolution V or higher. One is the 2_V^{5-1} design, an example of which is illustrated in Table 5.3. This design requires 16 experimental conditions, a number that has proven manageable in many situations. For example, Cook et al. (2016) successfully implemented a 16-condition factorial experiment in healthcare settings (this was a 2^4 design rather than a 2^{5-1}). The other recommended design is a 2_{VI}^{6-1}, which requires 32 experimental conditions. An example is described in Collins et al. (2011). Piper et al. (2016) successfully implemented this design in healthcare settings.

The recommendation here is to consider a 2_V^{5-1} design or a 2_{VI}^{6-1} design, not to use the exact design in one of the articles cited above. There are many fractional factorial designs, some of which may be better suited to a particular situation than others. For this reason, it is rarely appropriate to select a fractional factorial design by taking one described in the scientific literature, say in this book or the Collins et al. (2011) article, and simply inserting it into a new study. Instead, a much better and more productive approach is to use software like PROC FACTEX to search for a design. For example, suppose an investigator wishes to conduct an experiment with six factors. Based on this book, the investigator knows there is a 2_{VI}^{6-1} design. The investigator can use PROC FACTEX to search for the most appropriate 2_{VI}^{6-1} design for the situation at hand.

5.13.3 Resources for Further Study

This chapter has provided a brief introduction to fractional factorial designs and offered recommendations for some standard designs that are likely to be useful for conducting optimization trials. Because there are so many variations of fractional factorial designs, a comprehensive treatment is outside the scope of this book. In particular, there are some additional considerations pertinent to designs that bundle more than two effects (i.e., that are smaller than half-fraction). Readers are referred to the introductory treatment in Montgomery (2009) and the more advanced treatment in Wu and Hamada (2009).

5.14 Unbalanced Reduced Factorial Designs, Interactions, and Aliasing

In Chap. 1 it was stated that MOST does not require any particular approach to experimentation in the optimization phase, only that the approach selected is the best one according to the resource management principle, that is, it is the most efficient way to obtain the information needed for optimization. Factorial experiments have been emphasized in this book, because they tend to be the most efficient in most circumstances. This emphasis is not meant to rule out other possible approaches to examination of intervention components that have been used by intervention scientists (Collins et al., 2009; West & Aiken, 1997; West, Aiken, & Todd, 1993). These include conducting an individual experiment to investigate each component or conducting a multi-arm comparative experiment (MACE; Collins, Dziak, Kugler, & Trail, 2014) as discussed below.

Neither the individual experiments nor MACE approaches enable investigation of interactions between components. For an investigator who may see interactions as more of a nuisance than a revelation, the idea of avoiding factorial experiments and instead selecting an experimental design that does not involve interactions may be appealing, at least on the surface. However, although there may be good reasons for considering an approach other than a standard factorial or fractional factorial design for the optimization phase of MOST (some of which are discussed in Chaps. 6 and 8), avoidance of interactions is not one of them.

As will be shown in this section, an investigator can run from interactions but cannot hide. There are experimental designs that do not permit estimation of interaction terms, but there are no experimental designs that make interactions vanish. If interactions are present and sizeable, they will have an impact on results. As explained in this section, in experimental designs that permit examination of multiple components but not examination of interactions, the interactions become aliased with other effects (Collins et al., 2009).

5.14.1 Familiar Experimental Designs as Unbalanced Reduced Factorial Designs

As has been pointed out by West et al. (1993), West and Aiken (1997), and Collins et al. (2009), some familiar experimental designs, such as individual experiments and the MACE, are made up of a subset of the conditions from a factorial experiment and therefore can be considered reduced factorial designs.

Suppose Dr. B is interested in examining the three candidate intervention components *MI*, *PEER*, and *TEXT*, but, instead of conducting a factorial experiment like the one shown in Table 5.4, decides to conduct three individual treatment/control experiments. The first experiment would compare *MI* to a control; the second experiment would compare *PEER* to a control; and the third would compare *TEXT* to a control. Then the first experiment consists of Conditions 1 and 5 from Table 5.4; Condition 1 is the control and Condition 5, in which subjects are provided with motivational interviewing only, is the treatment condition. Similarly, the second experiment would consist of Conditions 1 and 3 from Table 5.4, and the third experiment would consist of Conditions 1 and 2. This is summarized in Table 5.14.

5.14 Unbalanced Reduced Factorial Designs, Interactions, and Aliasing

Table 5.14 Variations on the RCT used to examine individual intervention components

Experimental condition from Table 5.2	MI	PEER	TEXT
	Three individual experiments *MI* vs control		
1	No[a]	No	No
5	Yes	No	No
	PEER vs control		
1	No	No	No
3	No	Yes	No
	TEXT vs control		
1	No	No	No
2	No	No	Yes
	One-at-a-time comparative experiment		
1	No	No	No
2	No	No	Yes
3	No	Yes	No
5	Yes	No	No
	Constructive experiment		
1	No	No	No
5	Yes	No	No
7	Yes	Yes	No
8	Yes	Yes	Yes

[a]No means not included in intervention; yes means included in intervention

Now suppose instead of these three individual experiments, Dr. B conducts a MACE. MACEs are a type of multi-arm RCT; they are sometimes called a single-factor experiment, where the factor has several levels. One variation on the MACE was called the comparative treatment strategy by West et al. (1993) and West and Aiken (1997); here this design will be called the one-at-a-time comparative experiment to distinguish it from other varieties of MACEs. The purpose of this experiment is to examine the performance of each candidate component by enabling it to be compared to a control group or, in some cases, directly to one or more other components. In the one-at-a-time comparative experiment, there is a treatment condition corresponding to each component, plus a single control condition. In each of the treatment conditions, a single component is set to yes or high, and all the others are set to no or low. If a one-at-a-time comparative experiment was used to examine the performance of *MI*, *PEER*, and *TEXT*, it would involve experimental Conditions 1, 2, 3, and 5 from Table 5.4.

The one-at-a-time comparative experiment is illustrated in Table 5.14, along with another variation on the MACE, the constructive experiment (West et al., 1993; West & Aiken, 1997). Here the purpose is not to examine the performance of each individual component, but rather to compare the performance of several different versions of an intervention. In a constructive experiment the components have been ranked by importance. In other words, it has been determined that if only one component were to be included, it would be component *X*; if one more component was to be added, it would be component *Y*; and so on until all the components are added. In the example, suppose it has been determined that motivational interviewing is the most important component, followed by peer mentoring, followed by text messaging. Then the following experimental conditions would be included in the design: *MI* set to yes, *PEER* and *TEXT* set to no; *MI* and *PEER* set to yes, *TEXT* set to no; and *MI*, *PEER*, and *TEXT* set to yes. Often a control condition is included, which in this case would consist of *MI*, *PEER*, and *TEXT* set to no, or a baseline condition representing a standard of care which is to be augmented by the components being examined.

There are other variations on the MACE, such as one called the dismantling strategy, which is similar to the constructive research strategy except components are taken away one by one rather than added. These variations have the same statistical properties as the one-at-a-time and constructive experiments, so to avoid redundancy, they will not be discussed further.

Because fractional factorial designs, individual experiments, and MACEs are all subsets of complete factorial designs, they can all be considered reduced factorial designs. However, individual experiments and MACEs are generally not considered fractional factorial designs. Preserving the balance property of the factorial design is a hallmark of fractional factorial designs, whereas individual experiments and MACEs do not preserve this balance property. In other words, individual experiments and MACEs are unbalanced reduced factorial designs. This lack of balance is why their sample size requirements usually exceed those of complete and fractional factorial designs, as will be discussed in Chap. 6.

5.14.2 Aliasing in Individual Experiments and MACEs

Above it was discussed how fractional factorial designs are subsets of complete factorial designs, and it was shown that effects are aliased in fractional factorial designs. If individual experiments and MACEs are subsets of complete factorial experiments, then it makes sense to examine whether there is aliasing in these kinds of designs as well. This can be accomplished using the same approach that was used to examine aliasing in fractional factorial designs—by looking at the effect codes from the factorial experiment.

For convenience, Table 5.15a repeats the effect codes for the 2^3 factorial experiment, and Table 5.15b, c show the effect codes for the corresponding set of three individual experiments and the corresponding one-at-a-time comparative experiment, respectively.

Let us begin by examining the effect codes for the individual experiment to examine the performance of *MI*, shown in Table 5.15b. This experiment includes

5.14 Unbalanced Reduced Factorial Designs, Interactions, and Aliasing 185

Table 5.15 Comparison of effect codes for several types of experiments

(a) Effect Codes for 2^3 factorial ANOVA

Experimental condition	Intercept	Main effects			Interactions			
		MI X_1	PEER X_2	TEXT X_3	MI × PEER X_4	MI × TEXT X_5	PEER × TEXT X_6	MI × PEER × TEXT X_7
1	+1	−1	−1	−1	+1	+1	+1	−1
2	+1	−1	−1	+1	+1	−1	−1	+1
3	+1	−1	+1	−1	−1	+1	−1	+1
4	+1	−1	+1	+1	−1	−1	+1	−1
5	+1	+1	−1	−1	−1	−1	+1	+1
6	+1	+1	−1	+1	−1	+1	−1	−1
7	+1	+1	+1	−1	+1	−1	−1	−1
8	+1	+1	+1	+1	+1	+1	+1	+1

(b) Effect codes for individual experiments
MI vs control
Main effect of *MI* is aliased[a] with *MI* × *PEER*, *MI* × *TEXT*, and *MI* × *PEER* × *TEXT* interactions

| 1 | +1 | −1 | −1 | −1 | +1 | +1 | +1 | −1 |
| 5 | +1 | +1 | −1 | −1 | −1 | −1 | +1 | +1 |

PEER vs control
Main effect of *PEER* is aliased with *MI* × *PEER*, *PEER* × *TEXT*, and *MI* × *PEER* × *TEXT* interactions

| 1 | +1 | −1 | −1 | −1 | +1 | +1 | +1 | −1 |
| 3 | +1 | −1 | +1 | −1 | −1 | +1 | −1 | +1 |

TEXT vs control
Main effect of *TEXT* is aliased with *MI* × *TEXT*, *PEER* × *TEXT*, and *MI* × *PEER* × *TEXT* interactions

| 1 | +1 | −1 | −1 | −1 | +1 | +1 | +1 | −1 |
| 2 | +1 | −1 | −1 | +1 | +1 | −1 | −1 | +1 |

(c) Effect codes for one-at-a-time multi-arm comparative experiment (MACE)
Aliasing is the same as for individual experiments
Example: *TEXT* vs control

1	+1	−1	−1	−1	+1	+1	+1	−1
2	+1	−1	−1	+1	+1	−1	−1	+1
3	+1	−1	+1	−1	−1	+1	−1	+1
5	+1	+1	−1	−1	−1	−1	+1	+1

[a]Aliased means code vectors are identical, or identical if one vector is multiplied by −1. Aliasing is indicated by shading

Conditions 1 and 5 from the complete factorial, so the effect of *MI* will be estimated by comparing the means of these two conditions. As Table 5.15b shows, the effect codes are identical (or identical if one vector is multiplied by −1) for the main effect of *MI* and the *MI* × *PEER*, *MI* × *TEXT*, and *MI* × *PEER* × *TEXT* interaction. The effect codes are shaded in the table to make this easier to see.

Thus, the result of the comparison of *MI* to control in an individual experiment is not the main effect of *MI* that would have been obtained in a factorial experiment involving *MI*, *PEER*, and *TEXT*. Instead, it is a combination that includes the main effect of *MI* along with the *MI* × *PEER*, *MI* × *TEXT*, and *MI* × *PEER* × *TEXT* interactions. In other words, in this design the main effect of *MI* is aliased with all of the interactions that involve *MI*. Table 5.15 also shows that the main effect of *PEER* is aliased with the *MI* × *PEER*, *PEER* × *TEXT*, and *MI* × *PEER* × *TEXT* interactions, and the main effect of *TEXT* is aliased with the *MI* × *TEXT*, *PEER* × *TEXT*, and *MI* × *PEER* × *TEXT* interactions.

Table 5.15c shows the conditions that make up the one-at-a-time comparative experiment. The conditions in this experiment are exactly the same as the conditions in the individual experiments approach. The only difference is that instead of repeating the control group three times, as is necessary in the individual experiments approach, the one-at-a-time comparative design involves a single control group. The one-at-a-time comparative experiment involves the same subset of experimental conditions from the complete factorial as are involved in the individual experiments approach, and therefore the aliasing in these two approaches is exactly the same. For example, Table 5.15b shows that the means of Conditions 1 and 2 are compared to estimate the effect of *TEXT* in both the individual experiment and the one-at-a-time comparative experiment. The shaded effect codes are the same for the effect of *TEXT* in both the individual experiment and the one-at-a-time comparative experiment; thus, the aliasing is also the same.

The constructive experiment involves a different subset of conditions from the factorial experiment. These conditions are shown in Table 5.16. Interestingly, even though the experimental conditions are different, the aliasing is essentially the same for the constructive experiment as it is for individual experiments and the one-at-a-time comparative experiment. For example, consider the effect of *TEXT*, which in the design in Table 5.16 would be obtained by comparing Conditions 7 and 8. The shading in the table points out that the vectors for *TEXT*, *MI* × *TEXT*, *PEER* × *TEXT*, and *MI* × *PEER* × *TEXT* are identical, so these effects are aliased.

It is instructive to categorize unbalanced reduced factorial designs in terms of resolution and bundling using the same criteria that are applied to fractional factorial designs. Because they alias main effects with two-way interactions, individual

Table 5.16 Effect codes for constructive multi-arm comparative experiment (MACE)[a]

Experimental condition	Intercept	Main effects			Interactions			
		MI	*PEER*	*TEXT*	*MI* × *PEER*	*MI* × *TEXT*	*PEER* × *TEXT*	*MI* × *PEER* × *TEXT*
		X_1	X_2	X_3	X_4	X_5	X_6	X_7
1	+1	−1	−1	−1	+1	+1	+1	−1
5	+1	+1	−1	−1	−1	−1	+1	+1
7	+1	+1	+1	−1	+1	−1	−1	−1
8	+1	+1	+1	+1	+1	+1	+1	+1

[a]Shading illustrates that the main effect of *TEXT* is aliased with *MI* × *TEXT*, *PEER* × *TEXT* and *MI* × *PEER* × *TEXT*. Aliased means code vectors are identical, or identical if one vector is multiplied by −1

experiments and MACEs would be classified as Resolution III. In individual experiments and MACEs, the size of the bundle depends on the number of components being examined; adding components will rapidly increase the number of effects in each bundle. For example, in a MACE examining three components, there are four effects in the bundle corresponding to the treatment-control comparison; in a MACE examining four components, there are eight effects in this bundle; in a MACE examining five components, there are 16 effects.

5.14.3 What Is the Effect of a Factor?

The present section focuses on estimation of the effect of a factor rather than how a factor may affect other factors or be affected by them. Let us return to one of the fundamental questions with which this chapter started: Does peer mentoring have a detectable effect on *adhere*? To address this research question, Dr. B will establish a factor, *PEER*, with two levels, yes and no, and manipulate this factor experimentally to estimate its effect. Several ways to manipulate *PEER* and the other factors experimentally have been reviewed. Dr. B can conduct a complete factorial experiment, or a balanced reduced factorial experiment (i.e., a fractional factorial experiment), or individual experiments, or an unbalanced reduced factorial experiment such as a MACE. There are subtle differences among these approaches in how the main effect of *PEER* is defined.

The objective of the complete factorial experiment is to yield an estimate of the classic main effect as defined in most statistics textbooks and in Chap. 3 in this book. The fractional factorial experiment has the same objective, but it requires assuming that one or more specific interactions are negligible in size to provide this estimate. This requirement is explicit, and any investigator who has the proper background to consider using a fractional factorial design is well aware of it.

The characterization of the objective of individual experiments or the MACE, that is, what is the intended effect estimate provided by these designs, is less straightforward. Here the focus is the MACE; everything said generalizes to individual experiments. Above it was said that in a MACE, the main effect of *PEER* is aliased with the *MI* × *PEER*, *PEER* × *TEXT*, and *MI* × *PEER* × *TEXT* interactions. Thus if these interactions can be assumed to be negligible in size, the MACE estimate is equivalent to the classical main effect. However, it is not typically considered necessary to make these assumptions to conduct a MACE. This implies that the effect of interest is in fact not the main effect of *PEER*, as it is in complete and fractional factorial designs, but the main effect of *PEER* combined with all the interactions involving *PEER*, whether they be negligible or sizeable.

A critical aspect of the classical definition of a main effect is that a main effect is averaged across all of the combinations of levels of all of the other factors in the experiment. By contrast, in a MACE, the effect of *PEER* is its effect at a single specific combination of levels of the other factors. This specific combination of levels varies depending on which variety of MACE design is used. For example, in a one-at-a-time comparative MACE (Table 5.14), the effect of *PEER* is defined as its effect when all of the other factors are set to the lower levels. In a constructive MACE (also shown in Table 5.14), the effect of *PEER* is defined as its effect when *MI* is set to the higher level and *TEXT* is set to the lower level.

This distinction is important for two reasons. First, as explained above, the effect of *PEER* as estimated in a MACE is not a main effect by the classical definition, although it may be of scientific interest. Instead, it is a conditional effect (also called a simple effect), that is, an effect conditioned on a particular combination of the levels of *MI* and *TEXT*. Second, main effects and conditional effects differ in how generalizable they are expected to be. As Fisher wrote:

> ...any conclusion...has a wider inductive basis when inferred from an experiment in which the quantities of other ingredients have been varied, than it would have from any amount of experimentation, in which these had been kept strictly constant.... (Fisher, 1971, p. 102)

In other words, a main effect has been shown to hold on average across a variety of combinations of levels of the other factors and therefore is, in a sense, more robust than a conditional effect based on a single specific combination of levels of the other factors.

This wider inductive basis can be important in optimization of interventions. One reason is that according to the continual optimization principle of MOST (see Chap. 1), the composition of any intervention is likely to change in the future as components are added or replaced to improve it. Thus as the intervention evolves, any given component is likely to be accompanied by different assortments of other components. Another reason that components with robust effects may be desirable is the reality of unplanned variability in the set of components making up an intervention. Such unplanned variability occurs when an individual does not participate in every component of an intervention. This may happen because ad hoc modifications removing some components were made, the individual chose not to participate in some components, an error was made, or some other reason. A robust component effect may be less likely to diminish under these changing circumstances.

Suppose a one-at-a-time MACE was used to establish that *PEER* has a conditional main effect when *MI* and *TEXT* are at the lower level. This says nothing about whether *PEER* will have an effect if either or both of the other factors are set to the higher level. Estimates of interactions would shed some light on this, but MACEs do not enable estimates of interactions.

This discussion is not intended to discourage the use of MACEs, which can be very efficient under certain circumstances (see Chap. 6). It is merely intended to make the point that although the objective of complete and fractional factorial experiments, individual experiments, and MACEs is to assess whether candidate components have a detectable effect, factorial designs on the one hand and individual experiments and MACEs on the other use subtly different definitions of the term "effect." Therefore, they address subtly different research questions.

5.15 A Few Points to Remember About Fractional Factorial Designs

This chapter has discussed reduced factorial designs and introduced fractional factorial designs, which are a special case of the factorial design. Fractional factorial designs involve a subset of the experimental conditions in a complete factorial. Here are some important points to remember about reduced factorial designs, particularly fractional factorial designs.

1. *Fractional factorial experiments are powered exactly the same as complete factorial experiments.* As compared to a complete factorial design with the same number of factors, a fractional factorial design has fewer experimental conditions. However, it requires exactly the same overall number of subjects (N) to achieve the same level of statistical power. This N is divided among fewer experimental conditions in a fractional factorial, so the per-condition n is larger than the per-condition n in a corresponding complete factorial that is identically powered.
2. *Fractional factorial designs require fewer experimental conditions than complete factorial designs, but they still require a lot more conditions than an RCT or a MACE.* Fractional factorial designs offer the most economy and tend to have the most desirable properties for applications in the behavioral, biobehavioral, and biomedical sciences when there are 5 or more factors. This means that a fractional factorial experiment typically will have at least 16 experimental conditions.
3. *The experimental conditions in a fractional factorial design are selected for inclusion or exclusion so as to preserve certain important statistical properties, NOT on a conceptual basis.* The investigator does not examine the list of experimental conditions, decide which ones are most interesting, and retain those in the fractional factorial design. Instead, the emphasis is on selecting a set of experimental conditions that enables the best estimation of important effects, given the resource constraints. The appropriate use of software makes it straightforward to select the experimental conditions to include in a fractional factorial design.
4. *All fractional factorial designs involve aliasing, or combining, of effects.* It is known which effects are aliased in every fractional factorial design, so the investigator can select a design strategically. The strategy is to select a design in which effects of scientific importance, typically main effects and lower-order interactions, are aliased with higher-order interactions that can be assumed to be negligible.
5. *All reduced factorial designs involve aliasing of effects.* Individual experiments and MACEs can be considered reduced factorial designs and therefore can be viewed as subsets of a corresponding complete factorial. However, they are not balanced. Like fractional factorial designs, unbalanced reduced factorial designs involve aliasing. In these designs interactions are aliased with each other and with main effects. Thus the use of approaches such as individual experiments and MACEs does not eliminate the need to consider interactions, merely the possibility of examining them.

5.16 What's Next

Chapters 3, 4, and 5 have reviewed complete factorial designs, balanced reduced (i.e., fractional) factorial designs, and unbalanced reduced factorial designs. Any of these designs may be used in the optimization phase of MOST. The investigator must select a design based on the resource management principle; in other words, the investigator must select the design that makes the best use of available resources. Chapter 6 discusses how to compare the efficiency and scientific yield of different experimental designs that may be under consideration for use in the optimization phase of MOST. Chapter 6 also reviews some practical considerations faced by investigators implementing factorial experiments.

References

Box, G. E. P., & Meyer, R. D. (1986). An analysis for unreplicated fractional factorials. *Technometrics, 28*, 11–18.

Chakraborty, B., Collins, L. M., Strecher, V., & Murphy, S. A. (2009). Developing multicomponent interventions using fractional factorial designs. *Statistics in Medicine, 28*, 2687–2708.

Collins, L. M., Baker, T. B., Mermelstein, R. J., Piper, M. E., Jorenby, D. E., Smith, S. S., . . . Fiore, M. C. (2011). The multiphase optimization strategy for engineering effective tobacco use interventions. *Annals of Behavioral Medicine, 41*, 208–226.

Collins, L. M., Dziak, J. R., & Li, R. (2009). Design of experiments with multiple independent variables: A resource management perspective on complete and reduced factorial designs. *Psychological Methods, 14*, 202–224.

Collins, L. M., Dziak, J. J., Kugler, K. C., & Trail, J. B. (2014). Factorial experiments: Efficient tools for evaluation of intervention components. *American Journal of Preventive Medicine, 47*, 498–504.

Collins, L. M., & Kugler, K. C. (Eds.) (2018). *Optimization of multicomponent behavioral, biobehavioral, and biomedical interventions: Advanced topics*. New York, NY: Springer.

Cook, J. W., Collins, L. M., Fiore, M. C., Smith, S. S., Fraser, D., Bolt, D. M., . . . Mermelstein, R. (2016). Comparative effectiveness of motivation phase intervention components for use with smokers unwilling to quit: A factorial screening experiment. *Addiction, 111*, 117–128.

Dziak, J. D., Nahum-Shani, I., & Collins, L. M. (2012). Multilevel factorial experiments for developing behavioral interventions: Power, sample size, and resource considerations. *Psychological Methods, 17*, 153–175.

Fisher, R. A. (1971). *The design of experiments*. New York, NY: Hafner Publishing.

Montgomery, D. C. (2009). *Design and analysis of experiments* (7th ed.). Hoboken, NJ: Wiley.

Nahum-Shani, I., & Dziak, J. J. (2018). Multilevel factorial designs in intervention development. In L. M. Collins & K. C. Kugler (Eds.), *Optimization of multicomponent behavioral, biobehavioral, and biomedical interventions: Advanced topics* (forthcoming). New York, NY: Springer.

Nahum-Shani, I., Dziak, J. J., & Collins, L. M. (2017). Multi-level factorial designs with experimentation-induced clustering. *Psychological Methods*. Advance online publication. https://doi.org/10.1037/met0000128

Piper, M. E., Fiore, M. C., Smith, S. S., Fraser, D., Bolt, D. M., Collins, L. M., . . . Baker, T. B. (2016). Identifying effective intervention components for smoking cessation: A factorial screening experiment. *Addiction, 111*, 129–141.

West, S. G., & Aiken, L. S. (1997). Toward understanding individual effects in multicomponent prevention programs: Design and analysis strategies. In K. J. Bryant, M. Windle, & S. G. West

(Eds.), *The science of prevention: Methodological advances from alcohol and substance abuse research* (pp. 167–209). Washington, DC: American Psychological Association.

West, S. G., Aiken, L. S., & Todd, M. (1993). Probing the effects of individual components in multiple component prevention programs. *American Journal of Community Psychology, 21*, 571–605.

Wu, C. F. J., & Chen, Y. (1992). A graph-aided method for planning two-level experiments when certain interactions are important. *Technometrics, 34*, 162–175.

Wu, C. F. J., & Hamada, M. S. (2009). *Experiments: Planning, analysis, and optimization* (2nd ed.). Hoboken, NJ: Wiley.

Chapter 6
Gathering Information for Decision-Making in the Optimization Phase: Resource Management and Practical Issues

Abstract It has been stated repeatedly in this book that the design of the optimization trial must be selected based on the resource management principle. This means that both the cost and scientific yield of a design must be evaluated and compared with that of other experimental designs under consideration. Every situation is different. In some situations, recruiting or retaining subjects may be very expensive, whereas in others, this is a relatively minor expense. Similarly, in some situations it is not difficult to implement a host of experimental conditions, whereas in others implementing a large number of conditions would be very costly. Experimental designs also differ with respect to the kind of resources they require. Some experimental designs require relatively few experimental conditions, but may require more subjects; others make very economical use of subjects, but require implementation of many experimental conditions. To add to the complexity, different experimental designs provide estimates of subtly different effects. This chapter attempts to provide the reader with the background needed to compare several different experimental designs that could be used in an optimization trial and to select the best one. Readers of this chapter should be familiar with the material in all previous chapters, particularly Chaps. 3 and 5.

Contents

6.1	Introduction	194
6.2	Selecting an Experimental Design Based on the Resource Management Principle	195
	6.2.1 Cost	196
	6.2.2 Scientific Yield	196
6.3	Number of Experimental Conditions and Number of Subjects Required by Individual Experiments, MACEs, Factorial Experiments, and Fractional Factorial Experiments	197
	6.3.1 Number of Experimental Conditions	197
	6.3.2 Sample Size	199
6.4	Costs of Conducting an Experiment	200
	6.4.1 Per-Subject Costs	201
	6.4.2 Per-Condition Overhead Costs	202
	6.4.3 Constant Overhead Costs	202
	6.4.4 Scenarios Illustrating Different Costs	203

© Springer International Publishing AG 2018
L. M. Collins, *Optimization of Behavioral, Biobehavioral, and Biomedical Interventions*, Statistics for Social and Behavioral Sciences,
https://doi.org/10.1007/978-3-319-72206-1_6

6.5	Identifying the Least Expensive Experimental Design When Exact Costs Are Unknown	207
6.6	Different Experiments Estimate Different Quantities: The Scientific Yield of an Experimental Design	209
	6.6.1 The Decision-Priority Perspective and the Resource Management Principle	211
6.7	Type of Outcome Variable	212
6.8	Conducting Random Assignment	213
	6.8.1 Simple Random Assignment	213
	6.8.2 Stratified Random Assignment	214
6.9	Avoiding and Dealing with Protocol Deviations	215
	6.9.1 Training and Supervising Staff	216
	6.9.2 Dealing with Unanticipated Disruptions to the Experimental Design	217
	6.9.3 The Robustness of Factorial Experiments	219
	6.9.4 The Aviation Approach to Experimentation	220
6.10	Avoiding Contamination Across Experimental Conditions	221
6.11	Registry of Optimization Trials	223
6.12	What's Next	224
References		224

6.1 Introduction

In the optimization phase of the multiphase optimization strategy (MOST), experimentation is necessary to obtain empirical information about the performance of the components that are under consideration for inclusion in an intervention. Selecting an experimental approach is one of the most important decisions made by an investigator using MOST. This decision determines the amount, kind, and quality of information the investigator will have available when selecting the components and component levels that make up the optimized intervention.

High-quality information is obtained by conducting a rigorous experiment. In MOST, any approach to experimentation is a possibility, so in a sense the toolbox is infinite; the only requirement is that the approach be selected based on the resource management principle. In other words, the investigator must seek the approach that makes the best use of available resources to yield the highest quality and most useful scientific information. This chapter shows how to compare the cost and scientific yield of several different experimental design approaches so that the resource management principle can be applied.

Which of the alternatives in the MOST toolbox will yield the most useful information depends on the type of information that is required, which, in turn, depends partly on the type of intervention that is to be optimized. The emphasis in this book is on fixed interventions. Optimization of fixed interventions usually requires approaches such as factorial designs, fractional factorial designs, individual experiments, and MACEs, which were reviewed in Chaps. 3, 4, and 5. This chapter is about comparing the cost and scientific yield of these approaches. Approaches that are primarily for optimization of time-varying adaptive interventions, such as the sequential, multiple assignment, randomized trial (SMART) and system identification, are not included in this chapter. More about adaptive interventions can be found in Chap. 8 of this book. In the companion volume (Collins & Kugler, 2018), Almirall, Nahum-Shani, Wang, and Kasari (2018) discuss SMARTs; and Rivera, Hekler, Savage, and Downs (2018) discuss system identification.

Some investigators might object to the kind of cost comparisons illustrated in this chapter, arguing that they are "apples to oranges" comparisons because an effect obtained in, for example, a factorial experiment has a different interpretation from an effect obtained in a multi-arm comparative experiment (MACE). This book takes the pragmatic position that to make an informed decision about the choice of experimental design for an optimization trial, an investigator may wish to know, given a fixed limit on available resources, which approach to experimentation will provide the most and highest-quality information. At times, an investigator may even adjust the scientific questions being posed to take advantage of the economy offered by a particular experimental design.

6.2 Selecting an Experimental Design Based on the Resource Management Principle

Let us return again to Dr. B, who, as first described in Chap. 1, wishes to develop an intervention to reduce viral load in HIV-positive individuals who are heavy drinkers. Dr. B has identified five intervention components (see conceptual model in Chap. 2) that are hypothesized to be critical in helping HIV-positive individuals to reduce their drinking and improve their adherence to antiretroviral therapy (ART). They are (a) motivational interviewing, to engage the participant in the process of examining his or her alcohol use and ART adherence; (b) peer mentoring, to provide contact with a positive role model; (c) text messaging, to provide access to a support system; (d) mindfulness meditation, to improve mental health; and (e) behavioral skills training, to improve behavioral skills for managing alcohol use and ART. Suppose there is currently no standard of care for the first four components, but physicians and HIV clinics routinely give HIV-positive individuals who drink heavily and have poor ART adherence a workbook to complete to help improve their behavioral skills. Dr. B is considering two levels for each of these components. Each of the first four components can be set either to no (not included in the intervention) or yes (included). The behavioral skills component can be at either a low level, consisting of the workbook only (i.e., standard of care), or a high level, consisting of the workbook plus training delivered by a behavioral skills counselor.

Dr. B is now in the optimization phase of MOST and would like to conduct an experiment to examine the performance of the five components described above. The question is, what kind of experiment? Should it be a complete or reduced factorial? If it is to be a reduced factorial, would it be best to conduct a separate experiment for each component? Would the best approach be a multi-arm comparative experiment (MACE)? Or should Dr. B conduct a fractional factorial experiment?

This book has stressed the necessity of following the resource management principle in MOST. This is critically important when selecting an experimental design for the optimization phase. To follow the resource management principle, the experimental design must be selected based on two criteria: cost and scientific yield. The objective is to balance these two criteria so as to make the best use of available resources.

6.2.1 Cost

This chapter will discuss two aspects of experimentation that help to determine the anticipated cost of implementing a particular experimental design. One aspect is the demands of the design in terms of the number of conditions the design involves and the number of subjects needed to maintain a desired level of statistical power. The other aspect is the anticipated cost of conditions and subjects. Different scenarios may involve different per-condition and per-subject costs, so that one design may be identified as the least costly in one scenario and a different design may be the least costly in another. This will be demonstrated later in this chapter.

6.2.2 Scientific Yield

The scientific yield of an experiment refers to the amount, quality, and type of information expected to be gained by conducting the experiment. The ideal experiment yields a lot of high-quality information of exactly the type that is needed. According to the resource management principle, it is important to select an experimental design that makes the best use of available resources to provide high-quality information of the type that is needed. It follows that it is important to avoid a mismatch between the research questions that have been identified as most important and the experimental design selected.

For example, an investigator may decide to obtain information about the effectiveness of individual intervention components by using the treatment package approach, that is, by assembling all the components into a package, conducting an RCT, and doing post hoc analyses to try to tease out component effects. A well-designed and well-implemented RCT can provide exactly the kind and quality of information that is needed to evaluate the effect of an intervention as a package, but it cannot provide very high-quality information about the effectiveness of individual intervention components. Post hoc analyses on data from an RCT rely on naturally occurring variation in participation in individual components, that is, on subject self-selection, rather than on random assignment to experimental conditions. Because any inferences about individual component effects would be weaker than inferences based on a randomized experiment, this is an example of a mismatch between the research questions and the experimental design selected.

As another example of a mismatch, consider an investigator who believes it is important to gather information on whether the presence or absence of one component has an impact on the performance of another. Suppose this investigator rigorously conducts a MACE. From this experiment the investigator may obtain a lot of high-quality information about the performance of a component at specific levels of the other components. However, as was discussed in Chap. 5, a MACE does not enable examination of interactions; instead, it aliases (combines) interactions with other effects. Because estimates of interactions are needed to address the research questions this investigator has designated as important, there would be a mismatch between the research questions and the selected experimental design.

In sum, different experimental designs estimate different effects. The type of effects each experimental design estimates, in other words, the scientific yield of various experimental designs, will be discussed further later in this chapter. First, cost is discussed.

6.3 Number of Experimental Conditions and Number of Subjects Required by Individual Experiments, MACEs, Factorial Experiments, and Fractional Factorial Experiments

6.3.1 Number of Experimental Conditions

Suppose Dr. B considers the following approaches: individual experiments on each component, a MACE, a complete factorial experiment, and a fractional factorial experiment. As discussed in Chap. 5, the experimental conditions in the individual experiments, MACE, and fractional factorial approaches are all subsets of the complete factorial design. These four designs differ in terms of the number of experimental conditions they require.

The individual experiments approach requires a two-condition experiment for each component. This is illustrated in Fig. 6.1. For Dr. B's study, a total of ten experimental conditions would be required. In general, if k components are to be examined using the individual experiments approach, and each component can take on two levels for the purposes of experimentation, then $2k$ experimental conditions will be required.

As was discussed in Chap. 5, there are several variations on the MACE, including the one-at-a-time comparative experiment, the constructive experiment, and the dismantling experiment. These all have identical statistical characteristics, so this chapter will discuss only one type of MACE, the one-at-a-time comparative experiment. As illustrated in Fig. 6.2, in this example, a MACE would have six conditions. One of these conditions is a control, in which all of the components are set to the

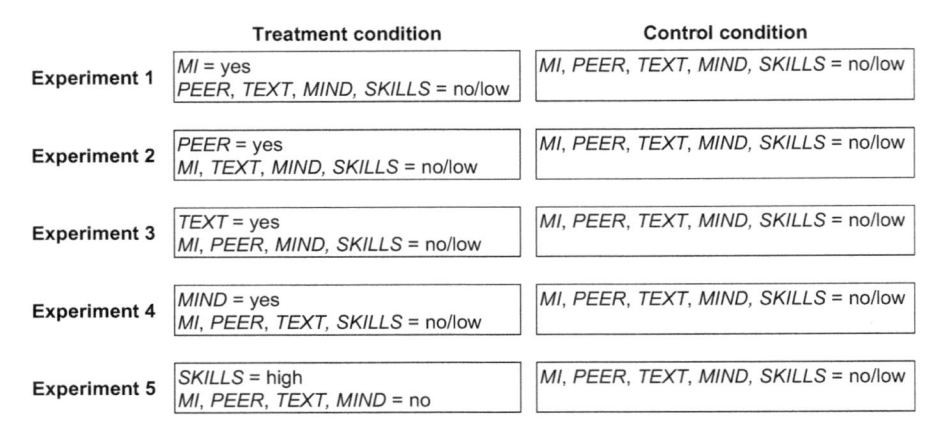

Fig. 6.1 Conditions in individual experiments approach

	Treatment condition	Control condition
Experiment 1	MI = yes PEER, TEXT, MIND, SKILLS = no/low	MI, PEER, TEXT, MIND, SKILLS = no/low
Experiment 2	PEER = yes MI, TEXT, MIND, SKILLS = no/low	MI, PEER, TEXT, MIND, SKILLS = no/low
Experiment 3	TEXT = yes MI, PEER, MIND, SKILLS = no/low	MI, PEER, TEXT, MIND, SKILLS = no/low
Experiment 4	MIND = yes MI, PEER, TEXT, SKILLS = no/low	MI, PEER, TEXT, MIND, SKILLS = no/low
Experiment 5	SKILLS = high MI, PEER, TEXT, MIND = no	MI, PEER, TEXT, MIND, SKILLS = no/low

Fig. 6.1 Conditions in individual experiments approach

Treatment conditions	MI = yes PEER, TEXT, MIND, SKILLS = no/low
	PEER = yes MI, TEXT, MIND, SKILLS = no/low
	TEXT = yes MI, PEER, MIND, SKILLS = no/low
	MIND = yes MI, PEER, TEXT, SKILLS = no/low
	SKILLS = high MI, PEER, TEXT, MIND = no
Control condition	MI, PEER, TEXT, MIND, SKILLS = no/low

Fig. 6.2 Conditions in a multi-arm comparative experiment (MACE)

lowest level. Each of the others is a treatment condition, in which one component is set to its higher level and the remaining components are set to their lower level. In general, if k components are to be examined in a MACE, $k + 1$ experimental conditions will be required.

Factorial experiments were discussed extensively in Chaps. 3 and 5. In a complete factorial experiment (Chap. 3) in which each factor has two levels, there are 2^k experimental conditions. In the example, there would be 2^5, or 32, conditions. A list of the conditions in the 2^5 design can be found in Table 5.2. In a balanced fractional factorial experiment, there would be 2^{k-1} conditions, that is, half the number of conditions in the corresponding complete factorial or fewer depending on the exact design selected. A list of the conditions in one possible fractional factorial design based on this example can be found in Table 5.3.

The number of experimental conditions in each of these approaches is summarized in Table 6.1.

Table 6.1 Comparison of number of experimental conditions and sample size requirements of different experimental designs that are subsets of a complete factorial with k 2-level factors

Design	Number of experimental conditions	Sample size requirement[a]
Individual experiments	$2k$	kN
Multi-arm comparative experiment (MACE)	$k + 1$	$(k + 1)N/2$
Complete factorial	2^k	N
Fractional factorial	2^{k-1} or fewer	N

[a] N = the sample size required to detect a regression weight of size β^* maintaining a desired level of power $1-\beta$ with a Type I error rate of α in a complete factorial design

6.3.2 Sample Size

It is possible to determine the absolute number of experimental conditions required by a particular design, as discussed above, because this number is always the same given a particular number k of components to examine, a particular number of levels to be examined in each component (in the example, two), and, if the design is a fractional factorial, the fraction. However, it is not possible to determine the absolute number of subjects required by a design, because in any experiment the required sample size is determined by a number of considerations: the desired level of power, the minimum effect size to be detected, and the alpha level to be used in hypothesis testing. (If there is a cluster structure in the data the design effect may be important also; see the Nahum-Shani and Dziak (2018) chapter in the companion volume. In the present chapter it is assumed there is no cluster structure.) However, it is possible to compare the relative sample size requirements by using one design as a baseline and examining the sample size requirements of the other designs in relation to this baseline.

The baseline for this comparison will be the complete factorial experiment. Suppose a particular α is selected for hypothesis testing, and power to detect a regression coefficient of size β^* (note that β^* refers to the standardized regression coefficient and β refers to the Type II error rate) is to be maintained at level $1-\beta$. For simplicity, it is assumed the effect size β^* is the same for all components. A power analysis demonstrates that in a complete factorial experiment, a sample size of N is required (see Chaps. 3 and 4 for how to power factorial experiments). Now let us compare the sample sizes that would be required to maintain this same desired level of power to detect an effect of the same size at the same α level if the other design alternatives are used (setting aside for the time being the important matter of what kind of effects each approach estimates).

This comparison is straightforward for fractional factorial designs. As was stated repeatedly in Chap. 5, fractional factorial designs have exactly the same sample size requirements as complete factorial experiments. Thus if N subjects are required to maintain a particular level of statistical power in a complete factorial design, N subjects are required by a balanced fractional factorial design comprised of a subset of conditions from this complete factorial.

The individual experiments approach involves conducting a separate experiment for each of the k factors. Consider the first of the k experiments. To detect a standardized regression weight of size β^*, a sample of size N is required. (This regression weight is, of course, conceptually not directly comparable to a regression weight that would be obtained based on data from a factorial experiment, for reasons discussed in Chap. 5. Nevertheless, it is instructive to make this comparison of resource requirements.) Now consider the second of the k experiments. Assuming the same expected standardized regression weight effect β^*, here too a sample of size N is required. Thus, across the k experiments a total sample size of kN is required.

The MACE involves k treatment-control comparisons. To maintain power at the desired level $1-\beta$, each of these comparisons must be based on a sample size of N. Consider the treatment-control comparison corresponding to the first of the k components. To ensure power at the desired level, it is necessary for the condition corresponding to the first component and the control condition each to have $N/2$ subjects, so that the comparison is made based on a total of N subjects. Similarly, to ensure power at level $1-\beta$ for each of the treatment-control comparisons, it is necessary for the corresponding treatment condition and the control condition each to have $N/2$ subjects. Thus, each of the k treatment conditions plus the single control condition must have $N/2$ subjects, for a total required sample size of $(k + 1)N/2$.

The sample size requirements for each design approach are summarized in Table 6.1.

6.4 Costs of Conducting an Experiment

It is helpful to divide the anticipated costs of conducting an experiment into three categories: per-subject costs, per-condition overhead costs, and constant overhead costs. Distinguishing among per-subject costs and per-condition overhead costs is important because, as was shown above, experimental design options vary in the resource demands they make in these two areas. Constant overhead costs do not enter into comparisons of the relative anticipated resource demands of various experimental designs, but a careful assessment of these costs can be important if the absolute resource demands of a design must be known for planning purposes. The three categories of costs are discussed in turn in the following sections.

As mentioned above, the resource management principle calls for considering both the cost of an experimental design and its scientific yield in selecting the one that makes the best use of available resources. In the next several sections, identification of the least expensive experimental design among several alternatives is discussed. It should be kept in mind that these designs vary in their scientific yield. This is discussed later in the chapter.

6.4.1 Per-Subject Costs

Per-subject costs are defined as the marginal cost of adding one subject to an experiment, *keeping the number of experimental conditions constant*. Per-subject costs can vary greatly across different commonly occurring scientific environments.

In some situations, subjects are free or nearly free of cost. Many universities maintain a human subjects pool, in which undergraduates participate in research as part of their course requirements. If undergraduates comprise a suitable population for the research being conducted, this is a highly economical alternative. In most other situations, there will be expenses associated with recruitment, screening, and retention of subjects.

Expenses associated with recruiting subjects may include advertising in media outlets or on public transportation (or this expense may belong in constant overhead costs, discussed below) and postage for mailing materials to potential subjects. In many intervention development studies, there is extensive screening of subjects, so that to obtain a single subject who is eligible to be randomized to an experimental condition, it is necessary to recruit and assess a number of potential candidates. For example, the list of exclusion criteria in Pellegrini, Hoffman, Collins, and Spring (2014) included numerous chronic medical conditions, the use of an assistive device for mobility, psychiatric hospitalization in the past 5 years, substance use dependence, and pregnancy, among others. Whenever there is a long list of exclusion criteria, many individuals who volunteer for a study will be ruled out. Thus a substantial amount of staff time may be required to weed out non-eligible volunteers and arrive at a single subject.

Costs may also be incurred for subject retention. Often subjects must be provided with an honorarium or compensated for expenses such as travel or childcare. Staff time may be required to ensure that the individual remains amenable to measurement until the study has been completed. Efforts may be devoted to keeping the subject engaged in the study, for example, by sending cards periodically to express the investigators' appreciation and keeping up to date on the individual's contact information. Such efforts are particularly required in longitudinal studies that follow subjects for months or years.

Unless all of the intervention components are computer-delivered, there are also likely to be expenses associated with delivering the components to each subject. As compared to a standard two-arm RCT, experiments that have more conditions can require correspondingly more resources to provide the level of training and supervision of staff needed to implement the experiment properly. This is discussed further in Sect. 6.9.

In some experiments, it is desirable to maintain a subject to staff ratio that falls within a particular range, so that adding, say, 20 subjects overall will require another 25% of a staff full-time effort (FTE) irrespective of how many experimental conditions comprise the experiment. This and similar expenses should be included in subject cost.

6.4.2 Per-Condition Overhead Costs

Per-condition overhead costs are defined as the marginal cost of adding one experimental condition to an experiment, *keeping the total number of subjects in the experiment constant*. This definition is the same even for designs such as MACEs, in which increasing the number of conditions would probably have to be accompanied by an increase in subjects to maintain power. Examples of expenses that may go in this category include salaries for additional staff hired solely because of the increased complexity of the design (e.g., to monitor adherence to study protocols and prevent protocol deviations), additional equipment or training manuals, and the cost of efforts devoted to training additional staff or to increasing the level of training of existing staff.

Careful thought is often needed to identify the costs that appropriately belong in the per-condition overhead category rather than in the subject cost category. For example, consider the cost of duplication of materials for subjects. Although with additional experimental conditions additional versions of these materials may be required, because the number of subjects is held constant, the total number of copies is the same. Thus any additional cost solely attributable to requiring more versions is likely to be minor. Now consider the cost of duplication of materials for staff. If each staff member needs a complete set of materials for all conditions, then the more conditions there are, the greater this expense is likely to be, irrespective of the number of subjects.

The overhead costs may vary across conditions. The costs associated with implementing a condition in which most or all of the factors are set to their lower levels may be less than those associated with implementing a condition in which most or all of the factors are set to their higher levels. Depending on how accurate an estimate of total costs is required, the investigator can either estimate an average per-condition overhead cost or compute the anticipated costs for each condition in a particular experimental design and sum them. Although the latter approach is more accurate, the former is probably all that is needed to compare the efficiency of a set of alternative designs under consideration. This is because, as will be demonstrated below, often the variability in resource demands among experimental designs is so large that fine-tuning the cost estimates would make little difference. If two designs appear very close in terms of resource demands based on an approximation, a follow-up computation using a more precise method can be used to decide between them.

6.4.3 Constant Overhead Costs

Constant overhead costs remain about the same irrespective of the number of subjects or conditions. For example, maybe 25% of the principal investigator's salary will be required no matter what experimental design is used. Another example of a cost that may not change in response to the choice of design is incentives given at the organizational level. For example, if an experiment is to be conducted in a large school district, a sum of money, say $10,000, may be set aside to reimburse the school district for administrative resources devoted to supporting the experiment.

If the purpose of computing anticipated costs is to compare the relative cost of several experimental designs that are under consideration to identify the least expensive one, it is not necessary to include constant overhead costs in the computations. The rank order of the designs will not change whether or not constant costs are included. In fact, as will be discussed below, all that is important for comparing the relative efficiency of various experimental designs is the ratio of per-condition overhead cost to per-condition subject cost, or vice-versa. However, if the overall costs are to be included in a report or in the budget of a grant proposal, then constant costs should be included. The computations in the present chapter do not include constant costs.

6.4.4 Scenarios Illustrating Different Costs

In this section two different research scenarios are presented. In each scenario a scientist is using MOST to develop and evaluate a multicomponent intervention and is in the process of deciding on what experimental design to use in the optimization phase.

In Scenario 1, a scientist is developing a multicomponent intervention to reduce anxiety in elderly individuals in assisted living. The subjects are all volunteers, and compensation is modest; each participant is provided with a $25 restaurant gift card. The program will be delivered by undergraduate human development majors, who will receive course credit for taking part in the research project. The investigator plans to hire hourly project manager staff to train and supervise the undergraduates, for the purpose of ensuring fidelity to the experimental procedures and preventing protocol deviations. Each condition increases the amount of training and monitoring of the undergraduates that is required of the project managers. It is estimated that each experimental condition requires additional project manager time, which will cost about $400.

In Scenario 2, a scientist is developing a multicomponent intervention to prevent excessive alcohol use in college students. The program is to be delivered at the very beginning of the fall of freshman year, and students will be asked to complete a pretest, an immediate posttest, and five survey occasions over the first 2 years of college. The program is to be delivered online, so once the experimental conditions and random assignment procedure are programmed, implementation of even a large number of experimental conditions does not require any additional management. Each experimental condition incurs an additional $25 in programming fees. The universities in which the experiment is to be conducted are requiring their students to participate in the intervention program, but of course cannot mandate participation in measurement, so it is important to incentivize the students to complete the surveys. Students are offered $50 in bookstore credit for completing each of the seven surveys, and a bonus of $50 for completing all seven surveys, for a total of $400 in bookstore credit.

Table 6.2 Comparison of hypothetical projected costs for various experimental designs, assuming per-condition overhead costs = $400, per-subject costs = $25

	Number of experimental conditions	N required	Condition costs	Subject costs	Total costs
Designs for examining four components					
Individual experiments	8	1408	$3,200	$35,200	$38,400
MACE	5	880	$2,000	$22,000	$22,000
Complete factorial	16	352[a]	$6,400	$8,800	$15,200
2_{IV}^{4-1} fractional factorial	8	352	$3,200	$8,800	$12,000
Designs for examining five components					
Individual experiments	10	1760	$4,000	$44,000	$48,000
MACE	6	1056	$2,400	$26,400	$28,800
Complete factorial	32	352	$12,800	$8,800	$21,600
2_V^{5-1} fractional factorial	16	352	$6,400	$8,800	$15,200
Designs for examining six components					
Individual experiments	12	2112	$4,800	$52,800	$57,600
MACE	7	1232	$2,800	$30,800	$33,600
Complete factorial	64	352	$25,600	$8,800	$34,400
2_{VI}^{6-1} fractional factorial	32	352	$12,800	$8,800	$21,600
2_{IV}^{6-2} fractional factorial	16	352	$6,400	$8,800	$15,200
Designs for examining seven components					
Individual experiments	14	2464	$5,600	$61,600	$67,200
MACE	8	1408	$3,200	$35,200	$38,400
Complete factorial	128	352	$51,200	$8,800	$60,000
2_{VII}^{7-1} fractional factorial	64	352	$25,600	$8,800	$34,400
2_{IV}^{7-2} fractional factorial	32	352	$12,800	$8,800	$21,600
2_{IV}^{7-3} fractional factorial	16	352	$6,400	$8,800	$15,200

[a]N required to maintain power ≥ 0.8 to detect a main effect corresponding to $\beta^* \geq 0.15$ at $\alpha = 0.05$

In Scenario 1 the per-subject costs are low in relation to the per-condition overhead costs, whereas in Scenario 2, the per-subject costs are high in relation to the per-condition overhead costs. Suppose in each scenario the investigator wishes to be able to detect a factor main effect corresponding to $\beta^* \geq 0.15$ with power ≥ 0.8 and $\alpha = 0.05$. In a factorial experiment, this would require $N = 352$. Using the factorial experiment as a baseline and the information provided in the brief description of Scenario 1, Table 6.2 compares the anticipated costs of individual experiments, MACEs, complete factorial, and fractional factorial designs to examine 4, 5, 6, and 7 components. For example, in Table 6.2, per-condition costs for the individual experiments approach = 8 × $400, and subject costs = 1408 × $25. Table 6.3 compares these same anticipated costs for Scenario 2.

Tables 6.2 and 6.3 illustrate several points.

First, even when used to examine the same number of components, different experimental designs can make very different resource demands and, as a consequence, can vary considerably in total costs. As an example, consider the designs for

6.4 Costs of Conducting an Experiment

Table 6.3 Comparison of hypothetical projected costs for various experimental designs, assuming per-condition overhead costs = $25, per-subject costs = $400

	Number of experimental conditions	N required	Condition costs	Subject costs	Total costs
Designs for examining four components					
Individual experiments	8	1408	$200	$563,200	$563,400
MACE	5	880	$125	$352,000	$352,125
Complete factorial	16	352[a]	$400	$140,800	$141,200
2_{IV}^{4-1} fractional factorial	8	352	$200	$140,800	$141,000
Designs for examining five components					
Individual experiments	10	1760	$250	$704,000	$704,250
MACE	6	1056	$150	$422,400	$422,550
Complete factorial	32	352	$800	$140,800	$141,600
2_V^{5-1} fractional factorial	16	352	$400	$140,800	$141,200
Designs for examining six components					
Individual experiments	12	2112	$300	$844,800	$845,100
MACE	7	1232	$175	$492,800	$492,975
Complete factorial	64	352	$1,600	$140,800	$142,400
2_{VI}^{6-1} fractional factorial	32	352	$800	$140,800	$141,600
2_{IV}^{6-2} fractional factorial	16	352	$400	$140,800	$141,200
Designs for examining seven components					
Individual experiments	14	2464	$350	$985,600	$985,950
MACE	8	1408	$200	$563,200	$563,400
Complete factorial	128	352	$3,200	$140,800	$144,000
2_{VII}^{7-1} fractional factorial	64	352	$1,600	$140,800	$142,400
2_{IV}^{7-2} fractional factorial	32	352	$800	$140,800	$141,600
2_{IV}^{7-3} fractional factorial	16	352	$400	$140,800	$141,200

[a]N required to maintain power ≥ 0.8 to detect a main effect corresponding to $\beta^* \geq 0.15$ at $\alpha = 0.05$

examining six components. In Scenario 1, the design options range from $15,200 to $57,600 in total costs; the cost of the most expensive design is nearly four times that of the least expensive design. In Scenario 2, the range is even wider, from $141,200 to $845,100. Here the cost of the most expensive approach is nearly six times that of the least expensive approach. This reinforces the importance of carefully following the resource management principle. Selecting an efficient design can make a huge difference in how much can be accomplished within a given budget, and can even make the difference between a study being feasible or infeasible.

Second, the individual experiments approach is usually the most expensive alternative and is always more expensive than a corresponding MACE. Tables 6.2 and 6.3 show that in both Scenarios 1 and 2, irrespective of whether the experiment were to examine 4, 5, 6, or 7 components, the individual experiments approach is always the most expensive. This is because although the individual experiments approach requires relatively few experimental conditions, it demands so many subjects that total subject costs tend to be high even when the per-subject cost is low.

MACEs always require fewer subjects and fewer experimental conditions than the corresponding individual experiments approach, so a MACE will always be less expensive than conducting individual experiments, no matter what the per-subject and per-condition overhead costs are. As Collins, Dziak, and Li, (2009) pointed out, in general conducting individual experiments is the least efficient way to collect information about the performance of intervention components.

Cost is not the only consideration. There can be a good scientific reason to conduct individual experiments. For example, it may be necessary to know the results of the first experiment before the second can be conducted, the second before the third can be conducted, and so on. But unless there is a similar compelling reason to conduct individual experiments, other approaches are usually strongly preferred.

Third, complete factorials are often less expensive than MACEs. In the four- and five-component examples in Table 6.2 and all of the examples in Table 6.3, the complete factorial design is cheaper than the MACE, even though it requires implementing many more experimental conditions. This is because of the dramatically lower sample size requirements of the factorial experiment as compared to the MACE. In the six-component example in Scenario 1, the MACE is more expensive than the complete factorial, but by only $1000. The only situation in Tables 6.2 and 6.3 in which the complete factorial experiment is greatly more expensive than the MACE is when there are seven components to be examined in Scenario 1. Here a complete factorial experiment would cost $60,000, as compared to a cost of $38,400 for a MACE. This is because a complete factorial with 7 factors has 128 conditions, as compared to 8 in a corresponding MACE, and in Scenario 1 the per-condition overhead cost is high in relation to the per-subject cost. In Scenario 2, in which the per-condition overhead cost is low in relation to the per-subject cost, even with 128 conditions the complete factorial is considerably less expensive than the MACE.

Fourth, fractional factorial designs are always cheaper than complete factorials and very often cheaper than MACEs. It is logical that a fractional factorial design will always be less expensive to implement than the corresponding complete factorial, because fractional factorials have exactly the same sample size requirements but involve half or fewer experimental conditions. Thus as compared to a complete factorial, a fractional factorial design is less expensive than a corresponding MACE in more situations. In both Tables 6.2 and 6.3, the fractional factorial designs are always less costly than not only the complete factorial but also the individual experiments and MACE approaches. This is true of even the half-fractions (the 2_V^{5-1}, 2_{VI}^{6-1}, and 2_{VII}^{7-1} designs), which, all else being equal, are the costliest of the fractional factorial designs in Tables 6.2 and 6.3, because they involve more experimental conditions.

Fractional factorial experiments will often be less expensive to implement than MACEs, but not always. This point will be discussed further below.

Fifth, information like that in Tables 6.2 and 6.3 *can be an important part of an argument justifying a design choice.* When preparing funding applications and journal articles, it can be helpful to include a smaller version of a table like Tables 6.2 and 6.3, including just the figures for the number of components under consideration in the proposed research. This can provide powerful evidence to support the argument for selection of a particular experimental design, especially if reviewer skepticism is expected concerning the utility and economy of factorial and fractional factorial designs.

6.5 Identifying the Least Expensive Experimental Design When Exact Costs Are Unknown

The anticipated costs displayed in Tables 6.2 and 6.3 can be used to rank the experimental designs from most to least expensive for a given number of components to be examined, assuming a factorial experiment would require $N = 352$. This rank order is determined not by the absolute costs per se, but by the ratio of per-condition overhead cost to per-subject cost (or, equivalently, the ratio of per-subject cost to per-condition overhead cost). In the remainder of this chapter, the ratio of per-condition overhead cost to per-subject cost will be called the cost ratio.

Above it was stated that in Scenario 1 the per-condition overhead cost is $400 and the per-subject cost is $25. Thus Scenario 1 represents a situation in which the cost ratio is $400/25 = 16$. This means that adding a single experimental condition would cost 16 times more than adding a single subject. Scenario 2 represents a situation in which the cost ratio is $25/400 = 0.0625$. This means that adding a single experimental condition would cost 0.0625 times what it would cost to add a single subject, or, put another way, adding a single subject would cost 16 times more than adding a single experimental condition.

Whenever the cost ratio is greater than one, adding a single experimental condition costs more than adding a single subject; whenever it is less than one, adding a single subject costs more than adding a single experimental condition. If the cost ratio is approximately one, the cost of adding a single experimental condition and the cost of adding a single subject are about the same. Given the variety of interventions and research settings, it would be possible to construct a plausible scenario for any of a wide range of cost ratios.

The results in Table 6.2 hold for any situation in which a factorial experiment requires $N = 352$ and the ratio of per-condition overhead costs to per-subject costs is 16. Similarly, the results in Table 6.3 hold for any situation in which a factorial experiment requires $N = 352$ and the ratio of per-condition overhead costs to per-subject costs is 0.0625. For example, consider Scenario 1. Suppose instead of $400 and $25, the per-condition overhead cost and per-subject cost were, say, $800 and $50. As compared to what is reported in Table 6.2, the costs would be double, but because the cost ratio would remain 16, the same designs would emerge as most and least expensive for a particular number of components. In other words, although it is necessary to know per-condition overhead cost and per-subject cost in order to determine the absolute amount of money required to implement an experiment, identifying the least costly design in relative terms requires only the cost ratio.

In fact, often even a rough estimate of the cost ratio, or a range within which the investigator is confident the cost ratio falls, is sufficient for decision-making purposes. For example, suppose an investigator wishes to examine five components and needs to know which will be less expensive: a MACE or a complete factorial design. Further suppose that as in Scenarios 1 and 2, a power analysis has indicated that a factorial experiment requires $N = 352$. The investigator is confident that the per-subject cost is $25 but is less confident about what the likely per-condition overhead cost will be.

Even with some uncertainty about the per-condition overhead cost, there may be a firm enough basis for establishing which of the two experimental designs is less costly. It can be shown by algebra that given that a factorial experiment requires $N = 352$, when five components are to be examined, a factorial design is less expensive than a MACE whenever the cost ratio is less than about 27, and a MACE is less expensive than a factorial design whenever the cost ratio is greater than about 27.

This was computed as follows: Let X represent the per-condition overhead cost. Suppose the per-subject cost is $1, so the cost ratio $= X/1 = X$.

The cost of a 2^5 factorial experiment is $32X + 352$.

The cost of a MACE, which requires 6 experimental conditions and 1056 subjects (see Table 6.1), is $6X + 1056$.

The value of X at which the costs are identical can be found by solving for X.

$$32X + 352 = 6X + 1056$$
$$X \approx 27$$

Thus a factorial experiment and a MACE cost about the same when the per-condition overhead cost is about 27 times the per-subject cost, or $27 \times \$25 = \$675$. If the investigator is reasonably confident that the per-condition overhead cost will be less than $675, then the complete factorial experiment will be less expensive than the MACE. If the investigator is reasonably confident that the per-condition overhead cost will be greater than $675, then the MACE will be the less expensive of the two designs.

Using the same logic, it can be shown that a half-fraction fractional factorial design will be less expensive than the MACE if the cost ratio is less than about 70. Thus if the per-subject cost is $25, a 2^{5-1} fractional factorial experiment will be less expensive than the corresponding MACE unless the per-condition overhead cost is about $1750 or greater.

Table 6.4 summarizes the minimum cost ratios that must be reached for a MACE to be less expensive than a complete or half-fraction factorial design for examination of 4, 5, 6, or 7 components when a factorial experiment requires $N = 352$. For example, the table shows that for a MACE to be less expensive than a complete factorial experiment to examine four components, the per-condition overhead cost must be at least 48 times the per-subject cost. Table 6.4 shows that the cost ratios are larger in the column for the half-fraction. This makes sense, because fractional factorial designs have fewer experimental conditions and therefore smaller total

Table 6.4 Approximate minimum ratio of per-condition overhead costs to subject costs required for MACE to be less expensive than alternative design when factorial design requires $N = 352$

Number of components	Alternative design Complete factorial	Half-fraction factorial
4	48	175
5	27	70
6	15	35
7	9	19

condition costs, all else being equal. Table 6.4 also shows that as the number of components increases, the cost ratio required for the MACE to be less expensive than the factorial and fractional factorial experiments becomes smaller. This is because as the number of components to be examined increases, the numbers of conditions in both the factorial and fractional experiments increase geometrically, while the number of conditions in the MACE increases linearly. Thus, all else being equal, as the number of components to be examined increases, total condition costs increase more rapidly for factorial and fractional factorial experiments than for MACEs.

6.6 Different Experiments Estimate Different Quantities: The Scientific Yield of an Experimental Design

In the optimization phase of MOST, the investigator typically wishes to assess the effectiveness of each of a set of components that are candidates for inclusion in an intervention. To gather the information needed to make this assessment, the investigator conducts an optimization trial. This chapter focuses primarily on three approaches to conducting an optimization trial: the factorial design, the fractional factorial design, and the MACE. As was discussed in Chap. 5, factorial and fractional factorial designs define the effect of a factor in the same way; the MACE defines it in a slightly different way. Therefore, whether an investigator selects a factorial or fractional factorial design on the one hand or a MACE on the other determines what is meant by the effectiveness of a candidate component.

Let us return to the example reviewed at the beginning of this chapter, involving five components that are candidates for inclusion in an intervention to increase ART adherence, and compare assessing the performance of one of these components, *MI*, within a 2^5 factorial experiment and the six-condition MACE depicted in Fig. 6.2. Suppose the data from the factorial experiment are analyzed using a factorial ANOVA with effect coding, and *MI* has a significant (however this is defined) main effect. This means that *MI* demonstrates an effect *on average* across all the possible combinations of the levels of *PEER*, *TEXT*, *MIND*, and *SKILLS*.

Now suppose instead of a factorial experiment, the MACE in Fig. 6.2 is conducted, the data are analyzed using one-way ANOVA, and the *MI* vs. control comparison is significant (again, however, this is defined). Here the presence of a statistically detectable effect means that there is an effect of *MI* at *one particular combination of levels of the remaining components*, in this case when *PEER*, *TEXT*, and *MIND* are all set to no and *SKILLS* is set to low. This is equivalent to a conditional effect or, as it is often called in the social and behavioral sciences, a simple effect.

Notice the difference between these two definitions of the performance of *MI*. The factorial experiment expresses performance as an average effect across all possible combinations of the levels of the other factors, whereas the MACE expresses performance as an effect at a single combination of levels of the other components. If a factorial experiment shows that there is a statistically detectable effect of *MI*, this does not imply that the effect of *MI* is present at any particular combination of levels of the factors. If a MACE shows that there is a statistically detectable effect of *MI*, this does not imply that the effect of *MI* averaged across all possible combinations of the levels of the remaining components is statistically detectable.

Which effect is more helpful in decision-making during the optimization phase of MOST? This depends on exactly what decision is to be made. Suppose the objective is to create a simple intervention by selecting only one of the five components to be set to the higher level. Thus the intervention could consist of either *SKILLS* set to high and *MI*, *PEER*, *TEXT*, and *MIND* all set to no, or any one of *MI*, *PEER*, *TEXT*, or *MIND* set to yes, and the remaining components set to no/low. In this case, the effect estimated using the MACE in Fig. 6.2 is exactly what is needed, because it evaluates the performance of each component with the others set to the lower level.

However, as has been discussed repeatedly in this book, the objective of experimentation in the optimization phase usually is not to select a single component, but rather to evaluate the performance of each of a set of candidate components in a highly efficient manner. It is implicit that none, one, some, or all of the components may be selected for the screened-in set, depending on their performance as demonstrated experimentally. It is also implicit that the final composition of the intervention may involve only a subset of the components in the screened-in set, with the final selection based on additional criteria such as cost or time (see Chap. 7 in this volume and the Dziak (2018) chapter in the companion volume). In other words, at the time the performance of, say, *MI* is being evaluated to determine whether or not it should be included in the screened-in set, it is unknown which other components and component levels will make up the final version of the intervention. Here the kind of information provided by a MACE—information about the performance of *MI* with the other components set to one particular configuration of levels—is of limited utility. It is more helpful to have a sense of the performance of the component more generally across many different configurations of components and levels.

As mentioned above, the presence of a main effect of, say, *MI* does not necessarily imply that *MI* has a detectable effect in each and every configuration of levels of the remaining factors. In a complete factorial experiment, each of these configurations is represented as an experimental condition, so it may be tempting to consider directly assessing *MI*'s performance with different configurations of levels of the remaining factors by conducting a series of comparisons of individual conditions to a "control" condition in which all the factors are set to the lower level. As discussed in Chap. 3, factorial experiments can be well-powered for detection of main effects and interactions even with per-condition ns in the single digits. However, few factorial experiments are powered for the purpose of comparing individual experimental conditions to each other. Even if resources were available to provide reasonable power for comparing the individual conditions in a factorial experiment, thereby turning the experiment into a giant MACE, it is questionable whether that would be the best use of those resources.

Complete and fractional factorial experiments are the only experiments discussed herein that can be used to estimate interactions. Examining interactions provides a sense of whether effects involving one or more factors vary depending on the level of one or more other factors. For example, if there is a substantial *MI* × *PEER* interaction, this means that the effect of *MI* varies depending on whether *PEER* is set to yes or no. Examining interactions can provide an excellent sense of whether the effect of a factor is likely to be robust across a variety of configurations of levels of the remaining factors. Thus although a factorial experiment does not enable direct comparison of individual conditions the way a MACE does, the ability to estimate interactions provides a good sense of the consistency of main effects. Chapter 4 covers interactions in detail, and Chap. 7 discusses the implications of interactions for deciding which components to select into the screened-in set.

6.6.1 The Decision-Priority Perspective and the Resource Management Principle

Chapter 3 discussed the difference between the conclusion-priority and decision-priority perspectives. From the conclusion-priority perspective, the top priority is to draw a scientific conclusion that will stand up when critically examined by other scientists. From the decision-priority perspective, the top priority is not to make scientific conclusions per se but to use rigorous scientific information to make the right decisions about which components and component levels are to be selected for inclusion in the screened-in set. The conclusion-priority perspective is more appropriate during the evaluation phase of MOST, and the decision-priority perspective is generally more appropriate during the optimization phase of MOST.

Taking the decision-priority perspective has important implications for selection of an experimental design using the resource management principle. From the conclusion-priority perspective, it may be sufficient to demonstrate that a variable has an effect, with little consideration of the context. By contrast, from the decision-priority perspective, it is critical to select an experimental design that provides the kind of information needed for selecting the components and component levels to be included in the screened-in set. Because the candidate components are ultimately to be combined into an intervention, usually this means that to make the best decisions, information must be obtained not only about the overall effect of each component but also about whether each component affects or is affected by other components. In other words, ideally information is needed about both main effects and interactions. However, implicit in the decision-priority perspective is the acknowledgement that decisions need to be based on the best information obtainable, even when that information is not perfect. Thus, although factorial experiments are frequently a good choice for optimization trials, there may be times when resources will be better managed with a different type of experiment. As was stated at the beginning of this chapter, the selection of the experimental design for the optimization phase is one of the most important decisions an investigator makes in MOST, and any experimental design is a possibility as long as it is selected based on the resource management principle.

6.7 Type of Outcome Variable

The outcome variable is selected during the preparation phase (see Chap. 2) of MOST, which, of course, is prior to the optimization phase. Everything discussed in this book is based on the assumption that the outcome variable is approximately normally distributed. It is possible to analyze the data from a factorial experiment or a MACE taking a generalized linear model (GLM) approach. In the example, the GLM approach would enable modeling, say, a simple binary outcome coded 1 if an individual is at least 85% adherent to the ART treatment regimen at the time of the posttest or 0 if the individual is not at least 85% adherent, or a time-to-event outcome such as the time to first lapse in adherence. This analysis is not difficult using standard statistical software. The X side of the equation would be the same as in the normal model. If data from a factorial experiment were being analyzed, the X side of the equation would include all of the vectors corresponding to the main effects and interactions, coded in the usual manner, along with any covariates. A GLM link function would be selected appropriate to the outcome variable. For example, logistic regression is another name for GLM using a logit link function. Ordinary linear regression is a special case of GLM.

Obtaining estimates of main effects and interactions via GLM analysis of non-normal outcome data from a factorial experiment is usually no more difficult than obtaining them using ordinary linear regression. However, it is critical to select an appropriate link function, and interpretation of the obtained results can be much less straightforward (e.g., Karaca-Mandic, Norton, & Dowd, 2012). The precise meaning of a main effect or, in particular, an interaction varies depending on which link function is used. Interpreting these quantities correctly and using them in making decisions about selection of components and component levels require extremely careful thought when the outcome variable is non-normal. This is an area where additional research and guidelines for scientists are badly needed.

The choice of outcome variable and link function also has an impact on statistical power. A power analysis based on one link function does not apply to another link function. In general, the sample size requirements when the outcome is non-normal tend to be greater than those required to maintain the same level of power in corresponding normal models. Therefore, to plan properly it is important to specify the appropriate distribution for the outcome variable and the corresponding link function when conducting a power analysis.

6.8 Conducting Random Assignment

Randomly assigning subjects to the conditions of a factorial experiment can be done following essentially the same procedures that are used to assign subjects in experiments with fewer experimental conditions. Of course, the sheer number of experimental conditions in a factorial experiment requires careful management of random assignment.

6.8.1 Simple Random Assignment

Suppose 352 subjects are to be randomly assigned to the conditions in a 2^4 factorial experiment, such as the one depicted in Table 6.5. As the table shows, a balanced design with 22 subjects in each of the 16 conditions is desired. The following procedure is straightforward and can help to achieve balance. Start by listing all the experimental conditions in the design and assigning a number to each condition, as shown in Table 6.5. Now consider random assignment of the first 16 subjects. To

Table 6.5 Experimental conditions and per-condition n in 2^4 factorial design with overall $N = 352$

Experimental condition	MI	PEER	TEXT	MIND	n
1	No[a]	No	No	No	22
2	No	No	No	Yes	22
3	No	No	Yes	No	22
4	No	No	Yes	Yes	22
5	No	Yes	No	No	22
6	No	Yes	No	Yes	22
7	No	Yes	Yes	No	22
8	No	Yes	Yes	Yes	22
9	Yes	No	No	No	22
10	Yes	No	No	Yes	22
11	Yes	No	Yes	No	22
12	Yes	No	Yes	Yes	22
13	Yes	Yes	No	No	22
14	Yes	Yes	No	Yes	22
15	Yes	Yes	Yes	No	22
16	Yes	Yes	Yes	Yes	22

[a]No means not included in intervention; yes means included in intervention.

proceed, it is necessary to sample random numbers without replacement from the uniform distribution of integers ranging from 1 to 16. Suppose this resulted in the following list of random numbers: 16, 12, 1, 11, 5, 6, 2, 10, 3, 9, 8, 7, 4, 13, 14, and 15. Then the first subject to be assigned would be placed in Condition 16, the second would be placed in Condition 12, the third would be placed in Condition 1, and so on. Once the first 16 subjects have been assigned to conditions, a new list of random numbers is used to assign the next 16 subjects to conditions in the same manner, and so on until all 352 subjects have been assigned. Assigning 352 subjects to 16 conditions in this manner will result in $n = 22$ per condition. This means that a total of 22 lists of random numbers is required. For convenience the lists can all be generated at once at the start of the assignment process.

6.8.2 Stratified Random Assignment

Sometimes it is desired to stratify random assignment to ensure that a certain characteristic is spread as evenly as possible across experimental conditions. For example, suppose the investigators conducting the experiment depicted in Table 6.5 wish to ensure that the percentage of males is roughly the same in all 16 experimental conditions. The percentage may not be 50; the investigators believe it is possible that fewer men than women will volunteer as subjects. Each experimental condition will be assigned roughly the same percentage of men if separate lists of random numbers are generated for men and women, and then the randomization is conducted separately. In other words, suppose the random numbers for women are in List W, and the random numbers for men are in List M. When a female subject arrives for randomization, the investigator would go to List W to find her experimental condition assignment; when a male subject arrives for randomization, the investigator would go to List M to find his. This stratified randomization procedure will ensure that the proportion of males and females in each experimental condition roughly reflects the corresponding proportions in the overall sample, whatever those proportions turn out to be.

It can be counterproductive to stratify on too many variables, particularly in complex factorial experiments involving several factors. It is a good idea to limit stratification of random assignment to one or, at the very most, two variables. Because factorial experiments are analyzed by combining experimental conditions, with means estimated based on several experimental conditions except in estimation of high-order interactions, some variability across individual conditions in characteristics such as gender is usually not a problem.

Now suppose the investigators wish to test some a priori hypotheses about gender as a moderator of the effects of several of the factors in the experiment. This will involve testing the significance of the interaction between gender and each factor gender is hypothesized to moderate (see Chap. 4). To maximize statistical power for these hypothesis tests, the investigators plan to recruit equal numbers of male and female subjects, and they wish to ensure that each condition is to be made up of half males and half females. The procedure for stratified randomization described above will accomplish this, provided that equal numbers of males and females are randomized.

6.9 Avoiding and Dealing with Protocol Deviations

A protocol deviation occurs when the experimenter errs by providing a subject a treatment other than what the assigned condition calls for. Sometimes this happens because the experimenter provides a component that was not supposed to be provided, or the experimenter may provide a higher level of a component when the assigned experimental condition called for a lower level. Sometimes protocol deviations occur because a component was withheld that the assigned condition called for, or a lower level of a component was provided when the condition called for a higher level.

It is critical to make sure that there are few—ideally, no—protocol deviations. There is a lot to keep track of when conducting an experiment in the field, particularly a factorial experiment involving several factors. In a factorial experiment, there are many more experimental conditions than there are in an RCT, and thus managing the experimental conditions can be a challenge. However, with careful planning and creative use of technology, it is feasible to implement many experimental conditions with few, or even no, protocol deviations. For example, Baker et al. (2016), Cook et al. (2016), Piper et al. (2016), and Schlam et al. (2016) describe a large study involving three factorial experiments implemented simultaneously in healthcare clinics. The three experiments, which involved a total of 80 experimental conditions, were implemented with very few protocol deviations. The investigators made extensive use of technology to manage the three experiments. For much more about the logistics of implementation, including recommendations for how to conduct a large factorial experiment in a field setting, see the Piper, Schlam, Fraser, Oguss, and Cook (2018) chapter in the companion volume.

Any protocol deviations should be carefully noted. If a protocol deviation is detected, it is important to review existing procedures with the staff conducting the experiment as soon as possible. It may be necessary to revise procedures with the objective of minimizing the probability that the deviation will be repeated.

Often a protocol deviation means that a subject assigned to one experimental condition, say Condition A, was in fact treated as if he or she was in a different condition in the experiment, say Condition B. It may be tempting to consider moving this subject after the fact from Condition A into Condition B. However, this should never be done. A subject must always be placed in the original randomly assigned experimental condition for data analysis purposes, even if he or she actually received a set of components and component levels that corresponds to a different experimental condition. If the investigator intervenes to change the subject's assigned condition, this is a nonrandom assignment; strictly speaking, the experiment could no longer be called completely randomized. It may be possible to conduct follow-up analyses to model the treatment that was in fact provided.

Every protocol deviation reduces the clarity with which it is possible to draw conclusions based on the results of an experiment. One or two protocol deviations probably do not constitute a disaster, but frequent protocol deviations can greatly distort results, typically by reducing observed effect sizes. Therefore, resources devoted to avoiding protocol deviations are typically well spent.

6.9.1 Training and Supervising Staff

Training and supervision resources must be sufficient to ensure staff have the necessary skills, background, and oversight to ensure that protocol deviations are kept to a minimum and also to maintain fidelity of delivery of each component. One determiner of how much training and oversight is required is how much the implementation of each level of each factor depends on humans. Any factors corresponding to components delivered electronically, such as the text messaging component in the example or the components described in the Kugler, Wyrick, Tanner, Milroy, Chambers, Ma, Guastaferro, and Collins (2018) chapter in the companion volume, are likely to require only a modest level of staff training resources. For these components, there will be up-front costs associated with programming the various experimental conditions. Once this programming is done, there is likely to be little need for training or supervision of staff, except in relation to how to answer questions and deal with technical issues. Because of the expense associated with training staff, both to conduct a factorial experiment and to implement the optimized intervention later when it goes to scale, investigators may wish to consider using technology in place of humans where it is reasonable to do so. In the example, perhaps mindfulness meditation training could be provided via a combination of video-recorded instruction and a mobile app.

Human-delivered components differ in how much staff training and supervision is required. If a component consists entirely of dispensing a pharmaceutical, such as the nicotine replacement components described in Piper et al., (2016), usually only a modest level of staff training and supervision is required, to ensure that subjects are provided with the correct type and dose of medication, along with adequate instructions. By contrast, components that involve counseling (e.g., Piper et al., 2016), motivational interviewing (e.g., Gwadz et al., 2017), coaching (e.g., Pellegrini et al., 2014; Pellegrini, Hoffman, Collins, & Spring, 2015), and the like may require a substantial amount of staff training to ensure fidelity of delivery.

As discussed above, protocol deviations occur when a subject is provided with the wrong treatment because of experimenter error. Whenever an experiment with multiple conditions is implemented, it is important to provide sufficient training to staff so that protocol deviations are avoided. It is also important to provide sufficient supervision and monitoring so that if a protocol deviation occurs, the cause can be determined, and future occurrences prevented.

Protocol deviations can occur when staff become confused about which experimental condition a particular subject is in or about what the factor levels are in a particular condition. If this happens, staff may deliver a component with perfect fidelity, but it is the wrong component for the assigned condition. Alternatively, a component that was supposed to be provided may be omitted. Training staff thoroughly, in addition to maintaining a level of implementation resources sufficient to ensure that staff are not rushed or overloaded, helps to prevent this. It is also prudent to prevent accidental protocol deviations by finding a way to maintain ongoing clarity about what treatment each subject is supposed to receive, so it becomes unlikely that staff will become confused. As mentioned above, the creative use of technology can be helpful; this is discussed in the Piper, Schlam, Fraser, Oguss, and Cook (2018) chapter in the companion volume.

Unfortunately, some protocol deviations may be deliberate. Staff may believe they have an ethical or moral obligation to withhold or, more likely, provide treatment counter to what is indicated by the assigned condition. Monitoring, such as audio or video recording of implementation or conducting spot checks, can help to prevent this. It is also a good idea to explain equipoise thoroughly to staff. In particular, it is important to convey that equipoise is not about how the investigators or staff feel about the likely effectiveness of a component, but rather about whether or not a convincing body of empirical scientific evidence has been accumulated concerning a component's effectiveness. It can be a delicate balance to convince staff that the components being examined are all exciting but that it is nevertheless ethical to withhold them because there is not yet enough scientific evidence to support making them the standard of care.

In some experiments, it may be possible to have staff specialize so that each individual implements a single component rather than entire conditions. In the example, this would mean that one staff member would do all the motivational interviewing, another would do all the behavioral skills training, and so on. This approach can help prevent protocol deviations, because it reduces both complexity and the temptation to add in a component that has not been assigned; each staff member needs to know only whether or not a subject is to receive one particular component, or the assigned level of that component, and has no responsibility for any other components. However, if only one or two staff members implement a particular component, the results may not generalize to other delivery staff.

6.9.2 Dealing with Unanticipated Disruptions to the Experimental Design

Disruptions to the design of a factorial optimization trial are serious and should be prevented whenever possible. However, if such disruptions do occur, they need not necessarily sink the experiment. In this section two examples of factorial experiments that suffered serious disruptions are discussed, along with the measures taken by the investigators to salvage the study.

In the first example, Strecher and colleagues (2008) wanted to examine six components that were candidates for inclusion in an online smoking cessation intervention. The original intended experimental design was a 2^{6-1} fractional factorial. This design has 32 experimental conditions. However, as Strecher et al. reported, "due to a programming error, 1 of the 32 arms was not filled." In other words, after the experiment was concluded, it was discovered that there was an experimental condition to which no subject had been assigned.

This was an unwelcome but, ultimately, not devastating development. Strecher et al. dealt with the empty cell by collapsing over one factor in statistical analysis of the data, essentially changing the original 2^{6-1} to a 2^{5-1} design with 16 experimental conditions. This new design did not have any empty cells. It was not completely balanced because one cell had approximately only half as many subjects as the other cells, but this was manageable in data analysis. The approach taken by Strecher et al. enabled straightforward analysis of the data, and the experiment produced meaningful and interesting results.

The second example of a serious disruption to an experiment is a study reported by Pellegrini et al. (2014, 2015). Pellegrini and colleagues wanted to examine five components that were candidates for inclusion in an adult weight loss intervention. The original intended experimental design was a 2^{5-1} fractional factorial with 16 experimental conditions. The experiment was to be conducted over approximately a 3-year period, with a target $N = 560$, or approximately $n = 35$ per experimental condition. Pellegrini et al. (2014) reported the original intended experimental design. Approximately one and one-half years into the experimentation period, the investigators discovered—to their dismay—that the experimental design they had reported, and were in the midst of implementing, was not the original intended design. The original intended design had been selected with the help of Proc FACTEX in SAS (see Chap. 5). There had been a clerical error in copying the list of experimental conditions produced by Proc FACTEX into word processing software during preparation of a grant proposal requesting funds to support the experiment. This clerical error produced the unintended design. When the grant was funded, the proposal was then used as a blueprint for the study, until the clerical error was discovered.

In a corrigendum to Pellegrini et al. (2014), Pellegrini et al. (2015) identified three possible courses of action they considered when the error was discovered. The first possible course of action was to stay with the unintended design and treat it as a 2^{5-1} in data analysis. This would have been reasonable, because the unintended design was balanced. However, its aliasing (see Chap. 5) was less desirable for the purposes of this research project than that of the original intended design. The second possible course of action was to stay with the unintended design but in data analysis collapse over one factor, making it a 2^4 complete factorial, which is essentially the approach taken by Strecher et al. However, the investigators needed to investigate all five components. The third possible course of action, the one that was ultimately selected, was to turn the unintended design into a new design, namely, a complete 2^5 factorial with 32 experimental conditions.

Fortunately, adding 16 conditions could be accomplished with minimal additional cost. The new design had the same target overall $N = 560$, but with double the experimental conditions, the new per-condition n was half of what it was in the 16-condition design, or approximately $n = 18$. At the time the error was detected, subject recruitment was approximately half complete. Once the original 16 conditions achieved $n = 18$, the investigators closed the books on them and made the transition to recruitment of subjects into the remaining 16 conditions in the new 32-condition design.

6.9.3 The Robustness of Factorial Experiments

In both the Strecher et al. and the Pellegrini et al. studies, a serious disruption to a factorial experimental design was suffered. The approaches to salvaging the study were different in the two cases. Strecher et al. chose to reduce the design by collapsing across one factor. Collapsing across one or more factors is nearly always a reasonable fallback position when something goes wrong with a balanced factorial experiment. If the problem is detected after the study has been concluded, as happened here, this may be the most viable option.

Because in the Pellegrini et al. study the error was detected early enough, the investigators had the option of revising their design by expanding it to include the experimental conditions needed to make a complete factorial. In the revised design, random assignment could not be done in the usual manner. Ordinarily, a 32-condition experiment would be conducted by randomly assigning to all 32 conditions at once. Because the design was changed while the study was in the midst of implementation, effectively 16 conditions were run first followed by the remaining 16 conditions. From one perspective, in the revised design time was confounded with some of the effects to be estimated. From another perspective, the revised design can be considered a 2_V^{6-1} fractional factorial (see Chap. 5) with *TIME* as a sixth factor having two levels: the time during which the first half of the conditions were run, and the time during which the second half of the conditions were run. Viewing the design this way makes it possible to determine exactly how time is aliased with other effects. In Pellegrini et al.'s design, each main effect is aliased with a four-way or five-way interaction involving the time factor. For example, the second factor is aliased with the four-way interaction involving time and the third, fourth, and fifth factors. In other words, the estimate of the main effect of the second factor would be approximately accurate unless this particular four-way interaction was large. As discussed in Chap. 5, Wu and Hamada (2009) call effects that are aliased only with four-way or higher-order interactions strongly clear. Viewing the new design as a fractional factorial, with time as a factor, lent clarity to the situation. This approach enabled Pellegrini et al. to weigh which was preferable, the aliasing in the unintended design or the aliasing in the revised design. Ultimately, they made the judgment call that the aliasing in the revised design was preferable.

Pellegrini et al. did not set out to use this approach, but it might have appeal to an investigator who wishes to conduct a 32-condition experiment in an environment in which only 16 conditions can properly be managed at one time. When not all conditions can be implemented simultaneously, confounding with time and interactions between time and experimental factors are often concerns. If time is considered an additional factor in a planful way, the investigator can select an experimental design so as to have manageable aliasing instead of unintended confounding. Proc FACTEX or similar software can be used to select a design that, like the one Pellegrini et al. used, keeps the main effects clear or strongly clear and otherwise has an acceptable alias structure.

These two examples illustrate the robustness of factorial experiments. Of course, the ideal is to avoid any disruptions to the design, but if disruptions do occur, they can often be dealt with in a manner that salvages the experiment and enables the investigators to meet most or all of the objectives of the study despite the unanticipated problem. If a disruption to a factorial experiment occurs, it is a good idea to seek the advice of a statistician who specializes in experimental design before making any decisions about what course of action to take.

6.9.4 The Aviation Approach to Experimentation

As long as humans have anything to do with experimentation, either as investigators or subjects, there will be a risk of protocol deviations, contamination, disruptions to design, and other implementation problems. One way to prevent these problems is to adopt the perspective taken in aviation and other fields in which errors are extremely costly. In these fields it is acknowledged that a certain amount of human error is inevitable, because humans can make errors even when they are doing their best to avoid them. The perspective taken in aviation and other fields is to (a) simplify procedures and make them explicit and unambiguous by making extensive use of clearly written checklists and (b) have more than one person, acting independently, review particularly important aspects of an endeavor before anything is finalized. For example, before each flight the pilot and copilot independently check all critical instruments. They must each verify that all the instruments are working properly before the plane can take off. The reason for having two people do this verification is not that it is a particularly difficult or complicated task for a qualified pilot to accomplish; rather, the redundancy is a part of standard procedures in aviation because the potential consequences of overlooking a malfunctioning instrument are so dire.

A checklist approach can help prevent protocol deviations. For example, a database can be maintained for use by the staff responsible for making sure subjects receive the appropriate treatment. The database can contain a record of the components and component levels that each subject is to receive according to his or her assigned experimental condition, along with a checklist to mark, perhaps on a touchscreen, as each component is provided to the subject. A similar approach can be taken without using an electronic database, for example, by maintaining a folder for each subject with a paper checklist corresponding to the components and component levels making up that subject's assigned condition.

The disruptions to the experimental design in the Strecher et al. and Pellegrini et al. studies might have been avoided if there had been independent reviews by two or more people before implementation began. In the Strecher et al. study, after the programming was completed and reviewed by the programmer, one or more of the investigators could also have reviewed the code before the online study was launched to ensure that the design was consistent in every detail with the design that had been selected for the experiment and that the random assignment procedure was programmed properly. Someone reviewing the code may have detected that there was a missing experimental condition or that random assignment had been set up overlooking one condition. Simply counting the conditions is not sufficient, because it is possible for a condition to be inadvertently duplicated while another is left out (although that is not what happened in Strecher et al.). It might also have helped to have one or more members of the research team act as pilot subjects. If randomization was done as described above, then the first round of random assignment would place each of the first 32 subjects in a different experimental condition; a second round of randomization would place the next 32 subjects in a different experimental condition; and so on. This procedure can be tested by conducting a test run in which research team members serve as pilot subjects to verify that each condition is assigned once and only once and that each condition is made up of the appropriate components and component levels.

In the Pellegrini et al. study, it might have been helpful to have one or more of the investigators check that the table that was included in the grant proposal included the correct set of experimental conditions based on the computer output. It also would have been a good idea to check that the experimental conditions that were to be implemented were consistent with the original computer results that were used to select the intended design.

Such recommendations seem obvious—in hindsight. Going forward it is difficult to prevent an error one has never seen before and therefore does not know to guard against. This is all the more reason to have several people double- and triple-check all procedures, looking vigilantly for any potential implementation problems. It can be difficult to find the time for double- and triple-checking when a complex research endeavor is in full swing, but in the long run, such an investment in time is always worthwhile.

6.10 Avoiding Contamination Across Experimental Conditions

Even when no protocol deviations occur, other threats to validity can occur in a field experiment. Shadish, Cook, and Campbell (2002) offer a comprehensive treatment of these threats to validity. Although Shadish et al. discussed primarily experimental designs that are variations on the RCT, for the most part the same threats to validity apply when conducting a factorial experiment. In this section one such threat to validity is considered, namely, contamination.

Contamination across experimental conditions occurs when subjects are provided the correct randomly assigned treatment by the experimenter, but then the treatment they receive or their reaction to treatment is influenced somehow as a direct or indirect result of contact with other subjects and/or learning about what treatment other subjects have been given. There are a variety of ways this can happen. One way is for a subject to seek out or receive some aspect of treatment because it has been assigned to another subject who recommends it. For example, consider two subjects, Mary and Bob, who are in the hypothetical experiment in Table 6.5. Suppose each has been participating in the experiment for several weeks. Mary is in experimental condition 12, in which *MIND* is set to yes, and Bob is in experimental Condition 11, in which *MIND* is set to no. Mary and Bob have shown up for assessment and are chatting in the waiting room. Imagine this dialog:

> Bob: "I'm still finding it difficult to keep up with the ART treatment—it seems overwhelming to me."
>
> Mary: "I've been struggling some too, but I am finding that the meditation we were taught is really helpful. Does it help you at all?"
>
> Bob: "What meditation? They haven't mentioned that to me."
>
> Mary: "Let me show you the meditation app they recommended. It's so easy!"
>
> Bob: "Wow, that looks great! Thank you! I will definitely download the app and try meditation."

As a result of his conversation with Mary, Bob is now receiving a version of a treatment intended for experimental Condition 12, instead of the appropriate treatment for his randomly assigned condition, Condition 11.

Contamination can weaken the effects of the factors. Imagine contamination taken to its logical extreme. If there were unlimited contamination, all subjects would receive the same treatment, and therefore none of the factors would demonstrate any effect. For this reason, measures should be put in place to prevent contamination. It is safest to limit unsupervised contact between subjects wherever possible. In situations where subjects may be acquainted, or become acquainted over the course of the study, they can be asked explicitly not to discuss the experiment until it is over, and not to deviate from their assigned treatment in any way. Such a strategy may (or may not) be successful with adults, but is probably futile with children or adolescents.

Close and sustained contact between subjects can be introduced by the experiment itself. Consider the hypothetical example. Suppose when *SKILLS* is set to yes, the behavioral skills training is delivered in a group setting in which discussion is encouraged. Subjects would be likely to discuss their experiences with working to maintain ART adherence, and it is almost inevitable that which treatment protocol each has been assigned would become a topic of conversation.

Sometimes subjects are together in naturally occurring clusters that make contamination nearly inevitable. It is probably not a good idea to assign people who are clustered in, for example, the same family, household, military barracks, elementary school classroom, or sports team to different experimental conditions. Even if the individuals were scrupulous about following the experimental protocol to the letter, they are in such close quarters and interact so frequently that the very fact that one has a particular treatment may influence the others.

When subjects are clustered—whether the clusters are experimentally produced, as when subjects who are otherwise unrelated are assigned to a group treatment setting in which they will interact, or naturally occurring, as when students are clustered in classrooms—and the clustering is likely to produce contamination, it may be necessary to assign entire clusters rather than individuals to experimental conditions. As discussed in previous chapters, this is called between-cluster randomization (Dziak, Nahum-Shani, & Collins, 2012; Nahum-Shani, Dziak, & Collins, 2017) or simply cluster randomization. Between-cluster randomization can greatly reduce contamination problems. Because everyone in a cluster is in the same experimental condition, interaction among the people in a cluster is no longer a problem. However, between-cluster randomization complicates statistical analysis and can have enormous implications for statistical power. Factorial experimentation when there is a cluster structure is discussed at length in the Nahum-Shani and Dziak (2018) chapter in the companion volume.

6.11 Registry of Optimization Trials

Many funders worldwide now require or strongly encourage investigators to register their trials. The objective is to promote transparency in the conduct of experimentation. The World Health Organization maintains a central data base of clinical trials (http://apps.who.int/trialsearch), which at this writing accepts data from 17 different registries. These include clinical trials registries maintained by the United States (clinicaltrials.gov), the European Union (www.clinicaltrialsregister.eu), the United Kingdom (www.isrctn.com), and other nations. At this writing, the Society for Research on Educational Effectiveness is in the process of establishing a comparable registry for educational trials (https://www.sree.org/pages/registry.php).

In general, optimization trials should be registered in whatever data base the investigator would register an RCT. This registration will include specification of the variables that will be used as outcomes for the purpose of optimization. Investigators may wish to note that today, most of these registries are set up primarily to accept two-arm or three-arm RCTs. Entering a more complex design, such as a factorial experiment, can be challenging, but it is possible. Perhaps as more optimization trials are registered, increased flexibility will be introduced into the trial registry systems. Some practical aspects of registry of optimization trials are discussed further in the Piper, Schlam, Fraser, Oguss, and Cook (2018) chapter in the companion volume.

6.12 What's Next

Chapter 2 discussed the preparation phase of MOST, in which the investigator prepares for optimizing an intervention. In this phase the investigator develops a detailed conceptual model, pilot tests as necessary, and selects an optimization criterion to be used when making decisions in the optimization phase. Chapters 3, 4, and 5 discussed experimental designs that may be used in the optimization phase of MOST. The present chapter reviewed how to select a design for the optimization trial based on the resource management principle and offered some practical recommendations for implementation of factorial experiments in field settings. The next chapter, Chap. 7, reviews what to do after the optimization trial has been conducted, and results have been obtained. Chapter 7 discusses how to make use of the information gathered via experimentation to make decisions about which components to select for inclusion in the screened-in set and how to select the levels of these components that will comprise the optimized intervention.

References

Almirall, D., Nahum-Shani, I., Wang, L., & Kasari, C. (2018). Experimental designs for research on adaptive interventions: Singly and sequentially randomized trials. In L. M. Collins & K. C. Kugler (Eds.), *Optimization of behavioral, biobehavioral, and biomedical interventions: Advanced topics* (forthcoming). New York, NY: Springer.

Baker, T. B., Collins, L. M., Mermelstein, R., Piper, M. E., Schlam, T. R., Cook, J. W., . . . Fiore, M. C. (2016). Enhancing the effectiveness of smoking treatment research: Conceptual bases and progress. *Addiction, 111*, 107–116.

Collins, L. M., Dziak, J. R., & Li, R. (2009). Design of experiments with multiple independent variables: A resource management perspective on complete and reduced factorial designs. *Psychological Methods, 14*, 202–224.

Collins, L. M., & Kugler, K. C. (Eds.). (2018). *Optimization of multicomponent behavioral, biobehavioral, and biomedical interventions: Advanced topics*. New York, NY: Springer.

Cook, J. W., Collins, L. M., Fiore, M. C., Smith, S. S., Fraser, D., Bolt, D. M., . . . Mermelstein, R. (2016). Comparative effectiveness of motivation phase intervention components for use with smokers unwilling to quit: A factorial screening experiment. *Addiction, 111*, 117–128.

Dziak, J. (2018). Optimizing the cost-effectiveness of a multicomponent intervention using data from a factorial experiment: Considerations, open questions, and tradeoffs among multiple outcomes. In L. M. Collins & K. C. Kugler (Eds.), *Optimization of multicomponent behavioral, biobehavioral, and biomedical interventions: Advanced topics* (forthcoming). New York, NY: Springer.

Dziak, J. D., Nahum-Shani, I., & Collins, L. M. (2012). Multilevel factorial experiments for developing behavioral interventions: Power, sample size, and resource considerations. *Psychological Methods, 17*, 153–175.

Gwadz, M. V., Collins, L. M., Cleland, C. M., Leonard, N. R., Wilton, L., Gandhi, M., . . . Ritchie, A. S. (2017). Using the multiphase optimization strategy (MOST) to optimize an HIV care continuum intervention for vulnerable populations: A study protocol. *BMC Public Health, 17*, 383.

Karaca-Mandic, P., Norton, E. C., & Dowd, B. (2012). Interaction terms in nonlinear models. *Health Services Research, 47*, 255–274.

References

Kugler, K. C., Wyrick, D. L., Tanner, A. E., Milroy, J. J., Chambers, B. D., Ma, A., ... & Collins, L. M. (2018). Using the multiphase optimization strategy (MOST) to develop an optimized online STI preventive intervention aimed at college students: Description of conceptual model and iterative approach to optimization. In L. M. Collins & K. C. Kugler (Eds.), *Optimization of multicomponent behavioral, biobehavioral, and biomedical interventions: Advanced topics* (forthcoming). New York, NY: Springer.

Nahum-Shani, I., & Dziak, J. J. (2018). Multilevel factorial designs in intervention development. In L. M. Collins & K. C. Kugler (Eds.), *Optimization of multicomponent behavioral, biobehavioral, and biomedical interventions: Advanced topics* (forthcoming). New York, NY: Springer.

Nahum-Shani, I., Dziak, J. J., & Collins, L. M. (2017). Multilevel factorial designs with experiment-induced clustering. *Psychological Methods*. Advance online publication. https://doi.org/10.1037/met0000128

Pellegrini, C. A., Hoffman, S. A., Collins, L. M., & Spring, B. (2014). Optimization of remotely delivered intensive lifestyle treatment for obesity using the multiphase optimization strategy: Opt-IN study protocol. *Contemporary Clinical Trials, 38*, 251–259.

Pellegrini, C. A., Hoffman, S. A., Collins, L. M., & Spring, B. (2015). Corrigendum to "optimization of remotely delivered intensive lifestyle treatment for obesity using the multiphase optimization strategy: Opt-IN study protocol". *Contemporary Clinical Trials, 45*, 468–469.

Piper, M. E., Fiore, M. C., Smith, S. S., Fraser, D., Bolt, D. M., Collins, L. M., ... Baker, T. B. (2016). Identifying effective intervention components for smoking cessation: A factorial screening experiment. *Addiction, 111*, 129–141.

Piper, M. E., Schlam, T. R., Fraser, D., Oguss, M., & Cook, J. W. (2018). Implementing factorial experiments in real-world settings: Lessons learned while engineering an optimized smoking cessation treatment. In L. M. Collins & K. C. Kugler (Eds.), *Optimization of multicomponent behavioral, biobehavioral, and biomedical interventions: Advanced topics* (forthcoming). New York, NY: Springer.

Rivera, D. E., Hekler, E. B., Savage, J. S., & Downs, D. S. (2018). Intensively adaptive interventions using control systems engineering: Two illustrative examples. In L. M. Collins & K. C. Kugler (Eds.), *Optimization of multicomponent behavioral, biobehavioral, and biomedical interventions: Advanced topics* (forthcoming). New York, NY: Springer.

Schlam, T. R., Fiore, M. C., Smith, S. S., Fraser, D., Bolt, D. M., Collins, L. M., ... Baker, T. B. (2016). Comparative effectiveness of intervention components for producing long-term abstinence from smoking: A factorial screening experiment. *Addiction, 111*, 142–155.

Shadish, W. R., Cook, T. D., & Campbell, D. T. (2002). *Experimental and quasi-experimental designs for generalized causal inference*. Houghton, Boston, MA: Mifflin and Company.

Strecher, V. J., McClure, J. B., Alexander, G. W., Chakraborty, B., Nair, V. N., Konkel, J. M., ... Pomerleau, O. F. (2008). Web-based smoking cessation programs: Results of a randomized trial. *American Journal of Preventive Medicine, 34*, 373–381.

Wu, C. F. J., & Hamada, M. S. (2009). *Experiments: Planning, analysis, and optimization* (2nd ed.). Hoboken, NJ: Wiley.

Chapter 7
The Completion of the Optimization Phase

Abstract This chapter starts with the premise that an investigator has completed an optimization trial and analyzed the data. Now what? How does an investigator take these experimental results and use them to create an optimized intervention? In other words, how does an investigator use empirical data from an optimization trial to make decisions about which components and component levels will constitute the optimized intervention? This chapter describes one approach to making the decisions needed to complete the optimization phase of the multiphase optimization strategy (MOST). Familiarity with the material in Chaps. 1, 2, 3, 4, 5, and 6 is assumed throughout the present chapter.

Contents

7.1	Introduction	228
7.2	Overview of the Decision-Making Process	231
7.3	Some Fundamentals	233
	7.3.1 Effect Hierarchy, Effect Sparsity, and Effect Heredity Revisited	233
	7.3.2 A Decision-Making Approach Based on Effect Coding	234
7.4	Step 1: Identify the Important Main Effects and Interactions	234
	7.4.1 Defining What Constitutes an Important Effect: Main Effect and Interaction Criteria	234
	7.4.2 Selecting the Criteria	236
	7.4.3 The Main Effect and Interaction Criteria and Statistical Power	237
	7.4.4 Example of Step 1	237
7.5	Step 2: Divide the Candidate Components Into the Screened-In and Screened-Out Sets	239
	7.5.1 When Provisional Decisions Are Likely to Be Reversed	240
	7.5.2 Reconsidering Provisional Decisions Based on Interactions Involving PEER	242
	7.5.3 Reconsidering Provisional Decisions Based on Interactions Involving SKILLS	245
	7.5.4 Reconsidering Provisional Decisions Based on Interactions Involving MI	245
	7.5.5 Summary of Selection of the Screened-In and Screened-Out Sets of Components	246
	7.5.6 A Note About the Lower Level of Components	247
	7.5.7 Orphan Interactions	247
7.6	Step 3: Apply the Optimization Criterion	248

© Springer International Publishing AG 2018
L. M. Collins, *Optimization of Behavioral, Biobehavioral, and Biomedical Interventions*, Statistics for Social and Behavioral Sciences,
https://doi.org/10.1007/978-3-319-72206-1_7

	7.6.1	The All Active Components Criterion	248
	7.6.2	The Need for Other Optimization Criteria	250
	7.6.3	Constraints on Money	252
	7.6.4	Constraints on Time	253
	7.6.5	A Note on Cost and the All Active Components Criterion	254
	7.6.6	Constraints on Multiple Resources	254
	7.6.7	Four Different Optimization Criteria, Four Different Interventions…Which One Is Best?	255
	7.6.8	When a Component From the Screened-In Set Must Be Omitted From the Optimized Intervention	257
	7.6.9	The Shortcomings of Ad Hoc Modifications	257
	7.6.10	Reusing the Results of an Optimization Trial With Different Constraints....	258
	7.6.11	A Note on the Estimation of Cost	259
	7.6.12	The Possibility of a Bayesian Approach	259
7.7	When There Is More Than One Outcome Variable		260
	7.7.1	When Measures of Mediators Are Used as Short-Term Outcomes	260
	7.7.2	When There Is More Than One Primary Outcome	261
7.8	Why Bother?		262
7.9	After the Decision-Making Is Complete		263
	7.9.1	Secondary Analysis and Qualitative Data	263
	7.9.2	The Evaluation Phase of MOST	263
7.10	What's Next		265
References			265

7.1 Introduction

Let us return once again to Dr. B. As described in Chaps. 1 and 2, Dr. B wishes to develop an intervention to reduce viral load in HIV-positive individuals who drink heavily. Based on prior scientific literature, clinical experience, and a well-specified conceptual model (reviewed in detail in Chap. 2), Dr. B has identified five intervention components that are hypothesized to be critical in helping HIV-positive individuals to reduce their drinking and improve their adherence to antiretroviral therapy (ART). They are (a) motivational interviewing, to engage the participant in the process of examining his or her alcohol use and ART adherence; (b) peer mentoring, to provide contact with a positive role model; (c) text messaging, to provide access to a support system; (d) mindfulness meditation, to improve mental health; and (e) behavioral skills training, to improve behavioral skills for managing alcohol use and ART. Suppose there is currently no standard of care for the first four components, but physicians and HIV clinics routinely give HIV-positive individuals who drink heavily and have poor ART adherence a workbook to complete to help improve their behavioral skills. Dr. B is considering two levels for each of these components. Each of the first four components can be set either to no, that is, not included in the intervention, or yes, that is, included. The behavioral skills component can be at either a low level, consisting of the workbook only (i.e., standard of care), or a high level, consisting of the workbook plus training delivered by a

7.1 Introduction

Table 7.1 Factors and levels in hypothetical 2^5 factorial experiment

Component	Factor name	Lower level	Higher level
Motivational interviewing	*MI*	No[a]	Yes
Peer mentoring	*PEER*	No	Yes
Text messaging	*TEXT*	No	Yes
Mindfulness meditation	*MIND*	No	Yes
Behavioral skills training	*SKILLS*	Low (workbook only)	High (workbook + training)

[a]No means not included in intervention; yes means included in intervention

behavioral skills counselor. The primary outcome variable is number of days of adherence to ART in the 30-day period following the conclusion of the intervention and will be abbreviated *adhere*.

Suppose, as described in Chap. 2, Dr. B has completed the preparation phase of MOST, that is, has prepared a detailed conceptual model, selected a set of candidate intervention components, pilot tested them, and selected an optimization criterion. Dr. B then moved on to the optimization phase and selected an experimental design for the optimization trial. Dr. B decided to conduct a 2^5 factorial experiment (see Chaps. 3, 4, 5, and 6) in which each of the five candidate components is represented by an experimental factor. The components, the corresponding experimental factors, and the levels of each factor are listed in Table 7.1. (For a list of the 32 experimental conditions included in the 2^5 factorial design, see Table 5.2.) After registering the optimization trial as appropriate (e.g., in clinicaltrials.gov), including specifying that *adhere* is the primary outcome variable, Dr. B conducted the experiment. Then, Dr. B analyzed the data using a factorial ANOVA with effect coding to obtain estimates of the main effect of each factor on *adhere* and interactions between factors.

At this point, Dr. B is ready to build an optimized intervention. This chapter describes a step-by-step approach for accomplishing this. It is assumed that a factorial optimization trial has been conducted, although the general approach may be useful if a different experimental design has been used.

It may be helpful at this point to reiterate the difference between components, which are referred to in lowercase roman letters (e.g., motivational interviewing), and experimental factors, which are referred to in uppercase italics using an abbreviated version of the name of the corresponding component (e.g., *MI*). A component is a part of an intervention. By contrast, a factor is a part of a factorial experiment. Dr. B has identified a set of candidate intervention components. Each could be set at a variety of different levels of intensity, but Dr. B has identified two levels per component to consider for inclusion in the optimized intervention and will select one for each component. A factor is experimentally manipulated by the investigator for the purpose of comparing two or more levels of a component and understanding how a component may affect the performance of other components. In other words, which level of each candidate component is selected for inclusion in the intervention is determined largely by the information provided by the corresponding experimental factor.

Ultimately, the investigator must weigh the evidence provided by the optimization trial to decide whether the more intensive level of each component, which is usually the costlier level, is justified. Outlined in this chapter is a procedure for selecting the components and levels of each component that will make up the optimized intervention, based on the results of the optimization trial. The procedure is an updated and expanded version of the one originally presented in Collins, Trail, Kugler, Baker, Piper, and Mermelstein (2014). The present chapter covers a variety of optimization criteria, including the all active components criterion and criteria that specify a fixed upper limit on some specific resource such as cost or time. The approach suggested here is intended to help the investigator to build an intervention that produces the best result that can be obtained within a particular implementation budget, participant time burden, or other key constraint. If the key constraints are taken into account properly during the optimization phase, an optimized intervention shown to be statistically and clinically effective in the evaluation phase will be immediately scalable. In other words, there will be little or no need for ad hoc modifications to the intervention to enable it to be implemented in the intended setting.

This chapter takes a decision-priority perspective on the use of empirical results (see Chap. 3). From this perspective, the results of an optimization trial may be used in any consistent, rational, and scientifically sound manner. It follows that decisions may or may not be based explicitly on hypothesis testing, and if they are to be based on hypothesis testing, the traditional Type I error rate of $\alpha = .05$ may or may not be used. It also follows that the approach outlined in this chapter should by no means be considered the last word on approaches to decision-making in the optimization phase of MOST, because a particular investigator may determine that a different approach would work better in a given situation.

The approach presented here has some important limitations. It is suitable primarily for the all active components optimization criterion and criteria that involve a fixed upper limit on resources. Alternative approaches may be more suitable depending on the objectives in a given setting. In particular, the use of cost-effectiveness as an optimization criterion requires a somewhat more complex approach (see the Dziak (2018) chapter in the companion volume; Collins & Kugler, 2018). The approach described in this chapter is also primarily suited for situations with a single outcome variable, but it can be applied in a limited way in some circumstances when there is more than one outcome variable, as is discussed later in this chapter.

For simplicity, *throughout this chapter it will be assumed that a larger value of the outcome variable is more desirable and a positive sign on each regression weight indicates an effect in the desired direction.* The approach described in this chapter can be used when a negative regression weight is consistent with the desired direction, but special care must be taken to ensure that effects are interpreted correctly. Interpretation of results, particularly interactions, can become particularly challenging if the desired direction is represented by different signs for different factors in the experiment. In most cases interpretation of the results of the factorial ANOVA is more straightforward if the coding is set up so that a positive regression weight sign always indicates an effect in the desired direction.

7.2 Overview of the Decision-Making Process

The three-step decision-making process discussed in this chapter is illustrated in Fig. 7.1. The general logic can be summarized as follows: *Identify the main effects and interactions that are large enough to be considered important in decision-making; next, based on the estimates of these important effects, decide which*

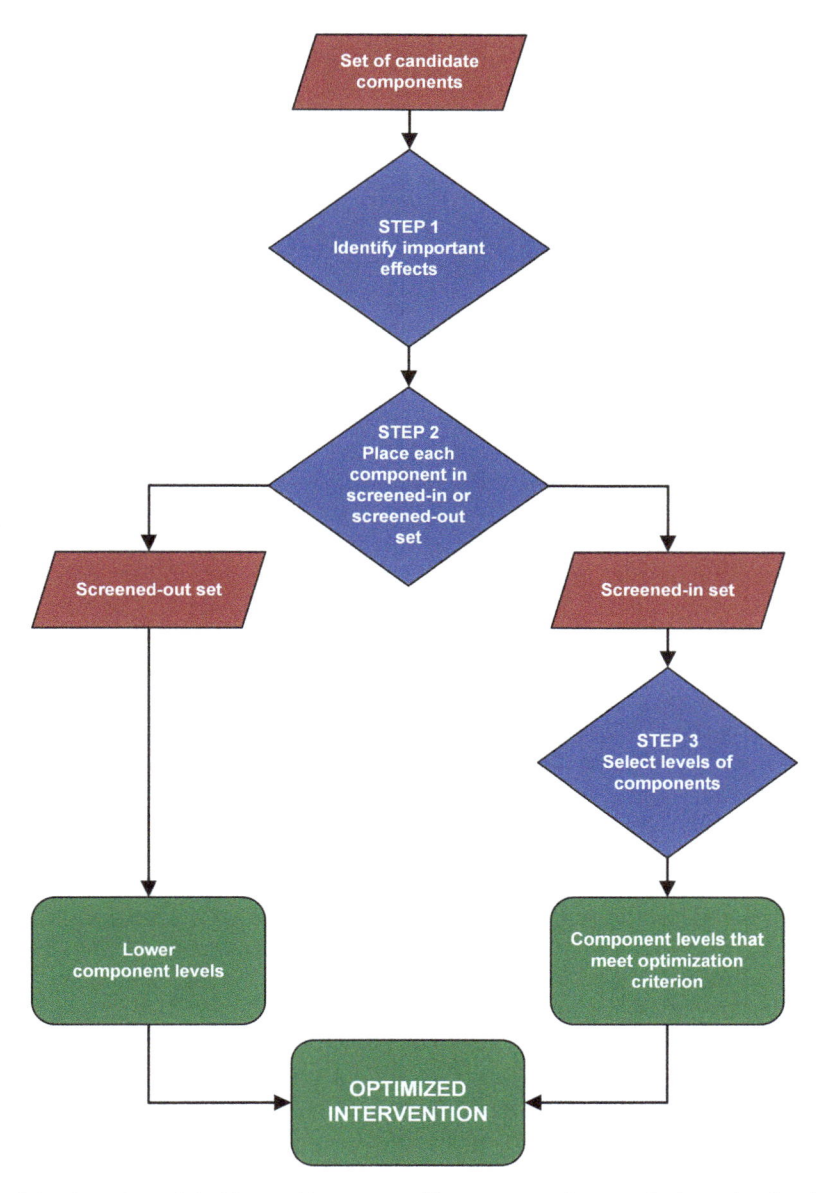

Fig. 7.1 Flow chart of decision-making process. If a component is set to the lower level and that level = no, the component is eliminated from the intervention by definition

components will be considered further for inclusion in the optimized intervention at their respective higher levels; finally, taking the optimization criterion (identified in the preparation phase of MOST) into account, decide which components will be included at the higher level in the optimized intervention. A more detailed overview follows.

Step 1. Identify the main effects and interactions that are considered large enough to be important in decision-making. The remaining effects are considered unimportant. The meaning of the term "important" will be discussed below.

Step 2. Divide the candidate components into the screened-in set and the screened-out set. When a component is placed in the screened-out set, the lower level of that component is automatically chosen and it is not considered further. The components in the screened-in set will be considered further for inclusion in the optimized intervention at their respective higher levels in Step 3, when the optimization criterion enters into the decision-making process.

Decisions about which components belong in the screened-in set and which belong in the screened-out set are made initially based on the important main effects and then reconsidered in the light of the important interactions. Recall the four desiderata: effectiveness, efficiency, economy, and scalability (Chap. 1). Step 2 of the decision-making process is focused primarily on the efficiency of the intervention being constructed—in other words, weeding out components and component levels that are not expected to contribute to effectiveness. Economy and scalability are considered in relation to effectiveness in Step 3.

Step 3. Make final decisions concerning selection of levels of the components in the screened-in set. These decisions are made based on the optimization criterion. If the optimization criterion is all active components, that is, economy and scalability per se are not to be taken into account in selecting components and component levels for the intervention, then at this point, decision-making is complete. The components and component levels that make up the optimized intervention have already been identified; the intervention will be made up of the higher levels of the components in the screened-in set and the lower levels of the components in the screened-out set. This intervention will be efficient by the definition in Chap. 1, because it will not contain components/component levels that do not contribute to effectiveness. However, it may or may not be economical or scalable.

Bringing an optimization criterion other than all active components into decision-making enables considerations of cost, broadly defined to include money, participant time, staff time, or any other resource, to play a role if desired. If an optimization criterion other than all active components has been selected, the objective is to develop an intervention that delivers the best outcome that is possible while taking resource constraints into account. Often, obtaining a better outcome requires more resources. Staying within a predetermined budget, in other words, developing a less resource-intensive intervention, may require selecting an intervention that does not include all of the higher levels of the components in the screened-in set and accepting a correspondingly less favorable expected outcome. In other words, in Step 3 the investigator can balance the desire for effectiveness against the imperatives of economy and/or scalability. Selecting the appropriate levels of the components in the screened-in set so as to meet this objective requires additional decision-making, explained below.

After decision-making is complete, the optimized intervention is assembled by combining the lower levels of the components in the screened-out set with the selected levels of the components in the screened-in set.

7.3 Some Fundamentals

7.3.1 Effect Hierarchy, Effect Sparsity, and Effect Heredity Revisited

The approach discussed in this chapter is guided by three principles: effect hierarchy, effect sparsity (also called the Pareto principle), and effect heredity. These principles, which are used widely in engineering, were discussed in Chap. 5 and will be discussed again briefly here; they are treated at greater length in Wu and Hamanda (2009). The effect hierarchy principle states that main effects are of primary scientific importance, followed by second-order interactions, then third-order interactions, and so on. The effect sparsity principle states that out of the many effects that can be estimated in a factorial ANOVA, only a few will be scientifically important. The effect heredity principle (as used in this book) states that an interaction is most likely to be scientifically important if at least one of its parent main effects is important; "orphan" interactions, that is, interactions that involve only factors that do not demonstrate important main effects, should be viewed with extreme caution unless they were designated scientifically interesting in the conceptual model.

The effect hierarchy principle suggests an approach in which initial provisional decisions about selection of components and component levels are made based on main effects unless there is a compelling reason to do otherwise, because, according to this principle, they are of primary scientific importance. Decisions can then be reconsidered in the light of important interactions, starting with two-way interactions, then three-way interactions, and so on. The effect sparsity principle suggests that decisions will be made based on the small subset of main effects and interactions that are scientifically important, rather than on all of the effects estimated in a factorial ANOVA. One of the steps in the decision-making process outlined here involves identifying the important effects. According to the effect heredity principle, it is unlikely (although not impossible) that an orphan interaction will be designated important. This is discussed further below.

The effect hierarchy, effect sparsity, and effect heredity principles should be considered rough guidelines that are useful in the absence of a priori predictions, rather than strict rules. In particular, they are always overruled by what the investigator has specified in the conceptual model.

7.3.2 A Decision-Making Approach Based on Effect Coding

The decision-making process outlined in this chapter is based on the assumption that effect $(-1, 1)$ coding has been used to conduct the factorial ANOVA. As will be seen below, the decision-making process involves examining the effects in a full regression model (i.e., a model containing all of the ANOVA effects) and then identifying which are important (by a priori criteria, discussed below). Decisions are made based only on the important effects. This means that a new regression model is needed, one that includes only the important effects. The new model will be referred to as the parsimonious prediction model.

When effect coding has been used, removal of one or more effects from the regression model does not change the estimates of the other effects when the experiment is balanced and changes them only minimally even in most unbalanced experiments. Thus, the investigator can be confident that the estimates that remain will be approximately the same in the parsimonious prediction model as they were in the original full model. By contrast, when dummy $(0, 1)$ coding is used, removal of even one effect can change the estimates of all of the other effects, and regression models based on the same data but made up of different subsets of effects are likely to be very different. This means that it would be difficult to arrive at the screened-in set and screened-out sets or to identify the levels of the components in the screened-in set that satisfy an optimization criterion, based on a factorial ANOVA that used dummy coding. (For a more detailed treatment of the interpretational and operational differences between effect-coded and dummy-coded effects in factorial ANOVA, see the Kugler, Dziak, and Trail (2018) chapter in the companion volume.)

7.4 Step 1: Identify the Important Main Effects and Interactions

The first several subsections provide an introduction to Step 1, followed by an example.

7.4.1 Defining What Constitutes an Important Effect: Main Effect and Interaction Criteria

Because the term "significant" may or may not refer exclusively to statistical significance at the traditional level and because what constitutes a significant effect varies across applications of MOST, in this book the term "important" will be used to refer to effects that meet the investigator's chosen criterion for significance. In this section some possible operational definitions of "important" are discussed.

7.4 Step 1: Identify the Important Main Effects and Interactions

A factorial ANOVA can yield many effect estimates; for example, a factorial ANOVA of data from a 2^5 factorial experiment can yield up to 31 main and interaction effect estimates. An approach is needed for sorting through all of these estimates and identifying the ones that are important, so that decision-making can be based primarily on these important effects. Any effects not designated important can be treated as effectively zero for decision-making purposes and therefore are set to zero in the parsimonious prediction model.

It is helpful to identify two a priori criteria at the start of Step 1 to serve as operational definitions of what is meant by "important": a main effect criterion and an interaction effect criterion. There are several ways to express these effect size criteria. One convenient effect size measure that has been used earlier in this book is the standardized regression coefficient:

$$\beta^* = \frac{b}{s_y}.$$

β^* represents the change in $\widehat{Y}$, in standard deviation units, associated with a one-standard deviation change in X. Another familiar effect size estimate is Cohen's d, which, in a factorial experiment with two levels per factor, represents the difference between $\widehat{Y}$ for the high and low levels of a factor, in standard deviation units. β^* and d can be expressed in terms of each other, so they are interchangeable; in a factorial experiment with dichotomous factors, $\beta^* = d/2$. Thus, the investigator may select whichever has the most intuitive appeal. Alternatively, effect sizes can be expressed in raw units, in terms of unstandardized regression weights or even simple differences between means. This may be a good approach if the outcome is in meaningful units. For example, if a weight loss intervention is being developed, an important effect could be expressed in terms of pounds (e.g., an important main effect could be defined as a difference of three pounds).

Hypothesis testing is not strictly necessary when the decision-priority perspective is taken (see discussions of this perspective in Chap. 3 and elsewhere in this book), but it can be a useful tool in selection of components and component levels and can also provide a helpful organizing framework. If hypothesis testing is used, the main effect and interaction criteria are expressed in terms of significance at a particular level, so that an important effect is any effect significant at $p <$ some designated value. This can work well as long as the role of hypothesis testing is that of servant rather than master. In other words, if hypothesis testing is to form the basis for the main effect and interaction criteria, it is important not to be overly doctrinaire about hewing to a conventional Type I error rate. Instead, as discussed in Chap. 3, care must be taken to establish Type I and Type II error rates that reflect the relative scientific costs associated with Type I and Type II errors. If overlooking a potentially useful component would be about as undesirable or costly as mistakenly including an inert component, then the Type I and Type II error rate should be about the same. If overlooking a potentially useful component is considered more undesirable or costly than including an inert component, the Type II error rate should be *smaller* than the Type I error rate. This is an example of how the resource management principle (Chap. 1) is used in MOST. Given a fixed sample size, this may mean that an $\alpha > .05$ must be chosen to keep the Type II error rate at a desired level. It should be noted that given a fixed N and a design consisting of two-level factors, selecting a particular effect size for a main or interaction effect criterion is equivalent to selecting a particular α for hypothesis testing.

7.4.2 Selecting the Criteria

The main effect criterion represents the smallest main effect that a factor must demonstrate for its corresponding component to be selected into the screened-in set. In considering what main effect criterion to select, the investigator may think in terms of whatever metric is most natural, whether this is the raw or standardized regression coefficient, Cohen's d, or any other expression of effect size. For example, perhaps to be included in the screened-in set, a component must show evidence of changing the outcome at least .2 of a standard deviation, which corresponds to a Cohen's d of .2 or a β^* of .1.

One approach is to think in terms of a target overall effect size for the optimized intervention and consider what the individual component effect sizes would have to be to achieve this. For example, suppose the investigator wishes to build an intervention that will have an overall effect of at least 0.5 of a standard deviation and expects to select as few as four components for inclusion in the intervention. At a minimum, the main effect criterion should be $d = .5/4 = .125$. However, as is discussed below, the effects may not be purely additive; there may be some synergistic or antagonistic interactions (see Chap. 4) that would mean the overall effect of the intervention as a package would be larger or smaller than would be expected based on the main effects alone. In particular, if there are antagonistic interactions, the combined effect of the components may be less than the sum of the main effects. If antagonistic interactions are a possibility, a larger main effect criterion may be called for, say $d = .2$ ($\beta^* = .1$) or even $d = .3$ ($\beta^* = .15$). (Please remember that these are effect-coded, not dummy-coded, effects.)

The interaction criterion is perhaps less intuitive. One way to think about the interaction criterion is in terms of the minimum size interaction that would trigger a reversal of a provisional decision about whether a component should be placed in the screened-in or screened-out set. In deciding what criterion to use for interactions, it is a good idea to be familiar with the material in Chap. 4. For simplicity, in the decision-making example presented in this chapter, the same criterion will be used for all main effects and interactions. In practice, different criteria could be used for main effects and interactions or for effects pertaining to different components.

7.4.3 The Main Effect and Interaction Criteria and Statistical Power

As was stated above, although hypothesis testing is not strictly necessary in decision-making in the optimization phase of MOST, it can provide a useful framework. One area in which it can provide a useful framework is in selection of the appropriate sample size for the optimization trial. Even if classical hypothesis testing is not going to be used, it makes sense to base the experiment on a sample size that is large enough to provide reasonable Type I and Type II error rates. This, in turn, will ensure that effects can be estimated with reasonable precision, providing a solid basis for decision-making.

The usual approach to selecting an effect size for a power analysis is to comb the prior scientific literature for evidence of the expected magnitude of the effect. This book advocates using the main or interaction effect criterion, whichever is smaller, as the target effect size in the power analysis (see Chaps. 3 and 4 for more about power). There are two reasons why this book takes this position. First, at this writing relatively little scientific evidence exists about the effect size of individual components of an intervention. Second, from the decision-priority perspective the primary purpose of the optimization trial is to determine whether a component's main effect does or does not meet the main effect criterion and whether there are any interactions that will impact decision-making. Thus it makes sense to power the optimization trial to detect effects at the size of the main or interaction effect criterion, whichever is smaller, at whatever Type I and Type II error rates are desired. For decision-making purposes, if the observed effect is smaller than the main/interaction effect criterion, the investigator does not care what its magnitude is, and therefore the resource management principle suggests that resources should not be devoted to detecting any smaller effects.

It is especially important to ensure that a decision to select the less costly level can be made confidently when the purpose of an experiment is to determine whether there is a difference between two putatively active levels of a component, such as one vs. four sessions of counseling or delivery by a registered nurse vs. a community volunteer. If there is no detectable difference between one vs. four sessions of counseling, one session will be given; if there is no detectable difference between delivery by a registered nurse or a community volunteer, the community volunteer would probably be selected because the volunteer is less expensive. Such comparisons, which must be viewed as equivalence or noninferiority trials (Kraemer, 2011; Walker & Nowacki, 2011), should be approached particularly thoughtfully.

7.4.4 Example of Step 1

Suppose Dr. B has conducted the 2^5 factorial experiment briefly described in Sect. 7.2, with 512 subjects. The data from the experiment have been analyzed, and the results are displayed in Table 7.2 (note that these results are based on

Table 7.2 Results of analysis of variance on data from 2^5 factorial experiment[a]

		b-weight	t	p
	Intercept	5.033	105.655	<.001
Main effects				
	MI	0.167	3.510	<.001
	PEER	0.217	4.556	<.001
	TEXT	0.030	0.623	0.534
	MIND	0.013	0.279	0.780
	SKILLS	0.213	4.468	<.001
Interactions				
	MI × PEER	−0.119	−2.504	0.013
	MI × TEXT	0.134	2.804	0.005
	MI × MIND	−0.013	−0.272	0.786
	MI × SKILLS	0.181	3.795	<.001
	PEER × TEXT	−0.015	−0.307	0.759
	PEER × MIND	0.042	0.884	0.377
	PEER × SKILLS	0.077	1.616	0.107
	TEXT × MIND	0.031	0.657	0.511
	TEXT × SKILLS	−0.010	−0.209	0.835
	MIND × SKILLS	−0.054	−1.128	0.260
	MI × PEER × TEXT	0.046	0.969	0.333
	MI × PEER × MIND	0.007	0.143	0.886
	MI × PEER × SKILLS	−0.038	−0.798	0.425
	MI × TEXT × MIND	−0.013	−0.263	0.793
	MI × TEXT × SKILLS	0.035	0.731	0.465
	MI × MIND × SKILLS	0.009	0.189	0.850
	PEER × TEXT × MIND	−0.035	−0.725	0.469
	PEER × TEXT × SKILLS	−0.005	−0.096	0.924
	PEER × MIND × SKILLS	−0.035	−0.726	0.468
	TEXT × MIND × SKILLS	0.015	0.310	0.757
	MI × PEER × TEXT × MIND	0.010	0.214	0.830
	MI × PEER × TEXT × SKILLS	−0.051	−1.078	0.282
	MI × PEER × MIND × SKILLS	−0.039	−0.826	0.409
	MI × TEXT × MIND × SKILLS	−0.028	−0.589	0.556
	PEER × TEXT × MIND × SKILLS	−0.012	−0.250	0.803
	MI × PEER × TEXT × MIND × SKILLS	0.008	0.160	0.873

[a]$N = 512$. Standard error (all effects) = .048. Results are based on artificial data. Shading indicates that the effect meets the main effect or interaction criterion. In this example both criteria are $p < .15$

artificial data and that unstandardized regression weights are displayed). This section describes Dr. B's approach for identifying which are the important effects and then establishing the parsimonious prediction model.

Suppose Dr. B has selected $\alpha = .15$ for both the main effect and interaction criteria. (Readers who are uncomfortable with the use of $\alpha = .15$ are asked to recall that a decision-priority perspective is taken in this chapter and encouraged to reread

Sect. 3.10.) The rows corresponding to effects that meet the main effect and interaction criteria are highlighted in the table.

The main effects of the factors *MI*, *PEER*, and *SKILLS* all meet the main effect criterion, that is, each of these main effects is considered important because each is significant at $p < .15$ (see the shaded rows in the main effect section of Table 7.2). The interactions *MI* × *PEER*, *MI* × *TEXT*, *MI* × *SKILLS*, and *PEER* × *SKILLS* meet the interaction criterion and thus are considered important (see the shaded lines in the interaction section of Table 7.2).

Once the important effects have been identified, the parsimonious prediction model can be established. In the parsimonious prediction model, the coefficients corresponding to all effects not designated important are set to zero, and therefore those effects drop out of the equation. In this example, the parsimonious model is

$$\widehat{Y}_{adhere} = 5.033 + .167 X_{MI} + .217 X_{PEER} + .213 X_{SKILLS}$$
$$- .119 X_{MI \times PEER} + .134 X_{MI \times TEXT} + .181 X_{MI \times SKILLS}$$
$$+ .077 X_{PEER \times SKILLS}.$$

In this case, because the model is perfectly balanced (a luxury of artificial data), these coefficients are simply taken from Table 7.2. This is appropriate because when effect coding has been used to conduct a factorial ANOVA of balanced data, removing effects does not change the size of the remaining coefficient estimates (see Chap. 3). If the per-condition ns are not exactly equal, as is common in empirical data, it is safest to re-estimate the coefficients by reanalyzing the data using only the predictors in the parsimonious model. The estimates would be expected to change very little unless the data were severely imbalanced. As will be seen below, the parsimonious prediction model is needed in Steps 2 and 3.

Note that effects in the undesired direction must be identified as important if they meet the appropriate effect criterion and included in the parsimonious model. For example, consider a main effect in the undesired direction. In the example in this book, such an effect would have a negative regression coefficient. If the absolute value of this effect met the main effect criterion, it would be included in the parsimonious model, because it is necessary to take into account that selecting the lower level of this factor (e.g., setting it to no) will produce a more favorable outcome (see the Dziak (2018) chapter in the companion volume).

7.5 Step 2: Divide the Candidate Components Into the Screened-In and Screened-Out Sets

Consistent with the effect hierarchy principle described above, the starting point for decision-making in Step 2 is the main effect results of the factorial ANOVA. If a factor's main effect is designated important and it is in the desired direction, then the corresponding component is provisionally placed in the screened-in set; otherwise, the component is provisionally placed in the screened-out set. In the example, therefore, the provisional decisions are as follows: include motivational interviewing, peer mentoring, and behavioral skills training in the screened-in set and place text messaging and mindfulness meditation in the screened-out set.

Next, the provisional decisions based on main effects are carefully and systematically reconsidered based on the interaction results of the factorial ANOVA. One way to think about interactions is in terms of what they add to the "story" of the outcome over and above the contributions of any main effects (see Chap. 4). For example, consider a regression equation including the main effects of *MI* and *SKILLS* and the *MI* × *SKILLS* interaction, and let $b_{MI \times SKILLS}$ represent the coefficient corresponding to the *MI* × *SKILLS* interaction. If $b_{MI \times SKILLS} = 0$, then the contribution of the factors *MI* and *SKILLS* to $\widehat{Y}_{adhere}$ is determined solely by their main effects, and there is no need to reconsider any decisions made based on these main effects. If $b_{MI \times SKILLS} > 0$, the interaction is synergistic. This means that $\widehat{Y}_{adhere}$ will be larger than would be expected based solely on the main effects of *MI* and *SKILLS*. If $b_{MI \times SKILLS} < 0$, the interaction is antagonistic. This means that $\widehat{Y}_{adhere}$ is smaller than would be expected based solely on the main effects of *MI* and *SKILLS*. (Remember that this chapter has made the assumption that all of the factors are coded so that a main effect in the desired direction has a positive sign. If either *MI* or *SKILLS* were coded in the opposite direction, the signs of the coefficients corresponding to synergistic and antagonistic interactions would be different.)

The presence of important synergistic and antagonistic interactions sometimes demands a reconsideration of the provisional decisions based on the main effects and, depending on the exact results, can prompt a reversal of those decisions. A plot of any important interactions should always be examined before making any final decisions. (Where important and unexpected interactions are found, it may be a good idea to consider refining the conceptual model (Chap. 2) to include the interaction during the preparation phase of the next cycle of MOST and to build a replication of the finding of those interactions into the next optimization trial.)

Collins et al. (2014) recommended reviewing the provisional decisions by starting with the factor that demonstrated the largest main effect, then considering the important two-way interactions involving this factor, and then considering any important higher-order interactions involving this factor. This would be followed by consideration of the factor with the next largest main effect and so on. Thus the factors with important main effects will be considered in this order: first *PEER*, then *SKILLS*, and then *MI*.

7.5.1 When Provisional Decisions Are Likely to Be Reversed

Before reviewing the provisional decisions made above, let us consider under what circumstances provisional decisions are likely to be reversed. Consider four possible situations involving two factors, *A* and *B* (for simplicity assume any important main effects are in the desired direction and that the levels of the components are no and yes): (i) both *A* and *B* have important main effects, and the *A* × *B* interaction is synergistic; (ii) both A and B have important main effects, and the $A \times B$ interaction is antagonistic; (iii) only A has an important main effect, and the $A \times B$ interaction is synergistic; and (iv) only A has an important main effect, and the $A \times B$ interaction is antagonistic.

In situation (i), in which both A and B have important main effects, and they interact synergistically, it is unlikely that the provisional decision will be reversed. Both of the corresponding components will have been placed in the screened-in set, because the main effects suggest that each one will have a positive effect on the outcome. The parsimonious prediction model would look like this:

$$\widehat{Y} = b_0 + b_1 X_A + b_2 X_B + b_3 X_{A \times B},$$

where b_1, b_2, and b_3 are the coefficients corresponding to the main effect of A, the main effect of B, and the $A \times B$ interaction, respectively. The synergistic interaction suggests that the combination of A and B produces an additional positive effect, so this reinforces the provisional decision.

In situation (iv), in which only A has an important main effect, and the $A \times B$ interaction is antagonistic, it is also unlikely that the provisional decision will be reversed. Only the component corresponding to A will have been placed in the screened-in set. The parsimonious prediction model would look like this:

$$\widehat{Y} = b_0 + b_1 X_A + 0 X_B - b_3 X_{A \times B}.$$

The antagonistic $A \times B$ interaction suggests that adding the second component would have a negative net effect on the outcome. This also reinforces the provisional decision to place the second component in the screened-out set.

However, an antagonistic interaction between two factors does not necessarily mean that having both factors set to the higher level will result in a net negative effect on the outcome. In (ii), both components will have been placed in the screened-in set. The presence of an antagonistic $A \times B$ interaction means that the parsimonious prediction model will look like this:

$$\widehat{Y} = b_0 + b_1 X_A + b_2 X_B - b_3 X_{A \times B}.$$

Both main effects are moving the outcome in the desired direction, but the antagonistic interaction is moving it the opposite, undesired direction. Should the provisional decision be reversed to move one of the components from the screened-in set to the screened-out set to prevent this? As the above equation demonstrates and as will be illustrated below, whether one of the components should be moved from the screened-in set to the screened-out set depends on the size of b_3 in relation to b_1 and b_2. If b_3 is smaller than both b_1 and b_2, there is still a net gain associated with having both factors set to yes despite their antagonistic interaction, and therefore it is possible that both of the components should remain in the screened-in set. This situation demands careful thought and often requires the investigator to make a judgment call.

Situation (iii) also requires careful thought and possibly a judgment call. Here only the component corresponding to factor A has been placed in the screened-in set, but there is a synergistic $A \times B$ interaction:

$$\widehat{Y} = b_0 + b_1 X_A + 0 X_B + b_3 X_{A \times B}.$$

Factor B does not have an important main effect, but when it is set to yes, there is nevertheless a net gain because of the synergistic $A \times B$ interaction. Whether the component corresponding to B should be moved into the screened-in set depends primarily on whether the size of the interaction is sufficient to justify reversing the provisional decision.

If an optimization criterion other than all active components is being used, in Step 3 cost (broadly defined) will be a consideration in determining whether ultimately both components should be set to the higher level in the optimized intervention. Therefore, it is reasonable not to include cost at this point in the decision-making process. However, if the all active components criterion is being used, cost is not a formal consideration in decision-making, so it may be helpful to bring it in at this point in an informal way. For example, in situation (iii), the question is whether or not to move the component corresponding to Factor B into the screened-in set. The investigator may be less inclined to move this component into the screened-in set if it is very expensive, and more inclined to move it if it is relatively inexpensive.

7.5.2 Reconsidering Provisional Decisions Based on Interactions Involving PEER

It was stated above that the reconsideration of provisional decisions would start with the factor demonstrating the largest main effect, which in this case is *PEER*. *PEER* is involved in two interactions, *PEER* × *SKILLS* and *MI* × *PEER*.

The *PEER* × *SKILLS* interaction is synergistic, so this is an example of situation (i). The interaction, plotted using predicted cell means obtained from the parsimonious prediction model, is depicted in Fig. 7.2. (For more about how to plot and interpret graphs of interactions, refer to Chap. 4.) Thus according to the parsimonious prediction model, the *PEER* × *SKILLS* interaction produces a value of the outcome variable, $\widehat{Y}_{adhere}$, that is larger—in this case, better—than would be expected based on the main effects of these two factors alone. It is unlikely that this interaction will lead to a reversal of the provisional decisions to include peer mentoring and behavioral skills training in the screened-in set. Nevertheless, the plot of this interaction will be examined, because interactions should always be considered carefully.

The main effect of *SKILLS* is evident in the figure; there is a difference between $\widehat{Y}_{adhere}$ when *SKILLS* is set to low and $\widehat{Y}_{adhere}$ when *SKILLS* is set to high. Similarly, the main effect of *PEER* is evident; there is a difference between $\widehat{Y}_{adhere}$ at the no and yes levels of *PEER*. The figure illustrates how the combination of *PEER* = yes and

7.5 Step 2: Divide the Candidate Components Into the Screened-In... 243

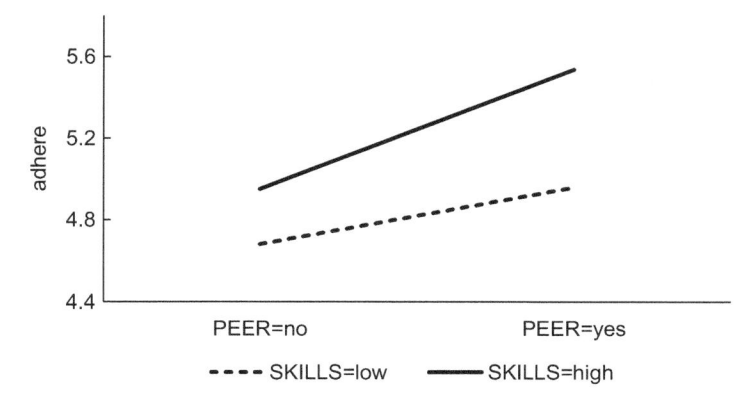

Fig. 7.2 Synergistic interaction between *PEER* and *SKILLS*. Artificial data

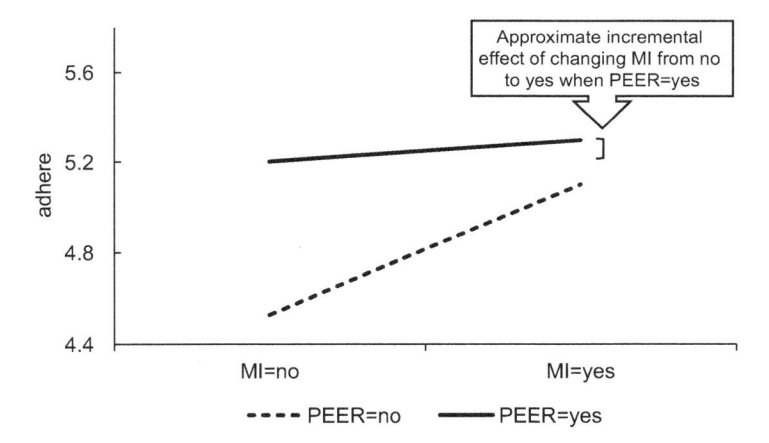

Fig. 7.3 Antagonistic interaction between *PEER* and *MI*. The bracket indicates the incremental effect of changing *MI* from no to yes when *PEER* is set to yes. Artificial data

$SKILLS$ = high seems to produce a boost in $\widehat{Y}_{adhere}$. Examination of this figure reinforces the decision to include the peer mentoring and behavioral skills training components in the screened-in set.

The $MI \times PEER$ interaction is depicted in Fig. 7.3. Because this is an antagonistic interaction, this is an example of situation (ii), so the provisional decision to include motivational interviewing in the screened-in set should be reconsidered particularly carefully. The decision to include motivational interviewing is being reconsidered rather than the decision to include peer mentoring because the *PEER* factor has a larger observed main effect size than the *MI* factor. If only one of the two components is to be selected for the screened-in set, it will be peer mentoring. Thus the working assumption is that peer mentoring is to be included in the screened-in set.

The investigator then needs to determine whether motivational interviewing should also be included.

Not every antagonistic interaction indicates a toxic combination of components or component levels (see Chap. 4). As discussed above, an antagonistic interaction means that when all the factors involved in the interaction are set to the higher level, the outcome is less favorable than would be expected based solely on the sum of the main effects. This does not necessarily mean that when all the factors involved in the interaction are set to the higher level, the outcome is less favorable than would be expected when only some of the factors are set to the higher level. For example, suppose Factor A and Factor B each have an important and positive main effect, and there is an important antagonistic $A \times B$ interaction. This may merely indicate that once one of the factors is set to yes, there are diminishing returns associated with setting the remaining factor to yes. Having Factor B set to yes when Factor A is set to yes will still move the outcome in the desired direction, if the main effect of Factor B is large enough to offset the reduction in the outcome produced by the antagonistic interaction. Thus *the presence of an antagonistic interaction involving two or more factors in an optimization trial does not necessarily mean that any of the corresponding components should be relegated to the screened-out set.*

A graph of the $MI \times PEER$ interaction appears in Fig. 7.3. When graphing cell means to determine whether a component should be moved from the screened-in set to the screened-out set or vice versa, it is usually helpful to place the factor corresponding to the component in question on the x-axis. Because the question is whether or not MI should be moved to the screened-out set, in Fig. 7.3 MI has been placed on the x-axis.

In the example, the coefficient for the main effect of MI is $b = 0.167$, and the coefficient for the $MI \times PEER$ interaction is $b = -0.119$. The evidence suggests that although the positive effect of including motivational interviewing in the intervention would be partially offset by the antagonistic interaction with peer mentoring, including both components nevertheless will move the outcome in the desired direction. It is evident from the figure that when $PEER$ is set to no, there is a large increment in $\widehat{Y}_{adhere}$ associated with changing MI from the no setting to the yes setting. By contrast, when $PEER$ is set to yes, the increment in $\widehat{Y}_{adhere}$ associated with changing MI from no to yes, indicated in the figure by a bracket, is much smaller.

Should the provisional decision to include motivational interviewing in the screened-in set be reversed? As discussed above, this is a judgment call. One reasonable perspective is the following: MI has shown an overall positive main effect, and when $PEER$ is set to the higher level, there is still a net gain associated with having MI set to the higher level. Therefore, it is expected that having both peer mentoring and motivational interviewing in the intervention would have a net positive effect, suggesting that both components should remain in the screened-in set. Later, in Step 3, the cost of MI will be introduced as a consideration.

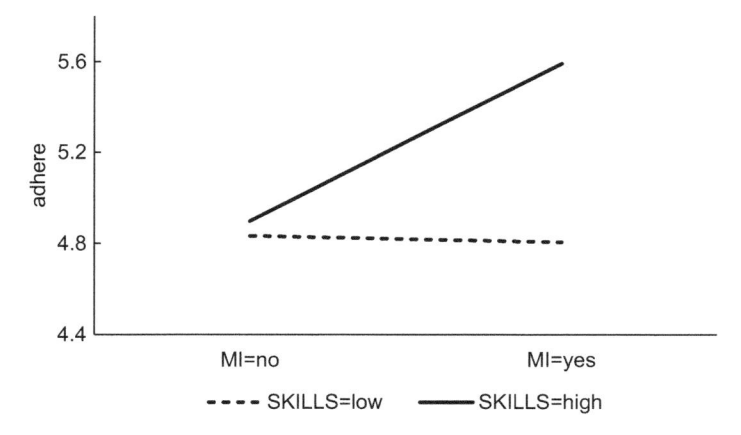

[Figure: plot with y-axis "adhere" from 4.4 to 5.6, x-axis with MI=no and MI=yes. Dashed line SKILLS=low stays near 4.8; solid line SKILLS=high rises from ~4.9 to ~5.6.]

Fig. 7.4 Synergistic interaction between *MI* and *SKILLS*. Artificial data

7.5.3 Reconsidering Provisional Decisions Based on Interactions Involving SKILLS

SKILLS, the factor showing the next largest main effect, is involved in two interactions that meet the interaction criterion: *PEER* × *SKILLS* and *MI* × *SKILLS*. *PEER* × *SKILLS* was discussed above.

The synergistic *MI* × *SKILLS* interaction is depicted in Fig. 7.4. Just as in Fig. 7.3, the main effect of each of the factors is evident. As Fig. 7.4 shows, when *SKILLS* and *MI* are both set to the high level, the combined effect is better than would be expected based solely on the main effects of *SKILLS* and *MI*. This confirms the decision to include motivational interviewing in the screened-in set. However, when *SKILLS* is set to the low level, on average there is no difference between having *MI* set to yes and set to no. These findings are particularly worth noting, because they suggest that if both components are not set to the higher level, it is possible that neither will be effective on average.

7.5.4 Reconsidering Provisional Decisions Based on Interactions Involving MI

MI, the factor showing the third-largest main effect, is involved in three interactions. Two of these, *MI* × *SKILLS* and *MI* × *PEER*, were examined above. The remaining interaction is *MI* × *TEXT*. This is an example of situation (iii). *TEXT* did not demonstrate an important main effect in the experiment, and therefore the corresponding component, text messaging, has provisionally been placed in the screened-out set. Because the empirical evidence suggests that *TEXT* has a synergistic interaction with *MI*, this provisional decision will now be reconsidered.

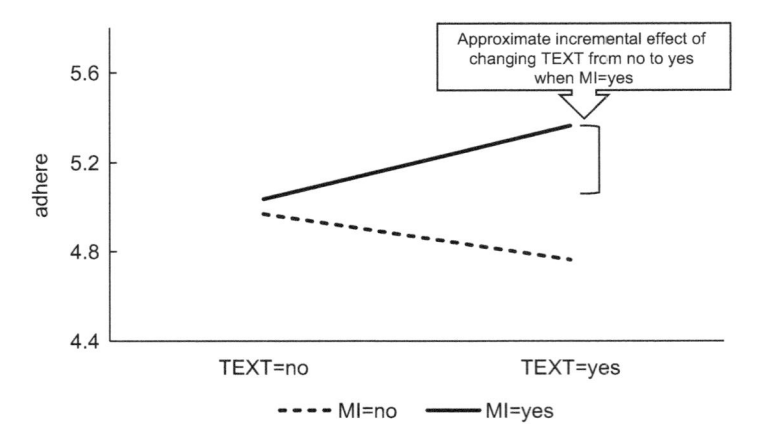

Fig. 7.5 Synergistic interaction between *MI* and *TEXT*. The bracket illustrates the incremental effect of changing the *TEXT* factor from no to yes when *MI* is set to yes. Artificial data

The *MI* × *TEXT* interaction is depicted in Fig. 7.5, with *TEXT* on the *x*-axis. As the figure shows, on average *TEXT* does not have an effect on the outcome variable. However, this is because its effect is quite different depending on the level of *MI*. When *MI* is set to no, *TEXT* appears to have an iatrogenic effect! (Reminder: These are artificial data.) However, when *MI* is set to yes, *TEXT* has a positive effect. Does this mean that the text messaging component should be included in the screened-in set? This depends in part on what the empirical results suggest about the expected size of the incremental gain in $\widehat{Y}_{adhere}$ produced by changing *TEXT* from no to yes when *MI* is set to yes. This incremental gain is indicated in Fig. 7.5 by a bracket. Whether this gain is large enough to justify moving *TEXT* into the screened-in set is a judgment call. Suppose in this case, the incremental gain appears large enough to Dr. B to justify including *TEXT* in the screened-in set. If an optimization criterion involving cost is to be used, this decision can be reconsidered again at that time. This will be demonstrated below.

7.5.5 Summary of Selection of the Screened-In and Screened-Out Sets of Components

The optimization trial examined five intervention components (motivational interviewing, peer mentoring, text messaging, mindfulness meditation, and behavioral skills training) by manipulating five corresponding factors (*MI*, *PEER*, *TEXT*, *MIND*, and *SKILLS*, respectively) in a factorial experiment. Based on the results of the experiment, it appears that the factors *MI*, *PEER*, and *SKILLS* have important main effects. Thus the corresponding components, namely, motivational interviewing, peer mentoring, and behavioral skills training, were provisionally selected for inclusion in the screened-in set. Text messaging and mindfulness meditation were provisionally placed in the screened-out set. When these provisional decisions were reconsidered based on the important interactions, it was decided that the text messaging component should be moved to the screened-in set. None of the other provisional decisions were reversed.

At this point Steps 1 and 2 of the decision-making procedure are complete, and the screened-in and screened-out sets of components have been identified. The screened-out set consists of mindfulness meditation. This component will be set to the lower level, in this case no, and will not be considered further. The screened-in set consists of motivational interviewing, peer mentoring, behavioral skills training, and text messaging. This set of components is eligible for further consideration in Step 3 of the decision-making process.

7.5.6 *A Note About the Lower Level of Components*

For the components that have been placed in the screened-out set, decision-making is complete at the end of Step 2. As Fig. 1 shows, no matter what optimization criterion has been selected, these components will be set to the lower level. If the lower level under consideration = no, the component is eliminated from the intervention by definition. If the lower level is something other than no, as in the behavioral skills component in the example, the component is included in the intervention, set to that lower level. Thus if the behavioral skills component was placed in the screened-out set, it would be included in the intervention at the low level, which consists of the workbook only.

For the components that have been placed in the screened-in set, decision-making continues in Step 3. These components may ultimately be set to either the high or the low level. The same reasoning about the low level applies here; in other words, a component set to the low level is eliminated from the intervention by definition if the low level = no.

7.5.7 *Orphan Interactions*

As mentioned above, it is possible for the results of a factorial ANOVA to include one or more orphan interactions, that is, unexpected sizeable interactions involving only factors that have not demonstrated important main effects. The heredity principle specifies that such interactions should be viewed with caution and that a synergistic orphan interaction involving a set of factors should not be considered immediate grounds for moving the corresponding components from the screened-out set into the screened-in set. First, it could be argued that an interaction made up of parent factors with unimportant main effects, particularly a higher-order interaction, is likely to be spurious. Second, if a factor does not demonstrate an important main effect but is involved in a synergistic interaction, this means the corresponding component works only when combined with certain other components and component levels. Such components are inherently less robust than those corresponding to factors that have demonstrated an important main effect in a factorial experiment, because by definition factors that demonstrate a main effect have been empirically demonstrated to work, on average, across a variety of combinations of levels of other components (see Chap. 3). In fact, from one perspective it is a worthwhile objective to build an intervention comprised as much as possible of components corresponding to factors that have demonstrated important main effects. If the components corresponding to the factors involved in, say, a higher-order orphan interaction are moved to the screened-in set, this makes several components that are of questionable robustness eligible to be included in the intervention or included at their higher levels.

If the results of an optimization trial include a sizeable synergistic interaction involving only factors that do not show main effects, this suggests that a good strategy may be to combine the corresponding components into a single component. This new component can then be examined in a future optimization trial, if it appears sufficiently promising.

It should be reiterated that the heredity principle applies only to unanticipated interactions—legitimate orphan interactions are possible. Theory or prior literature may predict that a set of, say, five intervention components, each of which is inactive on average, will be transformed by synergy into a powerful collective force if they are combined into an intervention. This would translate to an a priori prediction of an important five-way interaction with no important main effects on the parent factors. In MOST, such an interaction would be featured in the conceptual model. If a prediction about any interaction is made a priori and included in the conceptual model, the heredity principle, which is invoked only in the absence of a priori guidance, does not apply.

7.6 Step 3: Apply the Optimization Criterion

This section discusses how to apply the optimization criterion to the screened-in set to complete the decision-making process. Several different optimization criteria will be illustrated.

7.6.1 The All Active Components Criterion

This is the one case in which decision-making can stop as soon as the components have been sorted into the screened-in and screened-out sets. The investigator who uses the all active components criterion has decided to build an intervention by including the higher level of every component in the screened-in set, irrespective of cost, and the lower level of every component in the screened-out set. In the example,

7.6 Step 3: Apply the Optimization Criterion

the intervention would consist of providing participants with motivational interviewing, peer mentoring, behavioral skills training, and text messaging. Mindfulness meditation, which is in the screened-out set, would be set to no, so it would not be included in the intervention.

How successful was the decision-making procedure at selecting the set of components/component levels that produces the best outcome, regardless of cost? Although it may be desirable to identify the combination of components and component levels that produces the best outcome, there is no guarantee that the decision-making approach reviewed here will identify this combination in every case. A complicated set of important interactions can send mixed messages about which components should be included in the screened-in set, thereby making decision-making difficult. Moreover, the smaller the overall sample size N is, the less precise are the estimates of the regression weights, meaning that the regression weights are more likely to deviate substantially from the corresponding population parameters than they would be if the estimates were based on a larger sample size. Under these circumstances, the investigator may make decisions that are different from those that would have been made if better information were available. The aspirational goal of this decision-making approach is to select *one of the best* combinations, even when the available information is less than ideal, as it will be in most field research situations.

Ordinarily, an investigator who used an approach like the one described here to make decisions based on empirical data would be unable to know whether one of the best combinations of components had been selected, because the true model that generated the empirical data would be unknown. However, the above example used artificial data, so the true data generation model is known. It is

$$\mu_{adhere} = 5 + .2X_{MI} + .175X_{PEER} + .15X_{SKILLS} - .1X_{MI \times PEER} + .15X_{MI \times TEXT} + .125X_{MI \times SKILLS}.$$

Any regression weights not included in the above equation were set to zero. Because random error was added when the data were generated, the observed regression weights deviate from those in the equation.

Table 7.3 lists the combinations of components and component levels that are ranked fourth or better in order of the value of μ, the true model cell mean of the *adhere* outcome variable. Because of ties, more than four combinations are listed. In this example, the decision-making exercise identified one of two combinations that are tied for first place. The other top-ranked combination had all of the components set to the higher level. However, this combination includes mindfulness meditation, which in the data-generating model has a regression weight of zero, indicating it makes no contribution to the outcome. Even if cost in terms of money were a minor consideration, it would not be a good idea to include this component in the intervention because it would consume participant time and possibly other resources. It is possible that in the future these resources could be devoted to a component that would help to increase *adhere*. As discussed previously, one of the objectives of MOST is to arrive at an intervention that is efficient, in other words, eschews such components.

Table 7.3 Ten combinations of levels of factors with largest values of model mean outcome (μ)

MI	PEER	TEXT	MIND	SKILLS	μ	Rank
Yes[a]	Yes	Yes	Yes	High	5.70	1
Yes[b]	Yes	Yes	No	High	5.70	1
Yes	No	Yes	Yes	High	5.55	2
Yes	No	Yes	No	High	5.55	2
Yes	Yes	No	Yes	High	5.40	3
Yes	Yes	No	No	High	5.40	3
Yes	No	No	Yes	High	5.25	4
No	Yes	No	Yes	High	5.25	4
Yes	No	No	No	High	5.25	4
No	Yes	No	No	High	5.25	4

[a]Yes means included in intervention; no means not included in intervention
[b]Combination selected using the all active components optimization criterion

In the course of the decision-making, two difficult decisions arose: whether or not to include motivational interviewing and text messaging in the screened-in set. Ultimately, the decision was made to include both of these components. If different decisions had been made, one of the following screened-in sets would have been selected instead of the set that was ultimately chosen: (a) peer mentoring and behavioral skills training only or (b) motivational interviewing, peer mentoring, and behavioral skills training only. The use of artificial data provides a way of examining the results of these different decisions. As Table 7.3 shows, alternatives (a) and (b) are ranked fourth and third, respectively. Thus, if the decision-making approach outlined here had arrived at one of those two screened-sets, the set of components selected for the intervention would not have been as high-performing as the one chosen above, but nevertheless would have performed reasonably well. (A possible third alternative, namely, peer mentoring, text messaging, and behavioral skills training, is not on the table because, as noted above, the $MI \times TEXT$ interaction shows that $TEXT$ has an iatrogenic effect when MI is set to no.)

7.6.2 The Need for Other Optimization Criteria

Although the all active components criterion will produce an efficient intervention and is often an excellent choice of optimization criterion, it is not right for every situation. Sometimes key constraints must be included in decision-making to ensure economy or scalability. For example, suppose Dr. B needs to develop an intervention that costs less than, say, $500 per person to deliver. If the intervention that meets the all active components criterion costs $500 or more, then Dr. B will need to decide which components to select and which to jettison or set to the lower level to arrive at the best intervention that can be delivered within this budget constraint.

7.6 Step 3: Apply the Optimization Criterion

The optimization criterion provides a basis for making the decisions necessary to arrive at an intervention that provides the best outcome subject to constraints on cost, time, or any other resource. Depending on the optimization criterion, all, some, or even (rarely) none of the components in the screened-in set may be set to the higher level in the intervention. Thus, different optimization criteria may call for the inclusion of different components and component levels from the screened-in set in the intervention. This means that different investigators starting with the very same empirical results but different optimization criteria could, and often would, arrive at different optimized interventions. This is not a problem; on the contrary, it is appropriate, because different optimization criteria make different statements about the ultimate objectives of the intervention and about which constraints are critical. In a sense, they offer different visions for the optimized intervention.

Below several possible optimization criteria are reviewed. These are merely examples out of many possibilities.

The starting point for selecting the set of component levels that best meets the optimization criterion is the expected outcome for each possible combination of levels of the components in the screened-in set. Given the four components in the screened-in set, each with two possible levels, there are 16 different combinations of the levels of the corresponding factors. The combinations range from all of the factors set to the lower level to all of them set to the higher level. These are listed in Table 7.4, along with their expected outcomes, the $\widetilde{\widehat{Y}}_{adhere}$s, computed using the expression for the parsimonious prediction model. As a reminder that a decision has been made about a fifth component, *MIND* has been included, set to no in every combination.

Table 7.4 Parsimonious Model $\widehat{Y}$s for all possible combinations of factor levels corresponding to screened-in set

Combination number	MI	PEER	TEXT	MIND	SKILLS	$\widehat{Y}$ (parsimonious model)
1	No[a]	No	No	No	Low	4.708
2	No	No	No	No	High	4.618
3	No	No	Yes	No	Low	4.441
4	No	No	Yes	No	High	4.351
5	No	Yes	No	No	Low	5.227
6	No	Yes	No	No	High	5.445
7	No	Yes	Yes	No	Low	4.960
8	No	Yes	Yes	No	High	5.178
9	Yes	No	No	No	Low	4.652
10	Yes	No	No	No	High	5.286
11	Yes	No	Yes	No	Low	4.919
12	Yes	No	Yes	No	High	5.553
13	Yes	Yes	No	No	Low	4.694
14	Yes	Yes	No	No	High	5.635
15	Yes	Yes	Yes	No	Low	4.961
16	Yes	Yes	Yes	No	High	5.902

[a]No means not included in intervention; yes means included in intervention

7.6.3 Constraints on Money

As was mentioned above, suppose Dr. B needs to identify the best intervention that costs no more than $500 per person; in other words, Dr. B's optimization criterion is "Best expected outcome on *adhere* obtainable for no more than $500 per person." Suppose it costs $300 to supply motivational interviewing, $150 to supply peer mentoring, $50 for text messaging, and $25 for a behavioral skills workbook only, with an additional $200 for behavioral skills training, for a total of $225 for the higher level. Table 7.5 lists the 16 possible combinations of the levels of the components in the screened-in set, ordered by increasing cost.

Table 7.5 shows that Combinations 10, 12, 14, 15, and 16 exceed the $500 limit and therefore are ruled out immediately. Combination 16, which satisfied the all active components criterion, is the most expensive with a cost of $725. Of the combinations that cost $500 or less, the one with the largest value of $\widehat{Y}_{adhere}$ is Combination 6. Thus the intervention that meets the selected optimization criterion, that is, that provides the best expected outcome on *adhere* obtainable for no more than $500 per person, is made up of the peer mentoring component and behavioral skills training, including both workbook and training.

Table 7.5 Combinations of factor levels corresponding to components in screened-in set in order of cost in $US

Combination number	MI	PEER	TEXT	MIND	SKILLS	$\widehat{Y}$ (parsimonious model)	Cost (US$)
1	No[a]	No	No	No	Low	4.708	25
3	No	No	Yes	No	Low	4.441	75
5	No	Yes	No	No	Low	5.227	175
2	No	No	No	No	High	4.618	225
7	No	Yes	Yes	No	Low	4.960	225
4	No	No	Yes	No	High	4.351	275
9	Yes	No	No	No	Low	4.652	325
6	No	Yes	No	No	High	5.445	375
11	Yes	No	Yes	No	Low	4.919	375
8	No	Yes	Yes	No	High	5.178	425
13	Yes	Yes	No	No	Low	4.694	475
10	Yes	No	No	No	High	5.286	525
15	Yes	Yes	Yes	No	low	4.961	525
12	Yes	No	Yes	No	High	5.553	575
14	Yes	Yes	No	No	High	5.635	675
16	Yes	Yes	Yes	No	High	5.902	725

[a]No means not included in intervention; yes means included in intervention

7.6.4 Constraints on Time

Now suppose instead of money per se, Dr. B has identified participant burden, operationalized as time spent on the intervention, as a key constraint. Dr. B believes that participants will be most engaged in the intervention if it requires no more than 4 h, or 240 min, of their time. In this case the optimization criterion can be expressed as "Best expected outcome on *adhere* obtainable while requiring no more than 240 min of participant time."

Dr. B can take an approach to identifying the optimized intervention that is essentially the same as the one taken when the optimization criterion involved an upper limit on money. It is necessary to determine the anticipated amount of time that will be required for each combination of components. Any combinations that require more than 240 min of participant time are then ruled out immediately.

Suppose motivational interviewing is one 50-min session. Peer mentoring is six 30-min sessions for a total of 180 min. It is estimated that reading and thinking about the messages sent to provide text messaging will take no more than 10 min. It takes 30 min to complete the behavioral skills workbook, and the skills training takes another 90 min, for a total of 120 min.

The total expected amount of participant time required for each combination is shown in Table 7.6. The table shows that combinations 6, 8, 13, 14, 15, and 16 take more than 240 min to implement and therefore are ruled out immediately. Again,

Table 7.6 Combinations of factor levels corresponding to components in screened-in set in order of minutes of participant time required

Combination number	*MI*	*PEER*	*TEXT*	*MIND*	*SKILLS*	$\widehat{Y}$ (parsimonious model)	Minutes required
1	No[a]	No	No	No	Low	4.708	30
3	No	No	Yes	No	Low	4.441	40
9	Yes	No	No	No	Low	4.652	80
11	Yes	No	Yes	No	Low	4.919	90
2	No	No	No	No	High	4.618	120
4	No	No	Yes	No	High	4.351	130
10	Yes	No	No	No	High	5.286	170
12	Yes	No	Yes	No	High	5.553	180
5	No	Yes	No	No	Low	5.227	210
7	No	Yes	Yes	No	Low	4.960	220
13	Yes	Yes	No	No	Low	4.694	260
15	Yes	Yes	Yes	No	Low	4.961	270
6	No	Yes	No	No	High	5.445	300
8	No	Yes	Yes	No	High	5.178	310
14	Yes	Yes	No	No	High	5.635	350
16	Yes	Yes	Yes	No	High	5.902	360

[a]No means not included in intervention; yes means included in intervention

Combination 16, which meets the all active components criterion, is the most expensive at 360 min. Of the remaining combinations, Combination 12 is associated with the best outcome. Therefore, the intervention that meets the selected optimization criterion is made up of motivational interviewing, text messaging, and behavioral skills training, including both workbook and training.

7.6.5 A Note on Cost and the All Active Components Criterion

The all active components optimization criterion is used to achieve efficiency when constraints are not a consideration. Yet even under these enviable circumstances, if two interventions offer nearly the same expected outcomes, it may make sense to go with the less resource-intensive one. If reliable data on cost, time, or other resources are available, it can be helpful to go through the exercise of using the parsimonious model to predict the outcome of each combination of components/levels in the screened-in set. This information can be used to determine whether there are interventions that are as attractive or nearly as attractive as the one selected using the all active components criterion, but less costly, less burdensome, or otherwise less resource-intensive to implement. For example, Combination 12 is $150 cheaper to implement than Combination 16, takes only half as much participant time, and produces the third-best value of $\widehat{Y}_{adhere}$.

7.6.6 Constraints on Multiple Resources

A single optimization criterion can include constraints on more than one resource. Suppose Dr. B determines that the intervention can cost no more than $400 *and* must take no more than 210 min to implement. This would translate to an optimization criterion that can be stated as follows: "Best expected outcome on *adhere* obtainable for no more than $400 per person AND while demanding no more than 210 min of participant time."

Table 7.7 lists the combinations that exceed neither the cost (Table 7.5) nor time (Table 7.6) criteria, ordered by combination number. Of these combinations, Combination 5 is associated with the best outcome. Therefore, the intervention that meets this optimization criterion consists of peer mentoring and the behavioral skills training workbook only.

7.6 Step 3: Apply the Optimization Criterion

Table 7.7 Combinations of levels of components in screened-in set that cost ≤ $400 and require ≤ 210 min of participant time, in order of $\widehat{Y}$

Combination number	MI	PEER	TEXT	MIND	SKILLS	$\widehat{Y}$ (parsimonious model)	Cost (US$)	Cost (time)
1	No[a]	No	No	No	Low	4.708	25	30
2	No	No	No	No	High	4.618	225	120
3	No	No	Yes	No	Low	4.441	75	40
4	No	No	Yes	No	High	4.351	275	130
5	No	Yes	No	No	Low	5.227	175	210
9	Yes	No	No	No	Low	4.652	325	80
11	Yes	No	Yes	No	Low	4.919	375	90

[a]No means not included in intervention; yes means included in intervention

7.6.7 Four Different Optimization Criteria, Four Different Interventions...Which One Is Best?

In the preceding paragraphs, four different optimization criteria were used to build interventions based on the results of a single optimization trial. These optimization criteria defined the desired intervention as the one that (i) is made up of all the active components, irrespective of cost; (ii) produces the best expected outcome that can be obtained for no more than $500 per participant; (iii) produces the best expected outcome that can be obtained while requiring no more than 240 min of participant time; and (iv) produces the best expected outcome that can be obtained while costing no more than $400 per participant and requiring no more than 210 min of participant time. Given these different optimization criteria, it is not surprising that four different optimized interventions were produced. The optimization criteria and resulting interventions are summarized in Table 7.8.

The contrast across these four interventions helps to illustrate the difference between the concept of best and the concept of optimized. Recall the definition of optimization of an intervention from Chap. 1:

> Optimization of an intervention is the process of identifying an intervention that provides the best expected outcome obtainable within key constraints imposed by the need for efficiency, economy, and/or scalability.

Many scientists would say that the all active components intervention, Combination 16, is best because it is expected to produce the largest value of *adhere*. Here the definition of "best" is "most effective that can be obtained irrespective of cost." However, this definition of "best" may not be appropriate in every situation.

Would a scientist who must develop an ART adherence intervention that can be implemented for $500 per person or less consider Combination 16 "best," even though in his or her situation, it would not be scalable? What would be "best" to this scientist? The outstanding performance of Combination 16 may provide a useful standard, but otherwise matters little in practical terms, because this intervention is too expensive to go to scale. A more realistic definition of "best" for this scientist

Table 7.8 Comparison of interventions optimized according to different criteria

| Optimization criterion | Combination number | Components |||||| $\hat{Y}$ (parsimonious model) | Cost (US$) | Cost (minutes) |
		Motivational interviewing	Peer mentoring	Text messaging	Mindfulness meditation	Behavioral skills training			
All active components	16	*Included*	*Included*	*Included*	Not included	*Workbook + training*	5.902	725	360
Best outcome for ≤ $500 per participant	6	Not included	*Included*	Not included	Not included	*Workbook + training*	5.445	375	300
Best outcome for ≤ 240 min of participant time	12	*Included*	Not included	*Included*	Not included	*Workbook + training*	5.553	575	180
Best outcome for ≤ $400 per participant and ≤ 210 mins of participant time	5	Not included	*Included*	Not included	Not included	*Workbook only*	5.227	175	210

would be "as close as possible to the performance of Combination 16 without exceeding $500 per person." Out of the subset of combinations that can be implemented within the budget of $500 per person, Combination 6 shows the best performance. Combination 6 is not the best intervention in an absolute sense—it is outperformed by Combination 16—but it is the best *given this constraint on implementation costs* and therefore is the optimal intervention under this set of circumstances. Similarly, the other two optimization criteria listed in Table 7.8 produced two more interventions, Combination 5 and Combination 12. Neither of these outperforms Combination 16, but each is the optimal intervention according to its respective criterion and therefore can be considered best subject to a specific set of constraints. If the key constraints have been accurately specified in the optimization criterion, each of these interventions will be immediately scalable, without any need for ad hoc modifications, once it has been evaluated in an RCT.

7.6.8 *When a Component From the Screened-In Set Must Be Omitted From the Optimized Intervention*

Components in the screened-in set have been placed there because they contribute meaningfully toward moving the outcome in the desired direction. The above discussion has shown that nevertheless, depending on the optimization criterion in use, a component in the screened-in set may not be included in the optimized intervention at the higher level. For example, motivational interviewing is not included in two of the optimized interventions listed in Table 7.8.

Even if the higher level of a component from the screened-in set is ultimately not selected for inclusion in the optimized intervention, examining the component in the optimization trial has not been a waste of resources. Thanks to the optimization trial, there is now scientific evidence that there is a detectable difference between the higher and lower levels, which is adding to the body of knowledge about what works to effect change in the outcome. This is a valuable result in and of itself, and, assuming it is published in a peer-reviewed journal or otherwise made available to the scientific public, will be a building block in future intervention research.

7.6.9 *The Shortcomings of Ad Hoc Modifications*

The hypothetical example can be used to illustrate the inferiority of making ad hoc modifications to interventions to make them more scalable, as compared to using an approach like MOST to build an optimized intervention that is immediately scalable. Imagine Combination 16 has been evaluated in an RCT and shown to have a significant effect and now is to be implemented in healthcare settings. A healthcare clinic has been encouraged by the state health department to implement Combination

16 but cannot afford to implement it as is. Suppose the clinic does not have access to the results of the previous optimization trial. Without any information about the performance of individual components and interactions between components, the only way to arrive at an intervention that can be implemented is to make whatever ad hoc modifications to Combination 16 appear sensible.

Suppose the clinic decides that motivational interviewing and peer mentoring are too expensive and logistically complicated. Text messaging is automated and therefore logistically simple. The clinic is already using the behavioral skills workbooks and is readily able to provide behavioral skills training in addition to the workbook. Therefore, they plan to implement Combination 4. Although this may seem like a reasonable decision, Combination 4 has the worst observed outcome out of the 16 combinations that can be constructed from the screened-in set (see Table 7.4). The optimization trial demonstrated that there is an interaction between *MI* and *TEXT* indicating that when *MI* is set to no, *TEXT* has an iatrogenic effect. This suggests that the text messaging component should not be provided unless motivational interviewing is provided too. Combination 5 would provide a better expected outcome and would cost less; even an intervention consisting of all of the components set to the lower level—in other words, the current standard of care, which consists of only the behavioral skills workbook—would be expected to produce a better outcome.

This illustrates how ad hoc modifications to interventions may not identify the most efficient and economical alternative and can even backfire, producing an outcome considerably less favorable than could be obtained via a better use of the existing resources.

7.6.10 Reusing the Results of an Optimization Trial With Different Constraints

Now imagine a healthcare clinic has access to the results of Dr. B's optimization trial. This clinic has costs that are considerably different from those applied in the exercise above, because they enjoy the support of the National Foundation for the Promotion of Motivational Interviewing, which pays 80 percent of the costs of motivational interviewing. For this clinic, the cost of motivational interviewing is a small fraction of what was used in the cost analysis presented in Table 7.5. If it would be reasonable to assume the results of the original optimization trial would generalize to their setting and participant population, this clinic could select their own optimized intervention by identifying their desired optimization criterion and going through Steps 1, 2, and 3 of the decision-making process.

A similar exercise could be done if different constraints need to be applied in a different setting. For example, perhaps there is a skilled labor shortage in a particular area and staff time is at a premium. The results of an optimization trial can readily be reused based on a detailed set of results from the factorial ANOVA; it is not necessary to start with the original data, so there would be no risk of violating subject confidentiality. Of course, if the results of an optimization trial are intended for use in optimizing different interventions in a variety of settings, the investigator would be wise to take whatever measures are necessary to recruit a sample that is representative of the entire population of individuals for whom the intervention is intended.

Suppose an intervention is optimized based on the results of an optimization trial, and then a new intervention is optimized reusing the same optimization trial results but employing either a different optimization criterion or different data on cost. If the new optimized intervention is different from the previous one, and it is desired to evaluate the effectiveness of the new intervention as a package, it will be necessary to move to the evaluation phase of MOST and conduct a separate RCT.

7.6.11 *A Note on the Estimation of Cost*

To make effective use of an optimization criterion that involves cost in some form, in other words, any criterion other than all active components, the investigator has to have reasonably good estimates of the cost of implementing each combination of components. In this author's experience, most intervention scientists are well aware of the costs of conducting research, because they have experience writing grant proposals, where such knowledge is necessary. However, the cost of implementing an intervention in the intended clinical, educational, or community setting is often very different from the cost of conducting research. Moreover, implementation costs may vary within a nation and will definitely vary among nations. It would move intervention science, and possibly society, forward considerably if better communication about resource requirements were established between academic intervention scientists and those who are responsible for paying for and implementing the interventions developed by academic intervention scientists. Only then will intervention scientists be able to use the considerable power of controlled experimentation to ensure that society is obtaining the best return on its investment in interventions.

7.6.12 *The Possibility of a Bayesian Approach*

As mentioned in Chap. 3, in general, the decision-making described in this book has been made from a decision-priority rather than a conclusion-priority perspective. When the conclusion-priority perspective is taken, the decision-maker is guided by generally agreed-upon conventions, such as $p<.05$ as the definition of statistical significance. By contrast, when the decision-priority perspective is taken, there are no such guidelines. Instead, the investigator must make decisions based on all available information, taking uncertainty into account. This suggests that Bayesian methods are likely to be very helpful in the optimization phase of MOST, as they have been in the field of decision science. Guidelines for how to use Bayesian methods in making the decisions required by MOST, as well as applications of Bayesian methods in the experimentation conducted during the optimization phase of MOST, are needed.

7.7 When There Is More Than One Outcome Variable

In the examples of optimization criteria in the previous section, there was only one outcome, even when more than one type of constraint was specified. However, in some situations there may be more than one important outcome. This complicates matters considerably. Let us consider two related but distinct scenarios in which there may be more than one outcome variable.

7.7.1 When Measures of Mediators Are Used as Short-Term Outcomes

The first scenario is one where, as discussed in Chap. 2, the outcome of primary interest can occur in the distant future. An example is a school-based intervention program aimed at preventing drug abuse in adolescence, but delivered in middle school. If the primary outcome is drug use in, say, 10th grade, but the intervention is delivered in 6th grade, an investigator using drug use as the outcome for an optimization trial would have to wait 4 years to conclude the trial. As an alternative, measures of the target mediators can be used as short-term outcomes.

In this example, suppose the conceptual model specifies three candidate components, each targeting one of the following mediators: norms about drug use, expectations about the effects of drugs, and self-efficacy for resisting offers to use drugs. If the performance of each of these components was evaluated shortly after implementation of the optimization trial by using a measure of its respective target mediator, there would be three outcome variables, because there is a distinct mediator, and therefore a separate outcome variable, for each component. As explained in Chap. 2, the component targeting norms about drug abuse would be evaluated using a measure of norms, the component targeting expectations about the effects of drugs would be evaluated using a measure of expectations, and the component targeting self-efficacy for resisting offers to use drugs would be evaluated using a measure of self-efficacy.

When measures of mediators are used as short-term outcomes, particularly careful consideration should be given to selection of the main effect criterion. In general, a component's relation with the primary outcome will be smaller than its relation with the target mediator. Therefore, as compared to optimization based on the primary outcome, when optimization is based on target mediators, it may be advisable to use a larger main effect criterion.

Optimization based on mediators as short-term outcomes is most straightforward when the all active components optimization criterion is used. In the school-based drug abuse prevention example, there would be three ANOVAs, one for each outcome variable, and three sets of results. An adaptation of the decision-making procedure described above could be used to sort the components into the screened-in and screened-out sets. Based on the results, the investigator would establish, for example, which factors have an important and positive main effect on their own outcome variables. It would then be necessary to reconsider this preliminary decision not only on the basis of important interactions but also to take into account whether any factors have a negative main effect on any of the other outcome variables. Difficult decisions may arise if, for example, a factor has a positive main effect on its own outcome but has a negative main effect on a different outcome or is involved in a strong antagonistic interaction with another factor on a different outcome. However, each component would usually be designed primarily to affect its target mediator, so effects on other mediators would be expected to be minor.

In Chap. 6 it was mentioned that employing measures of mediators as short-term outcomes for the purpose of optimization can complicate or even rule out the use of some optimization criteria. The approach discussed in this chapter for incorporating cost into optimization involves using a single parsimonious prediction model. With different outcome variables used for decisions about different components, there would be more than one parsimonious model, and so it is unclear how they can be integrated to inform a decision about overall optimization involving cost. For this reason, currently the use of mediators as short-term outcomes is recommended with extreme caution for optimization criteria that involve cost. More research is needed in this area.

7.7.2 When There Is More Than One Primary Outcome

The second scenario is one where there is more than one primary outcome variable. Consider the hypothetical example used in this book, in which the primary outcome variable was *adhere*. Suppose in addition to *adhere*, there is a second primary outcome variable, *drinks*, number of alcoholic drinks consumed in the 30-day period following the conclusion of the intervention. The presence of *drinks* as a second outcome variable requires articulation of some important trade-offs. For example, are the two outcomes equally important, or should they be differentially weighted? If a component has an important effect on *drinks* but not on *adhere*, should it be placed in the screened-in set? What if a component has an important effect on *adhere* and a negative (iatrogenic) effect on *drinks*? The answers to questions like these must be decided upon by the research team, ideally during the preparation phase. This situation is in some ways more complicated than the first, because in the first scenario, each component is expected primarily to affect its own target mediator and to have at most only a small effect on other outcomes. By contrast, in this scenario, each component is expected to have an effect on both outcomes. Moreover, as in the first scenario, there will be more than one (in this case two) parsimonious prediction models. How can these be integrated when an optimization criterion involving cost is to be used? A brief discussion of some of the issues appears in Dziak (2018) in the companion volume, but much more research is needed, along with guidance for investigators.

7.8 Why Bother?

Why bother with the decision-making approach outlined here? It would be possible to use the saturated regression model developed while performing the factorial ANOVA, that is, the regression model based on all of the main effects and interactions corresponding to all of the factors, to create a $\widehat{Y}$ for each combination of levels of the factors in the experiment. This saturated model could then be used to select the combination of components and component levels that corresponds to the largest $\widehat{Y}$. Why not take this simple and expedient approach, instead of going through a complex decision-making process to identify the screened-in and screened-out sets?

Recall from Chap. 1 that one of the four desiderata for interventions is efficiency, which means that elimination of inactive components and unnecessarily high component levels from interventions is an important objective of the MOST approach. Suppose an investigator decides to examine the $\widehat{Y}$s and simply select the set of components and component levels corresponding to the largest one. There is nothing in this approach that weeds out inactive components and unnecessarily high component levels. Even a factor that demonstrates a positive effect so small that it may safely be attributed to random error will increase $\widehat{Y}$ by some amount. In fact, unless all of the effects are exactly zero, every factorial experiment will produce a largest $\widehat{Y}$ (which may correspond to several cells), even if all of the effects are small. Simply selecting a configuration of components and component levels that corresponds to the largest $\widehat{Y}$ could produce an intervention including, or even made up of, inactive components. As compared to an intervention made up of exclusively active components, the resulting intervention may be more likely to fail to show a detectable effect in an RCT. The approach outlined in this chapter is aimed at helping the investigator to arrive at an efficient optimized intervention made up exclusively of active components, in other words, an intervention that is likely to demonstrate both a clinically and statistically significant effect in an RCT.

7.9 After the Decision-Making Is Complete

7.9.1 Secondary Analysis and Qualitative Data

Secondary analysis of data from an RCT has a long history in intervention science. Secondary analysis of data from an optimization trial also can be extremely helpful. In particular, mediation and moderation analyses performed on data gathered during MOST, in both the optimization and evaluation phases, can be used in the preparation phase to inform development or refinement of the conceptual model.

There has been much research on mediation analyses based on data from studies in which there is a single treatment, such as an RCT (e.g., MacKinnon, 2008). It is also possible to conduct mediation analyses on data from a factorial screening experiment. This offers an opportunity to take a detailed look inside the "black box" of an intervention and test hypotheses about what variables mediate the observed main and interaction effects. Mediation analysis based on data from a factorial experiment is discussed in Smith, Coffman, and Zhu (2018) in the companion volume. Another area of interest may be moderation analysis to understand whether and how a component's effects vary depending on characteristics of the individual or the environment. Moderation analyses often involve examining whether there are sizeable interactions between individual or environmental characteristics and a component. Moderation analyses can be helpful in developing adaptive interventions, discussed in Chap. 8. Interactions in factorial optimization trials are discussed in Chap. 4.

Carefully collected and analyzed qualitative data have the potential to enrich any application of MOST tremendously. In particular, so little is currently known about interactions that qualitative data are likely to be essential in understanding interactions when they are detected. Qualitative data may also be helpful in understanding why a component did not work as anticipated and how it could be improved. Currently there is very little literature on the use of qualitative data in MOST. Examples of how qualitative data can enrich MOST would be helpful to intervention scientists.

7.9.2 The Evaluation Phase of MOST

At the conclusion of the optimization phase of MOST, the investigator moves to the evaluation phase, provided that the optimized intervention is expected to be sufficiently effective (see Chap. 1). In the evaluation phase, the optimized intervention is evaluated in a randomized controlled trial (RCT). This experimental design enables direct comparison of the optimized intervention against a suitable control or comparison treatment. What constitutes a suitable control treatment depends on the research question at hand and on what is the minimal treatment that it is ethical to provide for the control group subjects. Often the control group is provided with the current standard of care or is placed on a waiting list and provided with treatment after the experiment has been conducted. If the research question concerns whether a newly optimized intervention is detectably more effective than an existing intervention, the comparison treatment would be the existing intervention.

The purpose of optimization may be to start with an existing intervention and develop a new version of the intervention that is at least as effective, but more efficient, less costly, or more scalable. In this case an equivalency trial or noninferiority trial may be desired in the evaluation phase of MOST to demonstrate that the new version is at least as effective as the existing version.

Some readers may be wondering whether a separate evaluation phase is strictly necessary in MOST. These readers may feel there must be a way to combine the optimization trial with an RCT, so that only one experiment is necessary. After all, embedded in a factorial experiment are conditions corresponding both to a control (all of the factors are set to the lower or no level) and to the optimized intervention (each component set to the level selected for the optimized intervention). Perhaps the means of these two conditions could simply be compared in what would effectively be a two-arm RCT.

There are several reasons why this approach usually does not work well. First, the approach may not even be possible if a fractional factorial design was used for the optimization trial and a condition in which all the factors are set to the lower level or a condition corresponding to the optimized intervention is not included (see Chap. 5). Second, in general, even a factorial experiment that has a high degree of power for detection of main effects and interactions is unlikely to have sufficient power for detection of differences between means of two individual experimental conditions. To ensure enough power to compare two experimental conditions, it would be necessary to include as many subjects in each experimental condition of the factorial experiment as would be included in each arm of a fully powered RCT. This would typically be highly resource-intensive. As has been discussed, one motivation for using a factorial experiment is to estimate the effect of each component in an efficient way and thereby enable the investigator to determine which combinations of components and component levels will constitute the optimized intervention. The idea is that once the optimized intervention has been identified, the expected effect size may be substantial enough that an RCT with a relatively modest sample size will provide sufficient power.

A third reason has to do with the differences between the factorial experiment and the RCT in the concept of control. Briefly, each factor in a factorial experiment has its own control (this is discussed in detail in Chap. 3). This may be different from the type of control desired in an RCT, as described above. Cook et al. (2016), Piper et al. (2016), and Schlam et al. (2016) reported on three factorial experiments examining a total of 15 components that were candidates for inclusion in a comprehensive clinic-based smoking cessation intervention. Based on the results of the experiments, five components were selected for inclusion at the higher level (Piper et al., in press). In the evaluation phase of MOST, the investigators wanted to compare the performance of the optimized intervention against that of an intervention representing the standard of care that had evolved in the field since the experiments were begun some

4 years previously. Another example would be any experiment in which the lower level of the factors represents no treatment, which in some situations could be an ethical approach if at the conclusion of the experiment the subjects were offered any components they were not originally randomly assigned. The control for the RCT that would be directly analogous would be a wait-list control, but it might be desired to employ a standard-of-care comparison instead.

Investigators who are determining whether or not to proceed to the evaluation phase of MOST may wish to consider that the way the term "effective" is used in the optimization phase is subtly different from how it is used in the evaluation phase. In the optimization phase, effectiveness is expressed without direct reference to a control or comparison group; it is expressed in terms of the predicted outcome, $\widehat{Y}$, produced by the parsimonious model that was derived from the optimization trial. By contrast, in the evaluation phase, effectiveness is determined by means of an RCT and expressed by direct comparison of the mean outcome for the treatment group, who are assigned to receive the optimized intervention, to the mean outcome for the control group. As has been discussed earlier in this book, the optimized intervention—the one that is the most effective subject to constraints—may or may not demonstrate statistically detectable effectiveness in an RCT.

As has been stated repeatedly in this book, MOST is a framework, not an off-the-shelf procedure. Although the evaluation phase is a part of MOST, it may not be needed in each and every case. Every decision made in MOST must be guided by the resource management principle and tempered with common sense. There may be times when the results of the optimization trial are so clear there is little doubt the optimized intervention would be highly likely to have a statistically and clinically significant effect in an RCT. In this case, the investigator must turn to the resource management principle and determine whether the available resources would be better spent conducting an RCT to remove any remaining doubt about the ultimate effectiveness of the optimized intervention or strengthening the intervention further via another optimization trial.

7.10 What's Next

Chapters 1, 2, 3, 4, 5, 6, and 7 in this book have reviewed how to optimize a traditional fixed intervention. The next chapter discusses the MOST framework more broadly, to include adaptive interventions. Selection of an experimental design for optimization of various kinds of interventions is discussed.

References

Collins, L. M., & Kugler, K. C. (Eds.). (2018). *Optimization of multicomponent behavioral, biobehavioral, and biomedical interventions: Advanced topics*. New York, NY: Springer.

Collins, L., Trail, J., Kugler, K., Baker, T., Piper, M., & Mermelstein, R. (2014). Evaluating individual intervention components: Making decisions based on the results of a factorial screening experiment. *Translational Behavioral Medicine, 4*, 238–251.

Cook, J. W., Collins, L. M., Fiore, M. C., Smith, S. S., Fraser, D., Bolt, D. M., et al. (2016). Comparative effectiveness of motivation phase intervention components for use with smokers unwilling to quit: A factorial screening experiment. *Addiction, 111*, 117–128.

Dziak, J. (2018). Optimizing the cost-effectiveness of a multicomponent intervention using data from a factorial experiment: Considerations, open questions, and tradeoffs among multiple outcomes. In L. M. Collins & K. C. Kugler (Eds.), *Optimization of behavioral, biobehavioral, and biomedical interventions: Advanced topics*. (forthcoming) New York, NY: Springer.

Kraemer, H. C. (2011). Another point of view: Superiority, noninferiority, and the role of active comparators. *The Journal of Clinical Psychiatry, 72*, 1350–1352.

Kugler, K. C., Dziak, J. J., & Trail, J. (2018). Coding and interpretation of effects in analysis of data from a factorial experiment. In L. M. Collins & K. C. Kugler (Eds.), *Optimization of multicomponent behavioral, biobehavioral, and biomedical interventions: Advanced topics* (forthcoming). New York, NY: Springer.

MacKinnon, D. P. (2008). *Introduction to statistical mediation analysis*. New York: Lawrence Erlbaum Associates.

Piper, M. E., Cook, J. W., Schlam, T. R., Jorenby, D. E., Smith, S. S., Collins, L. M., ... Baker, T. B. (in press). A randomized controlled trial of an optimized smoking treatment delivered in primary care. *Annals of Behavioral Medicine*.

Piper, M. E., Fiore, M. C., Smith, S. S., Fraser, D., Bolt, D. M., Collins, L. M., et al. (2016). Identifying effective intervention components for smoking cessation: A factorial screening experiment. *Addiction, 111*, 129–141.

Schlam, T. R., Fiore, M. C., Smith, S. S., Fraser, S., Bolt, D. M., Collins, L. M., et al. (2016). Comparative effectiveness of intervention components for producing long-term abstinence from smoking: A factorial screening experiment. *Addiction, 111*, 142–155.

Smith, R. A., Coffman, D. L., & Zhu, X. (2018). Investigating an intervention's causal story: Mediation analysis using a factorial experiment and multiple mediators. In L. M. Collins & K. C. Kugler (Eds.), *Optimization of multicomponent behavioral, biobehavioral, and biomedical interventions: Advanced topics* (forthcoming). New York, NY: Springer.

Walker, E., & Nowacki, A. S. (2011). Understanding equivalence and noninferiority testing. *Journal of General Internal Medicine, 26*, 192–196.

Wu, C. F. J., & Hamada, M. S. (2009). *Experiments: Planning, analysis, and optimization* (2nd ed.). Hoboken, NJ: Wiley.

Chapter 8
Introduction to Adaptive Interventions

Abstract Up to this point, this book has focused primarily on optimization of fixed interventions, in which the intervention design calls for every participant to receive the same treatment. This chapter provides an introduction to adaptive interventions, in which the content, dose, or approach can be varied across participants and across time, with the objective of achieving or maintaining a good outcome for all participants. The multiphase optimization strategy (MOST) can be used to optimize adaptive interventions. This chapter briefly reviews alternative approaches for the optimization trial when an adaptive intervention is to be optimized. Familiarity with the material in all previous chapters, particularly 1–4, is assumed.

Contents

8.1	Adaptive Interventions: The Basics	268
	8.1.1 Rationale for Adaptive Interventions	268
	8.1.2 The Anatomy of an Adaptive Intervention	269
8.2	Intensity of Adaptation	272
8.3	Identifying the Components of an Adaptive Intervention and Selecting an Approach for the Optimization Trial	273
	8.3.1 The Sequential, Multiple Assignment, Randomized Trial (SMART)	276
	8.3.2 A Brief Note About Powering a SMART	278
	8.3.3 The Outcome Variable in SMARTs	279
	8.3.4 Two Persistent Sources of Confusion	279
	8.3.5 Optimization Trials for Higher-Intensity Adaptive Interventions	280
8.4	Summary of Selection From the MOST Optimization Phase Toolbox	281
8.5	Some Open Areas and Open Questions	283
	8.5.1 The Preparation Phase of MOST and Adaptive Interventions	283
	8.5.2 The Use of Optimization Criteria With Adaptive Interventions	284
	8.5.3 How Personally Tailored Should an Intervention Be?	284
	8.5.4 Robust Adaptive Interventions	285
8.6	What's Next	285
References		286

8.1 Adaptive Interventions: The Basics

8.1.1 Rationale for Adaptive Interventions

Recall Dr. B, who is developing an intervention to reduce viral load in HIV-positive individuals who drink heavily. The purpose of the intervention is to promote ART adherence and ultimately reduce viral load. Because this is a fixed intervention, the same components and component levels will be provided to all participants. (A more complete description of this hypothetical intervention can be found in Chap. 2.) Suppose Dr. B has completed one cycle of MOST using the approach described in the preceding chapters and determined that the optimized intervention is made up of peer mentoring, text messaging, and behavioral skills training. The outcome variable is *adhere*, the number of days of adherence to ART in the 30-day period following the conclusion of the intervention. Further suppose that in the evaluation phase of MOST, this intervention demonstrated a statistically and clinically significant effect.

Consistent with the continual optimization principle, Dr. B is now planning another cycle of MOST to improve the intervention. Based on the outcome data gathered during the evaluation phase of MOST, Dr. B knows that even though this intervention has been demonstrated to be effective overall, there is heterogeneity in response across participants. In other words, the intervention worked very well for some participants and not as well for others. Some participants developed good ART adherence and sustained this adherence; others adhered well for a while and then returned to previous habits of poor adherence; and still others did not develop good ART adherence. How can an intervention be developed that will result in good ART adherence for more participants?

One approach might be to develop a new higher-dose version of the fixed intervention, by adding components or increasing the dose or intensity of the existing components. Depending on the situation, this can be a reasonable strategy, but it has some potential drawbacks. One potential drawback is that even though the new intervention may be the right approach for some participants, for others it may lead to poorer outcomes than would have been achieved with the original intervention. For example, some participants who would have responded to the original lower-dose intervention may feel that the higher dose is too burdensome and disengage entirely. A second potential drawback is that even if the higher-dose version of the intervention does not lead to disengagement or other negative outcomes, it wastes resources to the extent that it provides some participants with more treatment than is needed to achieve the desired effect. For example, suppose the higher-dose version of the intervention involves motivational interviewing, which is relatively expensive to provide. The resources used to provide motivational interviewing to participants who would have developed good ART adherence without it are wasted.

A different approach would be to develop an adaptive intervention (also called by other names, such as dynamic treatment regimens (e.g., Collins, Nahum-Shani, & Almirall, 2014), stepped care (e.g., Sobell & Sobell, 2000), and treatment policies (e.g., Wahed & Tsiatis, 2004)). As mentioned in Chap. 1, the Almirall, Nahum-Shani, Wang, and Kasari (2018) chapter in the companion volume (Collins & Kugler, 2018) defines an adaptive intervention as follows:

> An adaptive intervention is a sequence of pre-specified decision rules that can be used to guide whether, how, or when—and based on which measures—to alter an intervention or intervention component (e.g., treatment type, duration, frequency or amount) at critical decision points during the course of care.

Thus, in an adaptive intervention design, the content, dose, or approach is varied in response to measured characteristics of participants or the environment. These measured characteristics are called tailoring variables. The purpose of an adaptive intervention design is to use finite resources strategically to enable a good outcome for all participants. In an adaptive intervention, heterogeneity in treatment is used to reduce heterogeneity in response. An adaptive intervention design aims to provide each participant with the dose or type of treatment that is sufficient to produce the desired effect for that individual and, importantly, with *no more* than what is sufficient. In an adaptive intervention, fewer resources are spent on those participants who respond well to a lighter touch, so that the resources saved can either be deployed to other participants who require more intervention to respond well or used to extend the reach of the intervention to serve more people. Adaptive intervention designs are potentially useful in any area in which interventions are used, and they are consistent with the ideas behind the precision medicine initiative (e.g., Collins & Varmus, 2015).

8.1.2 The Anatomy of an Adaptive Intervention

For illustration, let us reimagine Dr. B's hypothetical fixed intervention as an adaptive intervention. Suppose all participants initially are provided with behavioral skills training, a weekly meeting with a peer mentor for 4 weeks, and text messaging for the entire duration of the intervention. ART adherence is assessed at 45 days, via a report provided by a medication event monitoring system (MEMS®) cap on the participants' pill bottles. At this assessment, those who have been at least 80 percent adherent over the previous 14 days (i.e., Days 31–45) are considered adherent, whereas those who have not been at least 80 percent adherent over the previous 14 days are considered non-adherent. The intervention for adherent participants is reduced to daily text messaging only. The intervention for non-adherent participants also includes daily text messaging but in addition includes 4 more weeks of peer mentoring and one session of motivational interviewing. Figure 8.1 depicts the design of this adaptive intervention. (While reading this chapter, please be mindful of the difference between intervention design and experimental design; see Chap. 1 and the glossary.)

In the terminology used in the adaptive intervention field (e.g., Nahum-Shani et al., 2012a), behavioral skills training, peer mentoring, text messaging, and

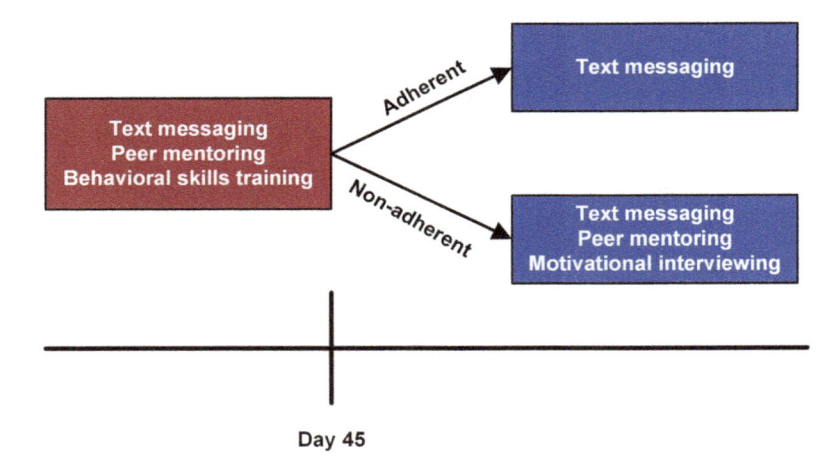

Fig. 8.1 Depiction of an adaptive intervention design. The intervention's objective is to encourage adherence to an HIV treatment protocol. After initial treatment, adherence is assessed at Day 45. Participants who are adherent continue with text messaging only. Those who are not adherent are provided with additional peer mentoring and a session of motivational interviewing in addition to text messaging

motivational interviewing are intervention options. Adaptive interventions have some features that are not found in fixed interventions, namely, decision points, tailoring variables, and decision rules. These features govern when and on what basis the intervention options are provided. At a particular decision point, a tailoring variable is assessed, and the decision rule determines what the course of action will be; that is, what intervention option will be selected in response to the value of the tailoring variable. In the example, all participants initially receive the same treatment. There is a decision point at 45 days. At this decision point, the tailoring variable, ART adherence over the last 14 days, is measured. The decision rules determine what intervention options will be provided if adherence is 80 percent or greater and what intervention options will be provided if adherence is less than 80 percent. The decision rules can be stated as follows: "If adherence is 80 percent or greater, continue with text messaging only. Otherwise, continue with text messaging plus provide four more weeks of peer mentoring and one session of motivational interviewing."

In any adaptive intervention, the tailoring variables, decision points, and decision rules are as important as the intervention options. In a sense, an adaptive intervention is a chain that is only as strong as its weakest link (Collins, Murphy, & Bierman, 2004). It is probably obvious that to have a highly effective adaptive intervention, it is necessary to have effective intervention options. It may be less obvious that appropriate and accurately measured tailoring variables, well-specified decision rules, and well-timed decision points are also essential.

To see the importance of accurate measurement of tailoring variables, imagine what might happen with Dr. B's adaptive intervention if it used a poorly measured

8.1 Adaptive Interventions: The Basics

tailoring variable. Suppose instead of the MEMS® cap report of adherence, the tailoring variable was a single self-report question, "How well do you feel you are adhering to the ART regimen?" with a five-point response scale ranging from *not at all* to *extremely well*. Those who select one of the two most positive responses will be considered adherent, and everyone else will be considered non-adherent. Self-report measures can make excellent tailoring variables, but a single vaguely worded question like this one is unlikely to be sufficiently valid and reliable to be useful for this purpose. Unreliability in a measure of a tailoring variable introduces random error into the assignment of treatments to participants. Taking this idea to its logical extreme, if a measure of a tailoring variable had a reliability of zero, using it to assign treatment would be equivalent to assigning treatment randomly. Invalidity in a measure of a tailoring variable can introduce bias into the assignment. In the example, suppose the measure is biased such that those who are less adherent are more likely to overstate how well they feel they are adhering. In this case, the decision rule will fail to provide increased intervention intensity to many individuals who would benefit from it. Of course, even the MEMS® cap report, although it is an excellent measure of adherence in most situations, is not perfect, because it records only when a pill bottle is opened, not whether the pill is taken after the bottle is opened.

In a highly effective adaptive intervention, the decision rule assigns the most appropriate of the available intervention options based on the observed value of the tailoring variable. Moreover, decision rules are specified a priori so that no ad hoc treatment decisions are made "on the fly." This means that decision rules must be sufficiently comprehensive to cover any situation that may arise and sufficiently clearly specified for intervention staff to be able to interpret them correctly. For instance, in the example discussed in this chapter, there should be explicit decision rules covering what to do if the participant stops using the MEMS® cap or if the MEMS® cap malfunctions.

Both the timing of decision points and, if there are multiple decision points, the interval between them (sometimes called the review interval) must be carefully selected in an adaptive intervention. In the example, if the decision point occurred before most participants could reasonably be expected to have developed the habits necessary to ensure good adherence, many individuals could inappropriately be placed in the non-adherent category. If the decision point occurred too late, say, 6 months after the initial treatment, a more intense additional treatment may be required to have an effect on the non-adherent participants.

The number of decision points featured in adaptive interventions varies. An adaptive intervention can have many decision points or as few as one, which may occur at the outset of the intervention before any treatment is provided or after an initial treatment has been provided. In the former case, the tailoring variable is usually based on measured characteristics of the individual, whereas in the latter case, the tailoring variable is usually a measure of how the individual responded to the initial treatment. The hypothetical example described earlier in this section is an example of an adaptive intervention with a single decision point that occurs after an initial treatment.

8.2 Intensity of Adaptation

For adaptive interventions that involve more than one decision point, the number of decision points and the length of the review interval determine the intensity of adaptation. If there are few decision points and the review interval is relatively long, then the adaptation of the intervention is low intensity. If there are many decision points or the review interval is relatively short, then the adaptation of the intervention is high intensity; such interventions are often referred to as intensively adaptive (Riley, Serrano, Nilsen, & Atienza, 2015).

The lowest-intensity adaptive interventions involve only a single decision point. One special case of low-intensity adaptive interventions is the matched intervention. In matched interventions, treatment is matched to individuals at a single decision point at the outset, usually based on a tailoring variable representing characteristics of the individual. This enables the treatment to differ depending on characteristics of the participant, environment, and so on, but not to vary over time within individuals. One example would be a drug abuse prevention program with different curricula for children in the sixth grade and those in the eighth grade. Another would be an intervention that allows for cultural adaptation by making different versions available for different ethnic groups. An adaptive intervention with a single decision point can also be time-varying. The example intervention illustrated in Fig. 8.1 is a low-intensity time-varying adaptive intervention. This intervention has only a single decision point, but it occurs after the outset of the intervention, enabling the treatment to vary over both individuals and over time within individuals.

Let us imagine an intervention similar to the one illustrated in Fig. 8.1 but with more intensive adaptation. Suppose over a 1-year period, the MEMS® cap report is transmitted wirelessly to an HIV clinic and reviewed every month on the first working day of that month by program staff. Any participant with a MEMS® cap report that indicates adherence at 80 percent or better over the previous month is sent an encouraging text message. Any participant with a MEMS® cap report that indicates less than 80 percent adherence is telephoned or contacted by text, and arrangements are made for a meeting either at the clinic or the individual's home. At this meeting, behavioral skills principles are reviewed. If a meeting is needed 2 months or more in a row, motivational interviewing is added. In this more intensively adaptive intervention, there are 12 decision points (i.e., monthly for 1 year), with a review interval of 1 month.

Intensity of adaptation is an important feature of an intervention that must be considered carefully. One reason is that cost, broadly defined to include participant time, usually increases along with intensity of adaptation. Another reason is that either insufficient intensity or too much intensity can lead to undesirable outcomes. An intervention with insufficiently intense adaptation, that is, an intervention that has a review interval that is too long, can let problems such as non-response go undetected until they become more difficult to deal with or even intractable. If the adaptation of the intervention is too intense, for example, if the tailoring variable is measured so frequently that meaningful change cannot reasonably be expected to take place between assessments, participants may find the frequent assessment and resulting adjustment of the intervention burdensome, annoying, or confusing. Another possibility is stigmatization of participants if the assessment is obtrusive enough for others to observe their frequent involvement with the intervention.

Improvements in technology, such as wearable sensors, smartphones with built-in global positioning systems (GPSs), and the like, are making it practical to develop interventions with increasingly intensive adaptation. This technology has made assessment of some tailoring variables less burdensome, less obtrusive, and more immediate and also has made it possible to use intervention designs that call for a more immediate response, often in real time or nearly real time. For example, consider an intervention that makes use of smartphones to reduce sedentary behavior. The activity monitor on the smartphone could be used to detect sedentary behavior, and a text message suggesting a brisk 2-min walk could be sent to any participant who has been sedentary for more than 1 h. Interventions like this one, in which the intensity of adaptation is very high, are often called just-in-time adaptive interventions (JITAIs; Klasnja et al., 2015; Nahum-Shani, Heckler, & Spruitz-Metz, 2015).

8.3 Identifying the Components of an Adaptive Intervention and Selecting an Approach for the Optimization Trial

Recall from the previous chapters in this book, particularly Chap. 2, that an intervention component is defined as any aspect of an intervention that can be separated out for study. There are many different appropriate and potentially useful ways to identify the components of any intervention. Different research questions can suggest very different approaches to conceptualizing components, and, by extension, different experimental designs for the optimization trial. This is particularly true with respect to the conceptualization of the components of adaptive interventions.

To illustrate this, let us take the intervention design depicted in Fig. 8.1 as a starting point and contrast the different approaches that could be taken by two scientists, Dr. B and Dr. C, to identify components in preparation for developing an optimized adaptive intervention. As will be seen below, because Dr. B and Dr. C have different research questions in mind, they have arrived at two different sets of components of essentially the same adaptive intervention. Both sets of components are appropriate; they reflect different research questions that are of interest to the two scientists. These different research questions require different choices of experimental design for the optimization trial.

Dr. B's research questions focus on the effectiveness of the intervention options as they are to be delivered in the adaptive intervention. From this perspective, it would be reasonable to identify four components: (1) text messaging, (2) peer mentoring, (3) behavioral skills training, and (4) motivational interviewing. Text messaging and behavioral skills training are fixed components. Text messaging is provided to all participants throughout the intervention, and behavioral skills training is provided to all participants at the outset of the intervention. Peer mentoring and motivational interviewing are adaptive. Four sessions of peer mentoring are initially provided to all participants; only those who are identified as non-adherent at Day 45 are provided with another four sessions. Initially motivational interviewing is not provided; only participants who have been identified as non-adherent at Day 45 are then provided with motivational interviewing.

Dr. B is interested in conducting an optimization trial to determine whether each of these components is effective, that is, the extent to which they move *adhere* in the desired direction. Given how Dr. B has conceptualized the components, consideration might be given to an optimization trial consisting of a 2^4 factorial experiment, for example, with the following factors: a factor corresponding to the text messaging component, labeled *TMS*, with levels no (text messaging not provided) and yes (provided); a factor corresponding to the peer mentoring component, labeled *PM*, with levels no (peer mentoring not provided) and yes (peer mentoring initially provided, followed by another four sessions if subject is non-adherent at Day 45); a factor corresponding to the behavioral skills training component, labeled *BST*, with levels no (behavioral skills training not provided) and yes (provided); and a factor corresponding to the motivational interviewing component, labeled *MI*, with levels no (motivational interviewing not provided to any participants) and yes (motivational interviewing provided if subject is non-adherent at Day 45).

Table 8.1 shows the conditions that would be included in this experiment. Note that the *PM* and *MI* factors, which correspond to adaptive components, are in a sense a bit different from the *TMS* and *BST* factors, which correspond to fixed components.

Table 8.1 Experimental conditions in 2^4 factorial design consistent with Dr. B's framing of components of adaptive intervention

Experimental condition	TMS	PM	BST	MI
1	No	No	No	No
2	No	No	No	Yes
3	No	No	Yes	No
4	No	No	Yes	Yes
5	No	Yes	No	No
6	No	Yes	No	Yes
7	No	Yes	Yes	No
8	No	Yes	Yes	Yes
9	Yes	No	No	No
10	Yes	No	No	Yes
11	Yes	No	Yes	No
12	Yes	No	Yes	Yes
13	Yes	Yes	No	No
14	Yes	Yes	No	Yes
15	Yes	Yes	Yes	No
16	Yes	Yes	Yes	Yes

8.3 Identifying the Components of an Adaptive Intervention and Selecting...

To illustrate why, let us contrast the *BST* and *MI* factors. Every experimental subject who is assigned to the yes level of *BST* is offered behavioral skills training. By contrast, an experimental subject who is assigned to the yes level of *MI* is provided with motivational interviewing only if identified as non-adherent at Day 45. For example, all subjects assigned to condition 4 will be provided with behavioral skills training, but only those who are non-adherent at Day 45 will be provided with motivational interviewing. Thus, the *PM* and *MI* factors call for subjects in the same experimental condition to receive different treatment options.

At first glance this may seem like a protocol deviation. However, upon reflection it becomes apparent that although subjects in a particular experimental condition may be provided with different treatment options, all are provided with the same adaptive treatment, because the same tailoring variable and decision rules are always used at the same decision point to determine which treatment option will be provided. In other words, Dr. B has chosen to view each of the two adaptive components as a mini-package consisting of the tailoring variable, decision rules, decision point, and intervention options. In Dr. B's framework, the tailoring variable, decision rules, and decision point are not separated out for individual examination. Instead, Dr. B has defined an adaptive component not only as the intervention option but also the tailoring variable, decision rule, and decision point. The motivational interviewing component, for example, is defined as not only the content of the motivational interview itself but also the decision point (Day 45), tailoring variable (adherence), and decision rule (motivational interviewing provided only to those who are non-adherent).

Now let us consider a different conceptualization of the components of this intervention. Dr. C is interested in a set of research questions subtly different from Dr. B's. Dr. C plans to provide all subjects with the same initial treatment as was used in Dr. B's study. The scientific questions of interest to Dr. C focus on the tailoring variable and decision rules that are used to assign the subsequent intervention options to participants. From Dr. C's perspective, there are three components to be investigated empirically: (1) the tailoring variable, (2) the decision rule for non-adherent participants, and (3) the decision rule for adherent participants. Here the research questions concern which alternative is best. With respect to the tailoring variable, Dr. C wants to know whether it is better to define non-adherence using (a) a strict criterion of 80 percent adherence overall or (b) a very strict criterion of 90 percent adherence overall, with no more than one day in the 14-day period without medication. With respect to the decision rule for adherent participants, Dr. C wants to know whether the decision rule should (a) continue with text messaging only or (b), in addition to text messaging, add four more weeks of peer mentoring. With respect to the decision rule for non-adherent participants, Dr. C wants to decide whether the decision rule should (a) continue with text messaging and add peer mentoring only or (b), in addition to text messaging and peer mentoring, provide a session of motivational interviewing. Note that in this particular framing of the components, the intervention options themselves are not separated out for individual examination. For example, Dr. C is not investigating whether text messaging has a detectable effect.

Whereas a factorial experiment is an appropriate choice for Dr. B's optimization trial, a traditional factorial experiment is not appropriate for Dr. C's optimization trial. This is due to Dr. C's framing of the components so as to prioritize examination of the tailoring variable and decision rules. Consider translating Dr. C's components into factors in a standard factorial design, as follows: the tailoring variable component is represented by a factor labeled *TAILOR* with two levels: strict and very strict. The component concerning the decision rule for non-adherent individuals is represented by a factor labeled *NON-ADH* with two levels: peer mentoring alone and peer mentoring plus additional motivational interviewing. The component concerning the decision rule for adherent individuals is represented by a factor labeled *ADH* with two levels: none and peer mentoring. A factorial experiment is not workable because *NON-ADH* and *ADH* cannot be crossed. For example, it is not possible for an individual subject to receive both peer mentoring as an adherent subject and peer mentoring plus motivational interviewing as a non-adherent subject simultaneously. The levels of the second factor pertain only to non-adherent individuals, and the levels of the third factor pertain only to adherent individuals.

Thus, an approach other than a traditional factorial experiment is needed for Dr. C's optimization trial. Fortunately, an appropriate alternative is available: the sequential, multiple assignment, randomized trial (SMART). The SMART will be briefly introduced below; a much more detailed treatment can be found in the Almirall, Nahum-Shani, Wang, and Kasari (2018) chapter in the companion volume.

8.3.1 The Sequential, Multiple Assignment, Randomized Trial (SMART)

Dr. C would probably choose to conduct an optimization trial to examine the identified components by using the abovementioned SMART experimental design. The SMART, which has been developed especially for optimization of time-varying interventions of low to moderate intensity of adaptation, is a special case of the factorial experiment, in which randomization of subjects occurs sequentially and may be restricted based on, for example, the outcome of a prior treatment.

Figure 8.2 illustrates a SMART that could be used to examine the three candidate components in this hypothetical example. This experimental design includes the three factors listed above: *TAILOR*, *NON-ADH*, and *ADH*. In Fig. 8.2 the first randomization is to one of the two levels of *TAILOR*, that is, to the strict or very strict definition of adherence. This first randomization is not restricted; all subjects would be involved. At 45 days there is a decision point. At this decision point, it is determined whether or not the subject was adherent (according to the definition specified by the level of *TAILOR* to which the subject was assigned). Immediately after the decision point, the next randomization occurs. This randomization is restricted according to whether the subject has been determined to be adherent or non-adherent. If the subject was adherent, he or she would be randomly assigned to a

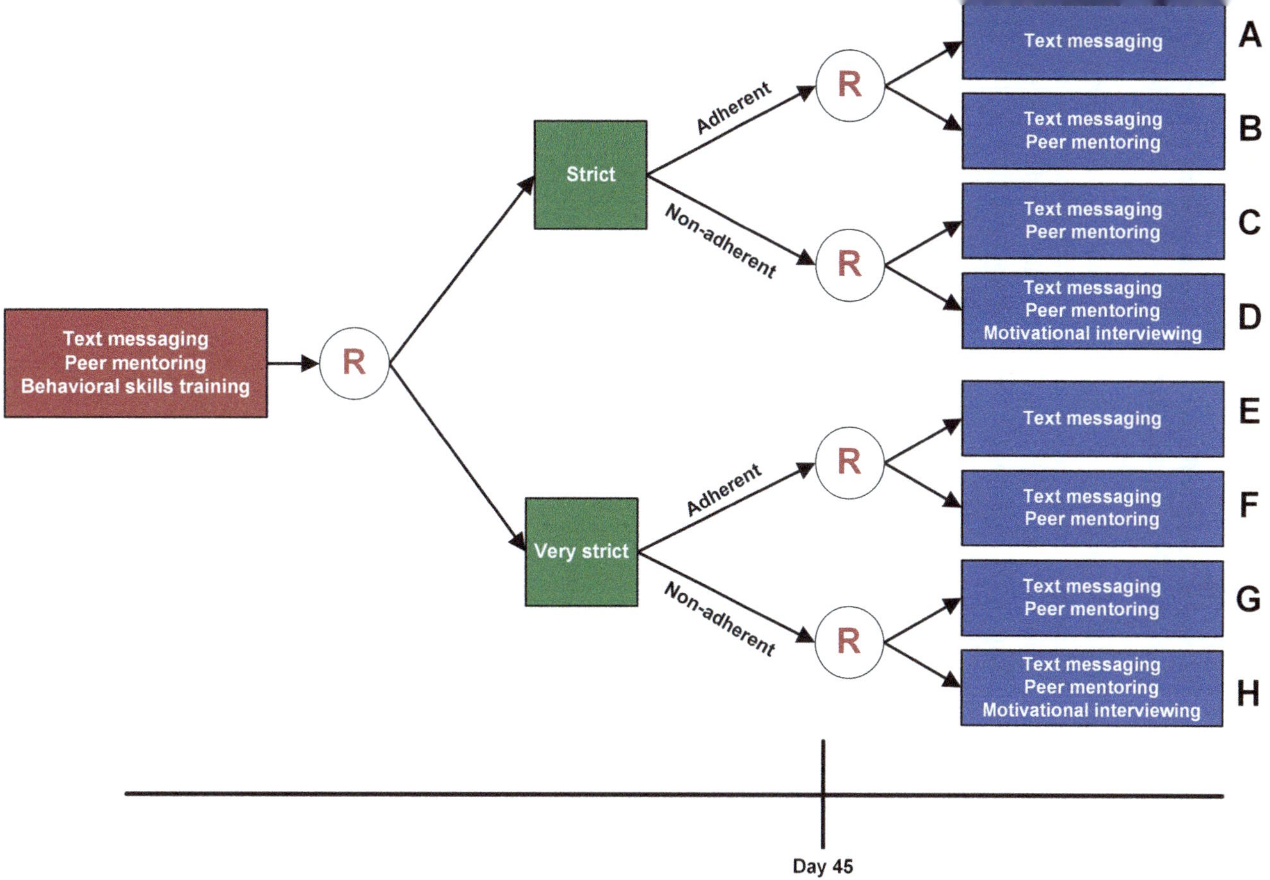

8.3 Identifying the Components of an Adaptive Intervention and Selecting... 277

Fig. 8.2 Depiction of a sequential, multiple assignment, randomized trial (SMART), an experimental design that can be used for optimization of certain adaptive interventions. An R within a circle indicates randomization. The first randomization is unrestricted, whereas the second randomization is restricted because it differs depending on whether the subject is classified as adherent or non-adherent by the tailoring variable. This SMART has eight experimental conditions, denoted by the letters at the far right

level of the *ADH* factor, that is, to receive either no additional treatment or 4 more weeks of peer mentoring. If the subject was non-adherent, he or she would be randomly assigned to a level of the *NON-ADH* factor, that is, the addition of peer mentoring only or peer mentoring plus one session of motivational interviewing. Figure 8.2 illustrates the sequential nature of the randomization; the second randomization could not have taken place before adherence was assessed at the decision point.

When data from a SMART are analyzed, the effects, consisting of main effects and, in some cases, interactions, can be estimated by combining experimental conditions (cells) in much the same way as is done in traditional factorial analysis of variance (ANOVA) (although depending on the research questions, this is not the only way to analyze data from a SMART; see Nahum-Shani et al., 2012b). As Fig. 8.2 shows, there are eight cells in this design. They are labeled A through H. The question about which is better, the strict or very strict definition of adherence, would be addressed by examining the main effect of *TAILOR*. The main effect of *TAILOR* is estimated by comparing the mean of A, B, C, and D to the mean of E, F, G, and H in Fig. 8.2. The question about which intervention option is better for adherent individuals, no additional treatment or 4 more weeks of peer mentoring, would be addressed by examining the main effect of *ADH*. This main effect would be computed by comparing the mean of A and E to the mean of B and F. Cells C, D, G, and H would not be included, because they contain the non-adherent individuals. The question about which intervention option is better for non-adherent individuals would be addressed by examining the main effect of *NON-ADH*. This main effect would be computed by comparing the mean of C and G to the mean of D and H. Cells A, B, E, and F would not be included, because they contain the adherent individuals.

8.3.2 A Brief Note About Powering a SMART

Statistical power must be considered carefully when planning a SMART. As discussed in Chaps. 3 and 4, a balanced 2^k factorial experiment provides the same statistical power for testing the significance of any regression coefficient of a given size, whether the coefficient corresponds to a main effect or an interaction (assuming effect coding is used). However, within a SMART available power may vary for different main effects of the same size, because the hypothesis tests may be based on different sample sizes. To see why, consider the SMART illustrated in Fig. 8.2. Suppose the main effect of *TAILOR* has been powered based on a particular N. The hypotheses concerning the main effects of *ADH* and *NON-ADH* inevitably will be tested based on a reduced sample size, because the original N is divided between the adherent and non-adherent subgroups. Therefore, there will be less power available for detection of these effects. The power for each of these main effects depends on how the original N ultimately is divided between adherent and non-adherent subjects, which can be difficult to predict before the experiment is conducted. A discussion of resources containing information about how to power SMARTs can be found in the Almirall, Nahum-Shani, Wang, and Kasari (2018) chapter in the companion volume.

8.3.3 The Outcome Variable in SMARTs

In the SMART described above and illustrated in Fig. 8.2, a natural choice for the outcome variable would be adherence as measured by the MEMs® cap, for example, number of days adherent in the 30-day period following initial treatment. Some readers may be concerned that basing the experimental outcome on the same measure as is used for the tailoring variable may capitalize on chance and increase the possibility of a spurious outcome. There may also be a concern about reactivity, specifically, that measurement of the tailoring variable itself will have an impact on the outcome. In the example, the very presence of the MEMs® cap may prompt participants to maintain better adherence.

Capitalization on chance would be a potential problem if the investigators changed the decision rules, tailoring variable, or decision points in an ad hoc fashion during the course of the experiment based on the outcome variable. For example, if, in the middle of the experiment, the investigators noted that subjects continued to be non-adherent when provided with one session of motivational interviewing and increased the number of sessions to two, any results could plausibly be attributed to capitalization on chance. Moreover, such a midcourse adjustment would be a clear violation of protocol in any controlled experiment. Administering intervention options that have been selected a priori, according to tailoring variables, decision rules, and decision points that have been determined a priori and are adhered to consistently throughout the experiment, does not constitute capitalization on chance.

Reactivity would be a potential problem if the MEMs® cap was included only for purposes of experimentation and removed in the later implementation of the optimized adaptive intervention. However, in the example, the tailoring variable, including its measurement, is an integral part of the adaptive intervention, as much as intervention options such as peer mentoring and motivational interviewing. In other words, the intervention design includes measurement of adherence via the MEMs® cap report; eventually when the intervention goes to scale, the MEMs® cap report will be included in the intervention design. Thus in using number of days adherent in the 30-day period following initial treatment as an outcome, the investigators are using a measure that was collected in the course of implementation of the intervention.

8.3.4 Two Persistent Sources of Confusion

This section attempts to shed some light on two persistent sources of confusion.

The first source of confusion is the distinction between an intervention design and an experimental design. The purpose of an intervention design is to administer treatment, whereas the purpose of an experimental design is to gather scientific information. This distinction sometimes becomes blurred, particularly in discussions about research to develop and evaluate adaptive interventions. In an adaptive intervention design, such as the one depicted in Fig. 8.1, tailoring variables, decision rules, decision points, and intervention options are specified a priori and constitute the intervention. Participants are typically not randomly assigned to these. By contrast, in an experimental design, such as the SMART depicted in Fig. 8.2, experimental subjects may be randomly assigned to tailoring variables, decision rules, decision points, or intervention options. This is done for the purpose of gathering scientific data and to identify the best treatment strategies—the experimental design is not used as a treatment strategy per se. In writing and in conversations at meetings, it is a good idea to prevent confusion by specifying intervention design or experimental design, rather than simply using the term design.

The second source of confusion is the distinction between an adaptive experimental design and a SMART. The adaptive experimental design is an approach to conducting a clinical trial, usually an RCT. The purpose of the clinical trial may be to evaluate any intervention, including a fixed or adaptive intervention. In an adaptive clinical trial, the experimental design itself may be modified, in a principled manner, during the course of the trial. The modification to the design is based on data collected during the course of the experiment (Kairalla, Coffey, Thomann, & Muller, 2012). For example, consider an adaptive clinical trial to examine four different intervention alternatives. Such a trial might monitor the outcome carefully in each experimental condition, and if one or more of the treatments do not look sufficiently promising, the randomization probabilities would be revised so that fewer, or even no, subjects would be assigned to the corresponding experimental condition. By contrast, although the purpose of a SMART is to obtain information that will be used to develop an adaptive intervention design, *the SMART is not an adaptive experimental design.* The SMART does not include the potential for revision of the design during the course of the experiment.

8.3.5 Optimization Trials for Higher-Intensity Adaptive Interventions

The SMART usually works best when the intensity of adaptation of the intervention being optimized is low to moderate. SMARTs can become unwieldy and inefficient when applied to higher-intensity interventions. When there are more than a few decision points, the number of experimental conditions in a SMART can multiply rapidly, and the N relevant to an individual research question can become small. A different approach is needed for intensively adaptive interventions. At this writing, optimization of intensively adaptive interventions, including JITAIs, is a very new area with enormous potential.

One approach to optimization of highly intensively adaptive interventions is the micro-randomized trial (Klasnja et al., 2015; Liao, Klasnja, Tewari, & Murphy, 2016), which is essentially a factorial experiment in which all experimental conditions are conducted repeatedly on individual subjects. This approach has roots in both classical factorial design and computer science. Another approach to optimization of highly intensively adaptive interventions is to view the adaptive intervention as a controller, much like a temperature control system in a home or cruise control in an automobile (Rivera, Pew, & Collins, 2007). From this control engineering perspective, the process being intervened on is a dynamical system, and system identification procedures can be used to model it. Based on the results, a controller can be built to provide the decision rules for the adaptive intervention. This approach is described in Lagoa, Bekiroglu, Lanza, and Murphy (2014) and in the Rivera, Hekler, Downs, and Savage (2018) chapter in the companion volume. Both micro-randomized trials and system identification experiments are essentially very sophisticated modern extensions of *N*-of-one approaches involving repeated measures experimentation performed on individuals. Assuming the sample size—in terms of both subjects and times of observation within subjects—is large enough, both offer the potential to aggregate the data in various ways to gain speed and efficiency in drawing inferences about both an individual subject and a population.

8.4 Summary of Selection From the MOST Optimization Phase Toolbox

Let us recall Fig. 1.1, which provided an overview of MOST. This figure listed a number of different approaches that may be used for the optimization trial, including factorial experiments, fractional factorial experiments, SMARTs, micro-randomized trials, and system identification experiments. All of these experimental designs (and potentially others) are part of the metaphorical MOST optimization phase toolbox. The investigator must select an appropriate tool from the toolbox for each optimization trial.

It has been emphasized throughout this book that different experimental designs are suitable for different research questions, and different research questions are likely to be raised in optimization of different kinds of interventions. Figure 8.3 summarizes the likely usefulness and relevance of several approaches to conducting optimization trials when applied to these different types of interventions. The leftmost part of the figure lists the different types of interventions that have been discussed in this book: fixed, in which there is no adaptation; matched, in which there is a single adaptation at the start of the intervention; and time-varying adaptive interventions, in which adaptation occurs over time. As the figure reflects, time-varying adaptive interventions range from less intense adaptation (fewer decision points, relatively wide spacing in time) to more intense adaptation (more decision points, relatively close spacing in time). The experimental design approaches

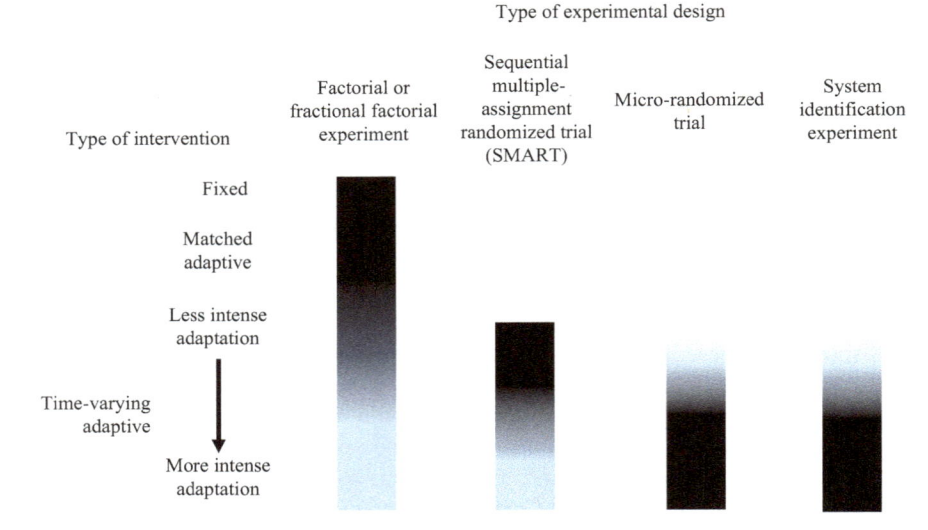

Fig. 8.3 Likely relevance of various experimental approaches for optimization of fixed interventions, matched adaptive interventions, and time-varying adaptive interventions with different intensities of adaptation. Lighter coloration in the blue bars signifies lower likely relevance; darker coloration signifies higher likely relevance

featured in Fig. 8.3 are factorial/fractional factorial experiments (combined in the figure because they are relevant for the same situations), SMARTs, micro-randomized trials, and system identification experiments. Under each category of experimental design approach, there is a shaded blue bar to signify the likely relevance of the approach for optimization of each type of intervention, with lighter shading indicating lower likely relevance and darker shading indicating higher likely relevance.

Figure 8.3 shows that for optimization of fixed interventions and matched adaptive interventions, factorial and fractional factorial designs are likely to be suitable, whereas SMARTs, micro-randomized trials, and system identification experiments are unlikely to be suitable. For optimization of time-varying adaptive interventions with less intense adaptation, factorial and fractional factorial designs remain relevant and merit consideration, but SMARTs are very often the most useful. As the intensity of adaptation increases, the relevance of the factorial/fractional factorial experiment and the SMART fades, but does not reduce to zero. For interventions involving more intense adaptation, such as JITAIs, the micro-randomized trial and the system identification experiment are usually the most relevant.

Figure 8.3 is provided with the caveat that it is intended to be only a rough guide, because the choice of experimental design for an optimization trial is unique to each situation and must be driven by the scientific questions at hand. As always, the choice of approach to the optimization trial is made based on the resource management principle (Chap. 1), which states that the investigator must seek the approach that makes the best and most efficient use of research resources. In other words, the objective in selecting a design for the optimization trial is to select the one that appropriately addresses the primary scientific questions in the most economical manner. In some cases, it will be appropriate to consider designs that are not included in Fig. 8.3.

8.5 Some Open Areas and Open Questions

8.5.1 The Preparation Phase of MOST and Adaptive Interventions

A detailed and clearly specified conceptual model is necessary when an adaptive intervention is being optimized, just as it is for optimization of fixed interventions. As discussed in Chap. 2, the conceptual model expresses all of what is known or hypothesized about how the intervention under development is to intervene on the behavioral, biobehavioral, or biomedical process.

The conceptual model provides the foundation for any strategy for dealing with participant heterogeneity in an adaptive intervention. If an intervention is to include different components or variations of components aimed at different people or at the same people at different times, the model must clearly specify that this is needed and detail which components or component levels are targeted to which subgroups, stages of development, or points in a time-varying process. Hypothesized or empirically observed moderation is often the motivation for including different components aimed at different people. For example, suppose gender moderates the effects of a particular component, so that it is more effective for females than for males. Assuming this component is satisfactorily effective in females, the investigator may want to retain it for females and develop a different version tailored for males—to develop an adaptive intervention that offers one version to males and one to females. The objective would be to produce equally positive effects in males and females.

Development of a conceptual model for an adaptive intervention can be particularly challenging because the conceptual model must address the reasons for the adaptation and the theoretical and practical justification for the strategy taken to deal with participant heterogeneity. If the adaptive intervention is time varying, the conceptual model must be informed by theories about change over time. However, many theories are not very specific about matters such as the rate of change over time, whether change is continuous or discontinuous, the anticipated time lag between cause and effect, and so on (Collins, 2006). All of these elements may come into play when selecting candidate tailoring variables, decision points, and decision rules.

Recall that according to the continual optimization principle (Chap. 1), it is important to keep returning to the conceptual model to improve and refine it based on the empirical results of the optimization and evaluation phases of MOST and any advancements in theory. Optimization trials can provide valuable data for the refinement of theory relevant to adaptive interventions. In this manner, as more research is conducted, the conceptual models underlying adaptive interventions will improve. As the underlying conceptual models are improved, adaptive interventions will become progressively more effective, efficient, economical, and scalable.

8.5.2 The Use of Optimization Criteria With Adaptive Interventions

As was mentioned in the Preface to this book, much of the work on adaptive interventions and optimization of interventions has occurred in parallel; these two research areas are just beginning to be integrated. One area where integration is needed is the application of optimization criteria, other than the all active components criterion, to adaptive interventions. In Chap. 2 a variety of optimization criteria were reviewed, and the use of these in decision-making to select the components and component levels making up an optimized fixed intervention was discussed in Chap. 7. At this writing, the optimization of adaptive interventions has been based on the all active components optimization criterion by default, because optimization has usually not been explicitly mentioned. It would be possible to use other optimization criteria, for example, to identify the most effective adaptive intervention that can be implemented for less than $500 per person. Before this can be done, research is needed on how to apply these kinds of optimization criteria based on empirical results from, for example, a SMART or a micro-randomized trial.

8.5.3 How Personally Tailored Should an Intervention Be?

A natural question that arises in connection with adaptive interventions is how far the effort to tailor an intervention to individual participants should go or, in other words, how elaborate the decision rules should be. More finely tuned adaptations may have the potential to produce better outcomes at a cost savings, because they come closer to distributing intervention resources so that each individual is provided exactly what is needed and no more than what is needed. In some settings elaborate decision rules incur very little cost. For example, in computer-delivered interventions, the bulk of the costs associated with decision rules are incurred in the initial programming. However, where an intervention must be partially or wholly delivered by humans, more elaborate decision rules can require additional staff training, more staff time spent reviewing tailoring variables and implementing changes in treatment dictated by the decision rules, and additional staff supervision to ensure that a complex set of decision rules is properly implemented.

On the one hand, the very personally tailored version may be more effective overall, but its reach and, therefore, its public health impact may be limited by the expense associated with the complexity of its decision rules. Sometimes an intervention with a simpler set of decision rules may be a bit less effective, but its ultimate public health impact may be greater if its economy means it has a wider reach. This trade-off between complexity and reach is directly related to the scalability of the adaptive intervention and, therefore, is an important consideration during the optimization phase of MOST. More research is needed on the cost of, and cost savings associated with, adaptation.

8.5.4 Robust Adaptive Interventions

Above it was mentioned that adaptive interventions and participant response to adaptive interventions can be viewed as a dynamical system, and an adaptive intervention can be viewed as a type of controller. From this engineering perspective comes the concept of robust control. As Ogunnaike and Ray (1994) put it, a robust controller "provides good performance and robust stability" (p. 1075) across "expected variations in the process parameters" (p. 1069). In other words, a robust controller is specifically designed to work well across a variety of environments and settings. The trade-off is that a robust controller may not work as well in any given setting as a controller specifically designed for that setting, and there may be variability in how well the controller works across settings. Whether or not this is an acceptable trade-off is highly situation-specific.

It may be a productive area of research to extend the concept of a robust controller to interventions. Imagine optimizing interventions using robustness as a part of the optimization criterion. The resulting interventions would be built to work well across a variety of settings such as different schools (low-income and high-income, public, charter, and private), communities (rural and urban), or hospitals (secular and religious, university and community); across participants with different characteristics (males and females, different ethnicities, different levels of literacy); and across intervention deliverers with different personalities and approaches.

8.6 What's Next

This chapter concludes the present book. The purpose of this book has been to provide a comprehensive introduction to MOST, with the objective of helping readers gain the perspective, background, and skills needed to optimize behavioral, biobehavioral, and biomedical interventions.

Readers who would like to learn more may be interested in the companion volume. This book includes eight chapters, each devoted to an advanced topic related to MOST. The Kugler, Wyrick, Tanner, Milroy, Chambers, Ma, Guastaferro, and Collins (2018) chapter provides a detailed example of the development of a conceptual model of an intervention as well as an example of how an iterative approach can be used in the optimization phase of MOST. Piper, Schlam, Fraser, Oguss, and Cook (2018) offer advice and lessons learned from a research team that has successfully implemented several large factorial experiments in field settings. Nahum-Shani and Dziak (2018) discuss experimental design and statistical analysis when there is a multilevel (cluster) structure in the data, including when the design of the optimization trial itself introduces the multilevel structure. The Almirall, Nahum-Shani, Wang, and Kasari (2018) chapter provides an introduction to optimization of adaptive interventions using the sequential, multiple assignment, randomized trial (SMART). Rivera, Hekler, Savage, and Downs (2018) discuss optimization of adaptive interventions from a control engineering perspective. Kugler, Dziak, and Trail (2018) compare and contrast effect coding and dummy coding of factorial experiments. Dziak (2018) discusses issues and open research areas related to cost-effectiveness and MOST. The final chapter, by Smith, Coffman, and Zhu (2018), reviews the possibilities that are opened up by mediation analysis of data when the treatment is a factorial experiment rather than a two-arm RCT.

What's next is for readers of this book to use MOST in their own work! Additional information helpful for investigators using MOST, as well as updates and supplemental material for this book and the companion volume, can be found at http://methodology.psu.edu/ra/MOST.

References

Almirall, D., Nahum-Shani, I., Wang, L., & Kasari, C. (2018). Experimental designs for research on adaptive interventions: Singly and sequentially randomized trials. In L. M. Collins & K. C. Kugler (Eds.), *Optimization of multicomponent behavioral, biobehavioral, and biomedical interventions: Advanced topics* (forthcoming). New York, NY: Springer.

Collins, L. M. (2006). Analysis of longitudinal data: The integration of theoretical model, temporal design and statistical model. *Annual Review of Psychology, 57*, 505–528.

Collins, L. M., & Kugler, K. C. (Eds.). (2018). *Optimization of multicomponent behavioral, biobehavioral, and biomedical interventions: Advanced topics*. New York, NY: Springer.

Collins, L. M., Murphy, S. A., & Bierman, K. L. (2004). A conceptual framework for adaptive preventive interventions. *Prevention Science, 5*(3), 185–196.

Collins, L. M., Nahum-Shani, I., & Almirall, D. (2014). Optimization of behavioral dynamic treatment regimens based on the sequential, multiple assignment, randomized trial (SMART). *Clinical Trials, 11*, 426–434.

Collins, F. S., & Varmus, H. (2015). A new initiative on precision medicine. *The New England Journal of Medicine, 372*, 793–795.

Dziak, J. J. (2018). Optimizing the cost-effectiveness of a multicomponent intervention using data from a factorial experiment: Considerations, open questions, and tradeoffs among multiple outcomes. In L. M. Collins & K. C. Kugler (Eds.), *Optimization of multicomponent behavioral, biobehavioral, and biomedical interventions: Advanced topics*. (forthcoming) New York, NY: Springer.

Kairalla, J. A., Coffey, C. S., Thomann, M. A., & Muller, K. E. (2012). Adaptive trial designs: A review of barriers and opportunities. *Trials, 13*(1), 145–154.

Klasnja, P., Hekler, E. B., Shiffman, S., Boruvka, A., Almirall, D., Tewari, A., & Murphy, S. A. (2015). Microrandomized trials: An experimental design for developing just-in-time adaptive interventions. *Health Psychology, 34*(S), 1220.

References

Kugler, K. C., Dziak, J. J., & Trail, J. (2018). Coding and interpretation of effects in analysis of data from a factorial experiment. In L. M. Collins & K. C. Kugler (Eds.), *Optimization of multicomponent behavioral, biobehavioral, and biomedical interventions: Advanced topics* (forthcoming). New York, NY: Springer.

Kugler, K. C., Wyrick, D. L., Tanner, A. E., Milroy, J. J., Chambers, B., Ma, A., & Collins, L. M. (2018). An iterative approach to building an optimized STI prevention intervention aimed at college students: The important of the conceptual model. In L. M. Collins & K. C. Kugler (Eds.), *Optimization of multicomponent behavioral, biobehavioral, and biomedical interventions: Advanced topics*. New York, NY: Springer.

Lagoa, C. M., Bekiroglu, K., Lanza, S. T., & Murphy, S. A. (2014). Designing adaptive intensive interventions using methods from engineering. *Journal of Consulting and Clinical Psychology*, *82*(5), 868–878.

Liao, P., Klasnja, P., Tewari, A., & Murphy, S. A. (2016). Sample size calculations for micro-randomized trials in mHealth. *Statistics in Medicine*, *35*, 1944–1971.

Nahum-Shani, I., & Dziak, J. J. (2018). Multilevel factorial designs in intervention development. In L. M. Collins & K. C. Kugler (Eds.), *Optimization of multicomponent behavioral, biobehavioral, and biomedical interventions: Advanced topics* (forthcoming). New York, NY: Springer.

Nahum-Shani, I., Hekler, E. B., & Spruijt-Metz, D. (2015). Building health behavior models to guide the development of just-in-time adaptive interventions: A pragmatic framework. *Health Psychology*, *34*(S), 1209–1219.

Nahum-Shani, I., Qian, M., Almirall, D., Pelham, W. E., Gnagy, B., Fabiano, G. A., . . . Murphy, S. A. (2012a). Experimental design and primary data analysis methods for comparing adaptive interventions. *Psychological Methods*, *17*(4), 457.

Nahum-Shani, I., Qian, M., Almirall, D., Pelham, W. E., Gnagy, B., Fabiano, G. A., . . . Murphy, S. A. (2012b). Q-learning: A data analysis method for constructing adaptive interventions. *Psychological Methods*, *17*(4), 478.

Ogunnaike, B. A., & Ray, W. H. (1994). *Process dynamics, modeling, and control*. Oxford: Oxford University Press.

Piper, M. E., Schlam, T. R., Fraser, D., Oguss, M., & Cook, J. W. (2018). Implementing factorial experiments in real-world settings: Lessons learned while engineering an optimized smoking cessation treatment. In L. M. Collins & K. C. Kugler (Eds.), *Optimization of multicomponent behavioral, biobehavioral, and biomedical interventions: Advanced topics* (forthcoming). New York, NY: Springer.

Riley, W. T., Serrano, K. J., Nilsen, W., & Atienza, A. A. (2015). Mobile and wireless technologies in health behavior and the potential for intensively adaptive interventions. *Current Opinion in Psychology*, *5*, 67–71.

Rivera, D. E., Hekler, E. B., Savage, J. S., & Downs, D. S. (2018). Intensively adaptive interventions using control systems engineering: Two illustrative examples. In L. M. Collins & K. C. Kugler (Eds.), *Optimization of multicomponent behavioral, biobehavioral, and biomedical interventions: Advanced topics* (forthcoming). New York, NY: Springer.

Rivera, D. E., Pew, M. D., & Collins, L. M. (2007). Using engineering control principles to inform the design of adaptive interventions. *Drug and Alcohol Dependence*, *88*, S31–S40.

Smith, R.A., Coffman, D. L., & Zhu, X. (2018). Investigating an intervention's causal story: Mediation analysis using a factorial experiment and multiple mediators. In L. M. Collins & K. C. Kugler (Eds.), *Optimization of multicomponent behavioral, biobehavioral, and biomedical interventions: Advanced topics* (forthcoming). New York, NY: Springer.

Sobell, M. B., & Sobell, L. C. (2000). Stepped care as a heuristic approach to the treatment of alcohol problems. *Journal of Consulting and Clinical Psychology*, *68*(4), 573.

Wahed, A. S., & Tsiatis, A. A. (2004). Optimal estimator for the survival distribution and related quantities for treatment policies in two-stage randomization designs in clinical trials. *Biometrics*, *60*(1), 124–133.

Glossary

Adaptive intervention design A sequence of prespecified decision rules that can be used to guide whether, how, or when—and based on which measures, called tailoring variables—to alter an intervention or intervention component (e.g., treatment type, duration, frequency or amount) at critical decision points during the course of care.

Aliasing The combining of effects that occurs as a result of deliberate experimental design decisions made by the investigator, particularly with respect to a fractional factorial design.

Alias structure Refers to how the aliasing is set up in a particular experimental design.

ANOVA Stands for analysis of variance.

Antagonistic interaction In factorial ANOVA, an interaction in which the combined effect of two or more factors is less favorable than would be expected based solely on the main effects.

Arm An experimental condition in an RCT.

Balance A highly desirable property of some factorial experiments. Balance is achieved when each level of each factor appears the same number of times, and every experimental condition has the same number of subjects.

Behavioral intervention An intervention that uses a strategy based on modification of affective, cognitive, or behavioral factors.

Biobehavioral intervention An intervention that uses a strategy based on both behavioral and biomedical approaches.

Biomedical intervention An intervention that uses a strategy based on pharmaceuticals, surgery, and the like.

Candidate intervention component A component from which a level (e.g., no or yes; low intensity or high intensity) is to be selected for inclusion in an intervention. This term may refer to any type of component.

Clear effect In a fractional factorial design, an effect that is aliased only with interactions involving three or more factors. See *strongly clear effect*.

© Springer International Publishing AG 2018
L. M. Collins, *Optimization of Behavioral, Biobehavioral, and Biomedical Interventions*, Statistics for Social and Behavioral Sciences,
https://doi.org/10.1007/978-3-319-72206-1

Conceptual model An expression of all of what is known or hypothesized about how the intervention under development is to intervene on the target behavioral, biobehavioral, or biomedical process. Similar to a logic model but more detailed about which individual components are aimed at which target mediators.

Content components Components that make up the content of an intervention. These are the parts of an intervention that are directly aimed at the behavioral or biological process being intervened on.

Component Any part of an intervention that can be separated out for study.

Constant component A component or set of components that is provided to all subjects in an optimization trial rather than experimentally manipulated.

Constant overhead costs In an experiment, costs that are expected to remain about the same irrespective of the number of subjects or experimental conditions.

Contamination Occurs when subjects in an experiment are provided the correct randomly assigned treatment, but then the treatment they receive or their reaction to treatment is influenced somehow as a direct or indirect result of contact with other subjects and/or learning about what treatments other subjects have been given.

Continual optimization principle The principle that optimization is an ongoing process of working toward a better intervention. This is one of the two fundamental principles of MOST. (See *resource management principle*.)

Cost ratio The ratio of per-condition overhead costs to per-subject costs, useful in determining which experimental design is most efficient in a given situation.

Decision points Designated times in an adaptive intervention design when the decision rules are to be implemented.

Decision rules In an adaptive intervention design, the rules that specify how the intervention will be varied in a systematic and principled manner based on the tailoring variables.

Design May refer to experimental design, intervention design, or research design.

Economy The degree to which an intervention produces a good outcome without exceeding budgetary constraints (where a budget may be placed not only on money per se but time or any other resource) on implementation and the degree to which it offers a high degree of effectiveness in exchange for the resources required to implement it.

Effect heredity As used in this book, the philosophy that an interaction is most likely to be scientifically important if at least one of its parent main effects is significant; otherwise, it should be viewed with extreme caution.

Effect hierarchy The philosophy that in the absence of a priori predictions to the contrary expressed in the conceptual model, lower-order effects in a factorial experiment are more likely to be scientifically important than higher-order effects.

Effect sparsity According to Wu and Hamada (2009), the expectation that "[t]he number of important effects in a factorial experiment is small" (p. 173). Also called the Pareto principle.

Effectiveness In this book effectiveness is used to refer in general to the degree to which an intervention produces a favorable outcome.

Efficiency The degree to which an intervention produces a good outcome while avoiding wasting money, time, or any other valuable resource.

Engagement/adherence components Components aimed at ensuring that participants remain engaged in the intervention for its entire duration, carefully follow its instructions, and otherwise comply with its requirements.

Experiment Manipulation of one or more independent variables for the purpose of empirically observing the effect on an outcome variable.

Experimental design In this book, refers to the design of an experiment to gather information needed to develop, optimize, or evaluate an intervention.

Factorial experiment An experimental design in which two or more independent variables are manipulated.

Fidelity components Components of an intervention aimed at maintaining a high level of fidelity of intervention delivery. Such components are usually aimed at those who deliver the intervention, or on the environment in which the intervention is to be delivered, rather than at the individuals who are the target of the intervention.

Fixed intervention design An intervention design in which all participants are offered the same set of intervention components in a uniform manner.

Fractional factorial design A reduced factorial design comprised of experimental conditions carefully selected to preserve the balance property and maintain a particular alias structure.

Fundamental principles of MOST The resource management principle and the continual optimization principle.

Intensity of adaptation The degree to which an adaptive intervention has many decision points and/or short review intervals.

Interaction effect criterion An a priori operational definition of what constitutes an important interaction. Used in the optimization phase of MOST when deciding which effects will be retained in the parsimonious prediction model.

Intervention A program with the objective of improving and maintaining human health and well-being, broadly defined.

Intervention component Any part of an intervention that can be separated out for study.

Intervention design The specific details of the approach taken by an intervention, including the intervention components, the settings of the components, any eligibility requirements for participants, etc.

Main effect (of a factor in a 2^k experiment) The difference between the mean response at one setting of the factor and the mean response at the other setting, collapsing over the levels of all remaining factors (e.g., Montgomery, 2009).

Main effect criterion An a priori operational definition of what constitutes an important main effect. Used in the optimization phase of MOST when deciding which effects will be retained in the parsimonious prediction model.

Matched intervention design An adaptive intervention design in which treatment is matched to individuals at a single decision point at the outset, usually based on a tailoring variable representing characteristics of the individual.

MOST Stands for multiphase optimization strategy.

Multicomponent intervention An intervention based on a strategy made up of more than one component.

Optimization criterion Operational definition of the optimized intervention, e.g., most effective that can be obtained for $< \$300$ per participant.

Optimization The process of identifying an intervention that provides the best expected outcome obtainable within key constraints imposed by the need for efficiency, economy, and/or scalability.

Optimization trial An experiment designed to collect the information needed to optimize an intervention.

Orphan interaction An interaction involving only factors that do not have detectable main effects.

Parsimonious prediction model A regression equation based on the results of the optimization trial but including only effects designated important according to the main effect criterion and the interaction criterion. Used in decision-making in the optimization phase of MOST.

Participant An individual who is taking part in the implementation of an intervention for clinical, as opposed to research, purposes. A participant does not provide research data. Contrast with *subject*.

Per-condition overhead costs The marginal cost of adding one experimental condition to an experiment, keeping the total number of subjects in the experiment constant.

Per-subject costs The marginal cost of adding one subject to an experiment, keeping the number of experimental conditions constant.

Pilot study A study aimed at examining feasibility in preparation for a more formal study and not intended for hypothesis testing. Also called pilot test.

Protocol deviation An error made in implementation of an experiment, occurring when the experimenter provides a subject who has been randomly assigned to a particular experimental condition a treatment other than what the assigned condition calls for.

Randomized clinical trial Same as randomized controlled trial.

Randomized control trial Same as randomized controlled trial.

Randomized controlled trial Although any true experiment could be called a randomized controlled trial, this term usually refers to a particular experiment design involving a limited number of experimental conditions, often only two, conducted for the purpose of direct comparison of the means of those experimental conditions.

RCT Stands for (i) randomized control trial; (ii) randomized controlled trial; (iii) randomized clinical trial.

Reduced factorial design An experimental design made up of a subset of the experimental conditions in a factorial design. A reduced factorial design may be balanced or unbalanced. See *balance*.

Research design The specific details of the procedures to be used in a study, such as selection and timing of measures, inclusion criteria for subjects, etc. If the study includes an experiment, then the experimental design is one aspect of the research design.

Resolution A shorthand method of expressing how clear the effects are in a fractional factorial design. A design's resolution is usually expressed as a Roman numeral, e.g., Resolution IV. A larger number indicates clearer effects.

Resource management principle The principle that an investigator using MOST must strive to make the best and most efficient use of available resources when obtaining scientific information. This is one of the two fundamental principles of MOST. (See *continual optimization principle*.)

Scalability The degree to which an intervention can be implemented widely in real-world settings exactly in the form in which it was evaluated.

Screened-in set (of components) In the optimization phase of MOST, the subset of candidate intervention components that remains after components have been eliminated based on the results of an optimization trial. These components may be included in the intervention at either the higher or lower level, depending on the optimization criterion.

Screened-out set (of components) In the optimization phase of MOST, the subset of candidate intervention components that have been eliminated based on the results of an optimization trial. These components will be set to their lower levels. If the lower level of a component = no, it will not be included in the intervention. If the lower level = lower intensity rather than no, the component will be included at the lower intensity.

Strongly clear effect In a fractional factorial design, an effect that is aliased only with interactions involving four or more factors.

Subject An individual who is taking part in and providing data for a research study. Contrast with *participant*.

Synergistic interaction In factorial ANOVA, an interaction in which the combined effect of two or more factors is more favorable than would be expected based solely on the main effects.

Tailoring variable In an adaptive intervention, characteristics of the participant or environment that form the basis for adaptation of the intervention, along with the decision rules, at particular decision points.

Target mediator A causal factor in a conceptual model that is specifically targeted by a particular intervention component.

Index

A

Adaptive experimental design, 280
Adaptive intervention, xi–xiv, xvii, 20, 21, 23, 26, 27, 38, 46, 142, 194, 263, 265, 268–286, 289–291, 293
Ad hoc modifications, x, 18, 188, 230, 257, 258
Aliasing, 152, 158, 159, 162, 165–168, 175, 178, 179, 189, 218–220, 289
Analysis of variance (ANOVA), 11, 12, 72, 73, 77, 79, 80, 88, 94–99, 109, 111, 116–119, 130–133, 137, 139, 140, 209, 229, 230, 233–235, 238–240, 247, 258, 261, 262, 278, 289, 293
Antagonistic interaction, 10, 123, 125–127, 236, 240, 241, 243, 244, 261, 289
Antiretroviral therapy (ART), xv, xix, 7–9, 13, 16, 21–23, 25, 28, 32, 39, 41–44, 47, 56, 59, 70, 80–83, 85, 88, 89, 117, 127, 129, 147, 195, 209, 212, 222, 228, 229, 255, 268–270
Arm, xi, 2, 3, 7, 102, 264, 289

B

Balance, 70, 76–79, 91, 98, 99, 105–107, 131, 137, 139, 140, 146, 196, 198, 199, 213, 217–219, 232, 234, 239, 278, 289, 291, 292
Bayesian, 90, 259, 260

C

Causal chain, 42, 43
Centering, 139, 140
Clustered data, 177
Clustering, xix, 85–87, 177, 199, 223, 286
Coefficient correction, 75, 76, 109–111, 118, 119, 134–136
Component, ix, 2, 68, 116, 147, 194, 228, 268
Conclusion-priority perspective, 87–93, 99, 100, 132, 157, 211, 259
Confounding, 6, 51, 155, 219, 220
Constant component, 81–83, 85, 290
Constant overhead costs, 200–202, 290
Contamination, 86, 220–223, 290
Continual optimization, 12–14, 16, 31, 45, 53, 91, 142, 188, 268, 283, 291, 293
Control, viii, xi, xii, 3, 6, 7, 9, 15, 25, 28, 30, 33, 44, 62, 72, 77, 79–81, 100, 112, 139, 151, 155, 173, 175, 182, 183, 185, 197, 200, 209, 211, 263–265, 281, 285, 286
Cost (broad definition), 2, 8, 60, 232, 242, 272, 291
Cost-effective, x, xi, xx, 29, 60, 61, 91
Cost ratio, 207, 208, 290

D

Decision-making, xvi, xvii, 11, 26, 46, 54, 57, 76, 89, 90, 92, 93, 119, 123, 130–133, 161, 180, 230–237, 239, 242, 247–250, 258, 259, 261–265, 284, 292
Decision-priority perspective, 87–93, 100, 132, 133, 157, 211, 230, 235, 237, 238, 259
Desiderata, 14–17, 232, 262
Dummy coding, 77, 97, 98, 101, 131, 139, 140, 180, 234, 236, 286

E

Effect coding, 77, 79, 80, 91, 94–99, 101, 131, 133, 134, 137, 139, 140, 150, 152, 167, 173, 180, 184, 209, 229, 234, 236, 239, 278, 286
Effect heredity, 130–132, 156, 233, 290
Effect hierarchy, 130–132, 156, 162, 233, 239, 290
Effect modification, 38, 51–53, 117, 134, 138–140, 142
Effect sparsity, 130–132, 156, 233, 290
Equipoise, 217

F

Factor, xv, 10, 37, 69, 116, 146, 198, 229, 274
Factorial, 62, 68–112, 146–190, 194, 229, 274, 293
Fixed intervention, xi, xvii, 21, 26, 27, 194, 265, 268–270, 282–284, 291
Fraction, 147, 149, 160, 161, 164, 170, 199, 208, 258
Fractional, xvii, xviii, 26, 77, 83, 86, 143, 146, 149–152, 155–162, 164–168, 170, 173–175, 178–181, 184, 186, 189, 194, 196–199, 204, 206–209, 211, 218, 219, 264, 281, 282, 289, 291, 293

G

The general linear model (GLM), 94–98, 212
Granularity, 49–51

H

Half-fraction, 149, 151, 152, 160, 164, 165, 168, 170, 180, 181, 206, 208, 209
Higher-order interactions, 53, 75, 76, 98, 128–130, 156, 159, 161, 162, 178, 180, 189, 219, 240
HIV, xv, xix, 7, 8, 14, 20, 21, 23, 38–41, 43, 44, 58, 81, 85, 86, 117, 147, 195, 228, 268, 270, 272

I

Iatrogenic, 11, 246, 250, 258, 261
Intensity of adaptation, 272, 273, 276, 280, 282, 291
Interaction, 5, 37, 72, 116, 152, 197, 229, 278
Intervention, 2, 36, 68, 116, 160, 194, 228, 267

J

Just-in-time adaptive intervention (JITAI), 273, 280, 282

L

Leaf springs, 3, 4, 6, 7, 25
Link function, 116, 212, 213
Logic model, 36, 37, 290
Log-transform, 116

M

Main effect, xvii, 11, 72–76, 116, 151, 204, 229, 278
Marginal, 119–121, 124–127, 201, 202, 292
Means, xix, 2, 36, 70, 118, 147, 230, 271
Mediation, xii, 3, 7, 19, 29, 43, 47, 263, 286
Mediator, 19, 27, 30, 37, 41–51, 54, 260–262, 290, 293
Meta-analysis, 51, 76
Minitab, 166
Model, xi, 7, 29, 85, 116, 147, 195, 228, 281
Moderation, 37, 39, 42, 51–53, 116, 120–122, 124–129, 141, 263, 283
Multi-arm comparative experiment (MACE), 182–184, 187–189, 194, 195, 197, 199, 200, 202, 204–206, 208–212
Multilevel data, *see* clustered data

O

Optimization criterion, xvi, xvii, 8, 9, 12, 25, 26, 29, 31–33, 55–62, 224, 229, 230, 232, 234, 242, 246–262, 284, 285, 292, 293
Optimization trial, xiii–xvii, 10, 13, 15–19, 21, 25, 26, 29, 30, 32, 33, 45, 46, 50, 51, 54–56, 59, 68, 116, 117, 130–132, 135, 180, 181, 195, 209, 212, 217, 223, 224, 229, 230, 237, 240, 244, 246, 248, 255, 257–260, 263–265, 273–278, 280–283, 286, 290, 292, 293
Orphan interactions, 233, 247, 248
Overhead, 105, 200–204, 206–208

P

Parent effect, 130, 233, 247, 248, 290
Pareto principle, 130, 156, 233, 290

Index 297

Per-condition overhead costs, 200, 202, 204, 207, 209, 290, 292
Per-subject costs, 196, 200, 201, 204, 207, 290, 292
Pilot test, xi, xvi, 7, 17, 18, 25, 54, 55, 224, 229, 292
Power, xii, 44, 79, 133, 146, 196, 237, 278
Predicted marginal means, 119
Proc factex, 166, 167, 170–172, 174, 175, 181, 218, 220
Protocol deviation, 179, 202, 203, 215, 216, 220, 221, 275, 292

R
Random assignment, 86, 139–141, 155, 177, 196, 203, 213, 214, 219, 221
Randomized controlled trial (RCT), viii, ix, xi, xii, 2, 4, 6, 7, 9, 12, 13, 15, 17–19, 27–33, 43–46, 51, 68, 69, 72, 76, 77, 79, 80, 91, 93, 96, 99, 102, 105, 111, 112, 138, 142, 150, 183, 189, 196, 201, 215, 221, 223, 257, 259, 262–265, 280, 286, 289, 292
Reduced factorial designs, xvii, 143, 146
Resolution, 160, 162, 164, 166, 168, 170, 171, 173, 174, 180, 186, 293
Resource management principle, 16–18, 26, 28, 29, 32, 46, 51, 62, 69, 90–93, 100, 105, 107–109, 133, 155, 162, 178, 181, 190, 194–196, 200, 205, 211, 224, 235, 237, 265, 282, 290, 291, 293
R (software), 166

S
SAS, 101, 166, 167, 171, 172, 218
Scientific yield, 178, 190, 194, 196, 200, 209
Screened-in set, 11–13, 16, 47, 56–59, 61, 88–93, 100, 161, 210, 211, 224, 232, 234, 236, 240–242, 244–255, 258, 261, 293
Screened-out set, 11, 57, 59, 89–91, 93, 232–234, 236, 239–248, 261, 262, 293
Sequential, multiple assignment, randomized trial (SMART), xiii, 26, 27, 194, 276–282, 284, 286
Standard care, xi, 3, 4, 7, 10, 28, 80, 147, 184, 195, 217, 228, 258, 264
Synergistic interaction, 10, 123, 130, 241, 243, 245, 247, 293
System identification experiment, 109, 281, 282

T
Target mediator, 37, 41–43, 45–51, 260–262, 290, 293
Time-varying, 38, 194, 272, 276, 281–283
Treatment, ix, xi, xviii, xix, 2, 6, 7, 9, 14, 19–21, 27, 28, 37, 42, 44, 61, 69–72, 79, 86, 97, 109, 111, 112, 127, 129, 139, 147, 162, 181, 182, 187, 196, 198, 200, 212, 215–217, 220, 222, 223, 263–265, 268–272, 275, 276, 278–280, 284, 286, 289–292

CPSIA information can be obtained
at www.ICGtesting.com
Printed in the USA
BVHW06*1237090518
515738BV00002B/3/P